EVOLUTIONARY ECOLOGY

BIOLOGY AND ENVIRONMENT
General Editor: Bernard Stonehouse

THE BIOLOGY OF PENGUINS
Edited by BERNARD STONEHOUSE

THE BIOLOGY OF MARSUPIALS
Edited by BERNARD STONEHOUSE
and DESMOND GILMORE

EVOLUTIONARY ECOLOGY
Edited by BERNARD STONEHOUSE
and CHRISTOPHER PERRINS

EVOLUTIONARY ECOLOGY

Edited by
BERNARD STONEHOUSE
School of Environmental Science
University of Bradford, England

and

CHRISTOPHER PERRINS
Edward Grey Institute of
Field Ornithology
University of Oxford

UNIVERSITY PARK PRESS
Baltimore • London • Tokyo

First published 1977 by
The Macmillan Press Ltd
London and Basingstoke

Published in North America by
UNIVERSITY PARK PRESS
Chamber of Commerce Building
Baltimore, Maryland 21202

Printed in Great Britain

Library of Congress Cataloging in Publication Data

Main entry under title:

Evolutionary ecology.

    (Biology and environment series)
    "Publications by David Lack": p.
    Includes indexes.
    1. Zoology—Ecology—Addresses, essays, lectures.
2. Evolution—Addresses, essays, lectures.
I. Stonehouse, Bernard. II. Perrins, Christopher M.
[DNLM: 1. Evolution. 2. Behavior, Animal.
3. Adaptation, Biological. 4. Ecology. QH546 E93]
QH541.145.E96 1977   591.5   76–30873
ISBN 0–8391–0885–0

Contributors

VAN BALEN, DR J. H. – Institut voor Ecologisch Onderzoek, Kemperbergerweg 11, Arnhem, The Netherlands

CAVÉ, DR A. J. – Institut voor Ecologisch Onderzoek, Kemperbergerweg 11, Arnhem, The Netherlands

CHITTY, PROFESSOR D. H. – Department of Zoology, University of British Columbia, Vancouver, Canada

FRY, DR C. H. – Department of Zoology, University of Aberdeen, Aberdeen

GIBB, DR J. – DSIR Ecology Division, Lower Hutt, New Zealand

HARRIS, DR M. P. – Institute of Terrestrial Ecology, Banchory Research Station, Glassel, Banchory, Kincardineshire

KLUYVER, PROFESSOR H. N. – Institut voor Ecologisch Onderzoek, Kemperbergerweg 11, Arnhem, The Netherlands

KREBS, DR J. R. – Edward Grey Institute of Field Ornithology, Department of Zoology, University of Oxford, Oxford

LEVIN, PROFESSOR D. A. – Department of Botany, University of Texas at Austin, Austin, Texas 78712

MEDWAY, LORD – Great Glemham House, Great Glemham, Saxmundham, Suffolk 1P17 1LP

NELSON, DR J. B. – Department of Zoology, University of Aberdeen, Aberdeen

NEWTON, DR I. – The Nature Conservancy Council, 12 Hope Terrace, Edinburgh 9

NISBET, DR I. C. T. – Massachusetts Audubon Society, Lincoln, Massachusetts 01773

ORIANS, PROFESSOR G. H. – Department of Zoology, University of Washington, Seattle 5, Washington

ORIANS, C. E. – Bowdoin College, Brunswick, Maine

ORIANS, K. J. – Overlake School, Redmond, Washington

OWEN, DR D. F. – 66 Scraptoft Lane, Leicester

PERRINS, DR C. M. – Edward Grey Institute of Field Ornithology, Department of Zoology, University of Oxford, Oxford

PYE, PROFESSOR J. D. – Department of Zoology and Comparative Physiology, Queen Mary College, Mile End Road, London E1 4NS

RICKLEFS, DR R. E. – Department of Biology, University of Pennsylvania, Philadelphia, Pennsylvania 19174

ROSENZWEIG, DR M. L. – Department of Ecology, University of Arizona, Tucson, Arizona 85712

SCHAFFER, DR W. M. – Department of Biological Sciences, College of Liberal Arts, University of Arizona, Tucson, Arizona 85721

SCHAFFER, M. V. – Department of Biological Sciences, College of Liberal Arts, University of Arizona, Tucson, Arizona 85721

SNOW, DR D. W. – Bird Room, British Museum (Natural History), Tring, Hertfordshire

TURNER, PROFESSOR B. L. – Department of Botany, University of Texas at Austin, Austin, Texas 78712

WATSON, DR A. – Institute of Terrestrial Ecology, Banchory Research Station, Glassel, Banchory, Kincardineshire

WYNNE-EDWARDS, PROFESSOR V. C. – c/o Department of Zoology, University of Aberdeen, Aberdeen

ZAHAVI, DR A. – University of Tel Aviv, Tel Aviv, Israel

Contents

1 Introduction

Dr B. Stonehouse and Dr C. M. Perrins

The twenty-one papers in this volume, centred on the theme of *Evolutionary Ecology*, are presented by biologists from a range of research fields. The factor common to all contributors is that their work has, in some important way, been influenced by David Lack. David Lambert Lack (1910–73), whose life and works are summarised in a group of obituary notes in *Ibis* Volume 115 part 3 (1973), was primarily an ornithologist. He was also an evolutionist of distinction whose main field of study was the ecology and evolution of birds, and whose writings appealed strongly to a very wide range of biologists in many fields of research. Among the contributors to this volume are several former students and colleagues, who knew Lack personally and benefited directly both from his teaching and from his warm personality. Other contributors represent the majority — biologists all over the world who knew him only from his writings, but gained a better understanding of their own research from his concepts and approach.

Son of a London surgeon, David Lack developed an early interest in ecology and the mechanisms of evolution through spare-time studies of birds, firstly as a schoolboy under the guidance of his father and later as an undergraduate escaping the aridity of a classical zoology course at Cambridge University. Influenced mainly by the semi-popular writings of W. P. Pyecraft, and later by Julian Huxley and J. B. S. Haldane, his earliest studies resolved themselves into a search for evolutionary answers to the ecological problems which his fieldwork disclosed. Though not uncommon today, this approach was novel in the early 1930s when his research career began. It was rare even among the few field biologists of the time who shared his interest in living animals and their responses to environmental pressures, and it offered only meagre opportunities for a professional career in science. So David Lack's first appointment on graduation was as a schoolmaster; it was as an amateur that he did the research on nightjars, robins and Galapagos finches which established him firmly as a scientist and writer. In 1945 he obtained his first and only professional appointment as a biologist — directorship of the Edward Grey Institute of Field Ornithology at Oxford University, which he held until his death.

The freshness of his approach is apparent in his early papers and books, and one of the reasons why they are still read with enthusiasm by successive generations of students. *The Life of the Robin* (1943), possibly his best-known book, can still be cited as an all-too-rare example of a readable biological monograph. The numerical data which he presented, from his pioneering banding study of robins, were among the first to be interpreted in life-table terms, and as such were regarded with suspicion and alarm by the scientific community. As he

recalled in an autobiographical note, when he announced at a meeting of the
British Ornithologists' Club that his banding returns indicated a three-fifths
mortality of adult robins each year, only one member of his audience – the
chairman, Landsborough Thomson – found it possible to believe him. In *Darwin's
Finches* (1947) he again broke new ground in relating species diversity and niche
utilisation, stressing the adaptive nature of a high proportion of the characters
which distinguish closely-related species and subspecies. His later books, including
The Natural Regulation of Animal Numbers (1954), *Population Studies of Birds*
(1966) and *Ecological Adaptations for Breeding in Birds* (1968) were daring
innovations of a different kind, bringing together his flair for field studies and his
wide knowledge of ornithological literature (helped by the proximity of the
Alexander Library of the Edward Grey Institute). Permeated as they were by his
unwavering belief that all life has been, and is being, shaped through the influence
of natural selection, and that natural selection operates exclusively through the
individual organism, these books form a lasting record of his clarity of thought,
vigorous sense of purpose and wit.

The papers presented in this volume fall into four categories, to all of which
Lack contributed significantly. In his own view, the two most serious contributions which he made to biology concerned the adaptations of closely related
species and the significance of reproductive rates. He noted with interest that
these concepts came to him within a few months of each other at a time (in 1943)
when he was seriously overworked and overtired. Although the first followed a
laborious assembling of facts, the second came long before he had gathered the
data which would justify it. The first concept is now fully integrated into
biological orthodoxy; few biologists working on isolating mechanisms and
adaptations of closely-related species are aware of their intellectual links with
David Lack's early studies. The second concept has remained controversial;
though both are represented in this volume, a much higher proportion of papers
relate to the wide-open question of how reproductive rates of plants and animals
are determined, and the nature of the selecting mechanisms which operate.

Lack's flair for guessing correct, simple answers to problems in evolutionary
ecology stayed with him throughout his working life, to the immense benefit of
his research students and the mystification – sometimes the exasperation – of
colleagues throughout the world. His diagnoses were not always right, though
right could often be found in them; those who could not subscribe to his simplistic
viewpoint nevertheless found it stimulating and sparred happily with him over
many years of fruitful controversy. We have included in this volume several contributions which would without doubt have evoked from David Lack a vigorous
reinterpretation of data, followed by a set of conclusions more in keeping with
his own way of thinking. He would never have forgiven us if we had not. We can
only regret, as he himself once regretted over a recently-deceased colleague, that
the term 'posthumous paper' does not mean precisely what it says.

We thank Mrs Heather French for her valuable editorial assistance, and
Miss Margaret Norris and the staff of the Alexander Library for their help in
compiling the list of David Lack's publications which appears at the end of this
book. Miss Hebe Jerrold compiled the index.

Section 1 Population regulation and the functions of territory

It is fitting that several biologists who often engaged in invigorating controversy with David Lack should open this symposium volume.

As Professor V. C. Wynne-Edwards notes generously in the first chaper of *Animal Dispersion in relation to Social Behaviour*, both the title of his book and his inspiration for writing it were drawn from Lack's *Natural Regulation of Animal Numbers*. His views, especially those concerned with social and group selection, often ran contrary to Lack's. They helped, however, to clarify the ideas which Lack expressed in later writings, notably in his appendix to *Population Studies of Birds* (which included a critical analysis of Wynne-Edwards's case) and in *Ecological Adaptations for Breeding in Birds* which Lack regarded as a reply to *Animal Dispersion* and '. . . an attempt to build a positive case on the other side'. In the opening chapter of this volume Professor Wynne-Edwards continues the argument, considering the impact of Hamilton's and Wilson's contributions to his thesis, and some of the implications of social selection — especially of altruism — in our own species.

In chapter 3 Dr Adam Watson, for long associated with Wynne-Edwards at Aberdeen, discusses aspects of his population study of Red grouse, drawing attention especially to the adaptive value of territorial behaviour in population limitation. Professor Dennis Chitty, who was a member of the Bureau of Animal Population during the many years when it shared a building with the Edward Grey Institute, reviews the development of a long-standing controversy with Lack over population regulation in cyclical breeders, with special reference to the microtine voles which were his main study species. Dr John Gibb, a former research student and colleague of Lack, in chapter 4 takes up the argument in relation to his own long-term studies of wild rabbits in New Zealand. Finally in this section, Dr John Krebs, who worked closely with Lack's research group on the breeding biology of titmice at Wytham Wood, near Oxford, contributes a chapter on his experimental investigation of the role of song in the establishment of territoriality in Great tits.

2 Society versus the individual in animal evolution

Professor V. C. Wynne-Edwards*

We were taught in our earliest childhood that we ought to respect the rights and feelings of others. By the same token, the oldest code of human morals that has come down to us, the Ten Commandments of Moses, is chiefly concerned with obligations that are to be observed between people. Only the first three commandments relate to something different, enjoining reverence for the deity. The last seven all have to do with proper relationships between individuals in society; keeping holy the sabbath day the same as everybody else, honouring one's father and mother, doing no murder, not committing adultery, not stealing nor bearing false witness nor coveting one's neighbour's possessions. To make the rules simple, Moses picked out a few particular obligations for mention, and many others, scarcely less important, are covered by the broader social virtues of being generous, temperate, just, honest, dutiful and brave.

We soon learnt that it is morally wrong to repudiate these obligations. If we did disobey them and were found out, we were faced with reprimand or punishment, and this in turn had a chastening effect on our conscience. Conscience no doubt depends partly on inborn neurological patterns which make us capable of feeling the emotions of guilt and virtue, and predispose us to comply with the social code; but training our conscience to work properly depends very much on learning the code, and on the reinforcement of good conduct that we need to receive from the people around us. Conscience tends to be strengthened by surroundings in which moral standards are high, and to be dulled where the prevailing morality is lax.

In many of our actions in daily life it is not left just to conscience whether we do right or wrong; it becomes also a question of keeping or breaking the law, and further reinforcement of good conduct then comes from a wish to avoid penalties that might be imposed by the courts. Society has evolved its legal institutions,

*Vero Copner Wynne-Edwards was born in Leeds on 4 July 1906. He studied zoology at Oxford, and in 1930 became Assistant Professor of Zoology to McGill University, Montreal. In 1946 he returned to Britain to become Regius Professor of Natural History at Aberdeen University. His main research field has been ecologically significant behaviour, especially in birds. His book *Animal Dispersion in relation to Social Behaviour* (1962) outlined an hypothesis of population homeostasis, that is, of populations that have become self-adjusting, through social and endocrine mechanisms; their controls have evolved to impose restraints on individuals for the common good, and above all for safeguarding the habitat from over-exploitation.

including codes of law and judicial procedures, to administer justice and protect
the common good in the many situations where people's interests clash, or where
it would be less than just, or ineffectual, to leave the conflicting parties or even
capricious bystanders to sort them out and impose a settlement. But conscience
and the law normally work together as a deterrent against anyone tempted to
seek selfish advantage at the expense of the community. They encourage us instead
to act responsibly and do our bit for the welfare of society as a whole. Though it is
not the whole story, it is true to say that, most of the time, what is right or wrong
boils down to a question of putting the common good above one's own selfish
gain.

The conflict between the interests of society on the one hand and those of the
individual on the other is not at all unique to mankind. It turns out to have deep
biological roots, and to affect in some degree a wide variety of living organisms.
Its evolutionary history must go back almost to the time when life began, more
than 2500 million years ago; but we can see it more clearly in the animal than in
the plant kingdom, and above all in the higher animals in which social relationships between individuals are well developed. Even such non-gregarious animals
as cats and eagles can have mutual social relationships with their neighbours just
as functional as those of gregarious antelopes or geese.

In broad terms the central objective of social development, and of society as a
biological institution, appears to be safe-guarding the long-term interests of the
social group, and securing its future survival. The central objective of the individual, on the other hand, is to stay alive until it has reproduced and thus made its
own contribution to the survival of the stock.

Darwin (1859, chapter 3) pointed out that individuals generally have to
struggle for existence, and not all are equally successful in achieving their central
objective. It is usually the fitter ones that live to breed, and thus, as parents, contribute the biggest share to the next and subsequent generations. The process of
natural selection is concerned with the characteristics that make individuals fit,
in so far as these characteristics can be inherited and passed on. It relates very
largely therefore to good and bad genes, enabling the better ones to be perpetuated
and squeezing the poorer ones out.

But what genes are good genes? Neo-Darwinists including my late friend
David Lack at Oxford and G. C. Williams at Princeton, have seen the answer as
being perfectly clear on logical grounds: good genes are the ones that promote
the fitness of their bearers, by maximising the contribution the bearers make to
the breeding stock of the next generation (*cf.* Lack, 1968, pp. 5–6; Williams,
1966). Authors like these conclude, in consequence, that all species must necessarily and quickly approach a condition where the individual's capacity to produce strong, viable offspring is raised as high as natural selection can make it.
They can see little possibility of genetic selection working in the reverse direction,
towards restraining fecundity, lengthening adolescence or shortening the lifespan,
if the effect of these would be to diminish the rate at which new breeding stock
was produced.

This may seem undeniable as a logical deduction, but it is not always possible
to equate it to the facts of nature as we observe them. We have learnt in the past
30 years that many of the higher animals have homeostatic adaptations for regulating their populations. This means that such animals can take the initiative them-

selves in keeping their numbers under control, and can compensate for the varying effectiveness of external checks on population growth, such as predation or disease. If these checks diminish, and population numbers consequently tend to rise and threaten overcrowding, the reproductive rate can be curbed, and, as I shall show later, numbers can usually be reduced. Conversely, when external causes of mortality intensify and the numbers in the population are heavily reduced, the survivors compensate by redoubling their productivity. Such homeostatic adaptations have repeatedly been shown to operate both in controlled laboratory experiments and in the wild; and they often include adjusting the fertility of individuals in response to a feedback from varying population density, or excluding a larger or smaller proportion of adults from breeding at all. Thus, under average conditions there can exist a reserve of reproductive capacity not being used, which can nevertheless be mobilised in an emergency, if the population is struggling to survive. Watson gives an illustration of this in chapter 3, on the Red grouse. Other classic examples were described by A. J. Nicholson (1955) and by Silliman and Gutsell (1958) and Silliman (1968).

There is commonly a prudential relationship between animal populations and their food supplies which is not always appreciated sufficiently. I can try to give a quick insight into this relationship by taking a striking illustration, the feeding ecology of the beaver (*cf.* Alesiuk, 1968; Bradt, 1938; Wynne-Edwards, 1970). Beavers depend for food, especially in winter, on the bark, twigs and buds of poplars and other trees, cutting down the trees with their chisel teeth, and floating most of the material away to store it in the beaver pond. Trees take years to grow, and if a stock of beavers is to remain in a particular habitat, the animals must not cut down in an average year more wood than is naturally replaced in the same area by annual growth. Beavers can fell small trees with relative ease, and they need to spend only a small fraction of their lives engaged in the process. If they were to produce more offspring and the population grew, more trees could certainly be felled each year, and if necessary transported to additional dams. But once the consumption rate had overtaken the rate of poplar regeneration, the biomass of trees would begin to decline. In a big watershed it might be 25 or even 100 years before the beaver population died out, after the last edible sapling had gone and the watershed was no longer habitable.

For their populations to be viable in the long term, beavers must clearly be adapted to live perpetually surrounded by a superabundance of food, without ever endangering the capital stock of trees that provides the annual increment and crop. This is what I mean by saying that their populations have to be curbed prudentially from within: it will always be too late if they let the population increase until hunger and want impose an ultimate limit to growth from without.

It may be asked whether or not beavers are prevented from increasing under natural conditions by the pressures of predation and disease. The evidence suggests otherwise. Over huge areas of Canada and the northern USA during the past 200 years, beavers were actually exterminated by trappers; but since the 1930s new legislation has brought in better management practices, and reduced the pressure of trapping. Reintroductions have been made to many areas where beavers had been exterminated, with the result that they have re-established themselves at a surprising rate. But they have only built up to a safe population density. Even in regions of human settlement where their natural predators have disappeared, beaver populations stabilise below the level at which they tend to destroy the habitat.

There is a parallel example in the lions and other carnivores in East Africa. They themselves have no serious predators, but they also succeed in living in equilibrium with their prey. Schaller (1972, p. 397) estimated that carnivores depending for food on the big herbivores in the Serengeti Park, collectively consume only about 9 - 10 per cent of the biomass of prey a year. The lions themselves (ibid. p. 122) are inactive on average for 10-21 hours out of the 24, suggesting again that lack of time and hunting success are not factors that might limit their rate of food consumption. Like the beavers' poplar trees, the ungulate prey takes time to grow and has a finite production rate, and it is likely that the predators are instinctively programmed to restrain their own consumption in the interests of prey-management, and the maintenance of a high crop yield. If it were not so, their numbers would top the threshold at which they began to deplete and eventually exterminate the stock of prey, as, in very similar circumstances, human whalers and trawlermen have done.

In chapter 3, Watson describes possibly the best-documented example there is at present, of a self-regulating vertebrate population living in the wild – the Red grouse; its regulatory process has been studied with great patience and thoroughness. Its population density is flexibly controlled by a system of individual territories that fill the habitat anew each autumn, and force out large numbers of unsuccessful surplus grouse, which are thereby doomed to die in the ensuing winter. Watson and his colleagues have shown experimentally that the changes in population density from place to place and year to year are related to the amount and quality of the food available in the habitat, in circumstances consistent with those I have just described. The chief homeostatic regulator of Red-grouse populations is the average size of the territory defended by the cock birds, which can vary severalfold.

The picture I am trying to build up is that of an observed world of nature, in which the vertebrate ecologist continually finds evidence that seems to contradict the conceptual world of the Darwinian purist, the world of compulsive short-term advantage and maximised reproductive output. What we see instead is a well-adjusted system of population control or homeostasis, which keeps populations around the level that the food resources can support without being overtaxed, and which, in short-lived species at least, allows population adjustments to be made from year to year and in some species even from month to month, as resource levels change. I developed this subject at length in my book *Animal Dispersion* (1962), and since that time the evidence for population homeostasis as a normal and widespread adaptation has grown unassailable.

It is a process that plays fast and loose with the fortunes of individuals, penalising or sacrificing them when necessary to achieve its aims. Its operation always appears to demand, to some degree, a social organisation within the population concerned. My own view is that this is a basic property of social organisations, and that sociality has evolved in part to enable the conflicts between members of a population, engaged in the struggle for existence, to be settled by conventional means, without escalating into violence; and thus to make it possible for individuals to come to terms and accept personal differences in rights and status. For instance, breeding populations of Red grouse and many other vertebrates are limited in number by the need to compete for or secure a territory of some kind, as a prerequisite for sexual maturation or mating. Obtaining or not obtaining a territory can be the most crucial single item among the achievements that determine the fitness of an individual. Yet the contest for territories is commonly won or lost without shedding blood, through some transient ritualised encounter; it may be a threat of menacing strength and weapons, or splendid adornments, or confident vocalising, or even overpowering aroma. Where the encounter is a one-to-one confrontation, each individual is ready at the moment of decision with alternative responses, to accept victory or defeat; and the contest is sometimes over in a matter of minutes or even seconds, because of the swiftness of appraisal and the readiness of the loser to concede the victory and signal his withdrawal. Territory winners, once established, for their part quickly accept each other's rights and boundaries.

Relationships among individuals living in a flock can be determined in a similar way, with a minimum of fighting, to produce a pecking order or hierarchy which is generally stable and accepted for the time being by the individuals concerned. Yet in times of food shortage, the ones at the tail of the hierarchy are the ones that starve to death; they are simply shed from the group in order that the group itself will survive. In the breeding season they may be excluded from reproducing, just like the outcasts in a territorial species that fail to establish territories.

Thus social competition typically results in giving rights to some individuals and withholding them from others; the paramount rights in question, which animal society confers on the successful, are the right to remain in the group, the right to food and other resources, and the right to reproduce. The effects of competition can have parallel, psychosomatic results. They are fed back into the individual's nervous and endocrine systems, with potentially powerful physiological and behavioural consequences, giving the stop or go in many instances to gameto-genesis or seeking a mate.

It is only slowly coming to light how drastic the effects of social competition can be. For instance, in chapter 3 Watson shows that, in an average year, less than 40 per cent of the Red grouse that enter the autumn competition each year emerge successfully, either by obtaining a territory if they are cock-birds or else by being accepted by established cocks if they are hens. The remaining sixty-odd per cent of both sexes become outcasts and, as winter comes on, they are gradually excluded from the right to belong and the right to feed, so that by the spring virtually all of them are dead.

This process of social selection is in fact by far the most intense selective agency to which the adult birds are subjected. Yet it depends essentially on an appraisal of one rival by another, and on their mutual decision as to which is to dominate and which is to give way. It is an appraisal that presumably takes into account the physical condition, the character and the confidence of the opponent; one can surmise that these qualities depend partly on nurture and diet and partly also on the general genetic make-up of the individual. Analogy with humans in a similar situation suggests that any good genotype stands a good chance, whatever its detailed make-up may be, and that the same good genes will not be present in every dominant individual. What we see is a social adaptation, which in the Red grouse creams off the best third or so of the population to become the parents of next year's generation, and gets rid of the rest, year in, year out.

In a number of animals, competing for territories or personal status involves actual physical assault, but it is typically ritualised, and the participants adhere to codes of behaviour which minimise serious injury. Although some species possess exceedingly lethal weapons in the form of horns, tusks, pincers, poison fangs and even electric organs, these are not used treacherously and without warning; the attacked individual is given a chance of defending itself, and though the fight may be a searching test of strength and determination, its purpose is not to kill. In a discussion of how ritualisation could have evolved, assuming that natural selection must act solely to promote Darwinian fitness, Geist (1971) accepts the premise that it would be advantageous to kill one's rivals and dispose of them for good, but that it is the difficulty of killing them quickly enough to avoid a swift and savage retaliation by the wounded party, with great risk of injury to the attacker, which has led to limited combat. Geist's long experience of the horned ungulates, especially the wild sheep, has shown that the dominant males do not seek to initiate attacks on rivals: they wait instead to be challenged. Thereby they minimise the energy expended in preserving their dominant status. He observes also that their defences, in the shape of thick hides and skulls, in general keep pace with their weapons; but nevertheless even ritual combat is dangerous for animals armed like these. Wounds are fairly common, and not a few are fatal.

Certainly the habitual and intentional killing of rivals is a rarity among animals. It occurs in the lion in East Africa, where field studies begun more than 20 years ago have shown that the adult males in the pride are supplanted every few years by younger adult males coming from outside, the progeny of another pride (Schaller, 1972). The incoming males often kill their predecessors. It may be significant here that the killing is incidental to a eugenic adaptation – male exogamy – which benefits the stock in the long term; additionally it rids the population of superannuated males who, if they remained alive, would be a useless drain on the food resources. More is thus gained by the killing than just the personal advantage of the incoming males. It is, of course typical of polygamous animals in general that there is normally an excess of males and unsuccessful adult males are expendable.

A complete contrast to the horned ungulates and lions can be found in the various species of birds in which the adult males compete in leks or tournaments – birds that include neotropical hummingbirds, cotingids and manakins, and the Australasian birds-of-paradise, and also the Blackcock and some other northern grouse. The local males gather daily at a traditional arena, just for a few hours in the Blackcock but in many tropical species for most of the day, and most of the year too, except for a 4 – 5 month break while they are moulting. None has any deadly weapons, and they do not fight; instead they confront each other with spectacular plumage and dramatic ceremonials and acrobatics. They include some of the most vividly adorned of living birds, such as the orange-and-black cocks-of-the-rock (*Rupicola*) and the sensational birds-of-paradise. Characteristically, they put great energy into their displays and Snow (1962; 1963) has shown that the tropical lek birds are without exception fruit or nectar eaters (the grouse are also largely vegetarian). The plants that produce fruit and nectar use the birds as agents for seed dispersal and pollination and the food they offer is maximally nutritious, plentiful and easy to see; a minimum of time is wasted, therefore, in the search for food. In contrast, insectivorous birds have to hunt laboriously for cunningly self-effacing foods. In few of these birds can the males be freed from duties at the nest; none has evolved a lek.

The vast majority of animals have equally innocuous competitive displays. Among these are the songs of birds, amphibia and insects; the visual displays of other, often highly coloured birds, insects and also many fish; the olfactory marking of trails and territories by mammals. It is notable that the lethal stings of ants, bees and wasps have evolved in females from ovipositors, and in none are the males armed, though they are often intensely competitive.

It is hard to resist the conclusion that killing rivals is not a rarity, for the primary reason that it is difficult to kill without getting hurt, as Geist suggests, but because it is potentially damaging to the society, especially in monogamous species. The inhibitions which make animals keep strictly to the rules in their intraspecific competition do indeed appear to find a functional counterpart in the Mosaic commandment to do no murder (*cf.* Lorenz, 1966).

It is only a step from inhibiting evil-doing to promoting a more constructive kind of behaviour, in which the individual is ready to sacrifice some of his energy and fitness, when required, for the benefit of the group. The worker bee is genetically a female, incapable of copulation and physiologically prevented from becoming sexually mature as long as there is a queen in the colony. Her Darwinian fitness is virtually nil and her energies are entirely devoted to serving the colony. In chapter 12 Fry describes analogous, if less self-sacrificing examples of this in birds. Evolutionists designate such behaviour as altruism, and it has always been a difficult problem to explain how altruism could have evolved, running as it does completely counter to the classical predictions of natural selection. The difficulty was first perceived, a little obliquely, by Darwin himself (1859, chapter 8), and was clearly appreciated by the founders of neo-Darwinism such as R. A. Fisher (1930) and J. B. S. Haldane (1932). It is a difficulty which applies not only to altruism but to any social adaptation that restricts the freedom of the individual to use all and every means in his power to maximise his own advantage.

I was much exercised by this in writing *Animal Dispersion.* The general hypothesis the book set out was the one I have been outlining here, that population homeostasis is a normal attribute of the higher animals, and that it is achieved through the medium of social interaction. Social adaptations, I showed, relate to the general good, and ultimately to the long-term survival of the group; it is their primary function to impose prudential curbs on the freedom of individuals, curbs which often result in lowering the fitness of the individuals. It has been clear to others besides myself, that groups in which nothing but individual advantage was promoted by selection would generally get into difficulties sooner or later, for instance by destroying the resources of the habitat, as I have illustrated with the beaver. What seemed to be required was a second, higher tier of selection, namely selection between groups: those groups that got into trouble would become extinct, and those that had efficient social adaptations, imposing prudential restraints on their members, would survive and spread. The evidence for the existence of local self-perpetuating groups in the animal kingdom is plentiful and widespread, although the effective size of the groups varies considerably. There are also a few notable exceptions of species whose habits are so vagrant that gene flow must occur rapidly on a continental scale, flooding the local populations out of any effective isolation and independence. Group selection, as it is called, had been discussed in a tentative and usually sceptical way by others, including both Haldane and Fisher, long before my time. And what was conspicuously absent from my promotion of it was a credible model or theory of how, in practice, group selection would take place.

In 1964 W. D. Hamilton published his two papers on the genetical evolution of social behaviour, drawing attention to an effect which Maynard Smith and later authors have called kin selection. The idea had been foreshadowed briefly by Haldane in 1932 (p. 209). Hamilton accepted that there are social adaptations which work to the disadvantage of many individuals possessing them, and he put forward a new model suggesting how they could have evolved. The simplest illustration of Hamilton's principle is that of the mother who expends energy nurturing her offspring. It is abundantly clear that by doing so she is increasing her own Darwinian fitness because she is promoting the likelihood that the offspring will survive, live to maturity and themselves successfully replicate her genes.

It is only slightly less obvious that an individual could also increase the replication of his genes by aiding the survival of his sibs, or of his nephews and nieces, because they all carry some fraction of the same genes as his own. The average fraction shared by sibs is the same as that shared by a parent with its offspring, namely a half. The probable fractions shared by an individual with other more remote relations are determined by simple arithmetic. Hamilton derived mathematically a quantity which he called an individual's inclusive fitness, which in effect sums the total inheritance the individual has invested, not only in his own genes but additionally in those he shares with his contemporary relatives. Hamilton showed that natural selection will tend to maximise this inclusive quantity also, just as it tends to maximise individual fitness; a positive selection value will consequently be attached to the activities of an individual which benefit his relations, and a negative one to activities that harm them.

Some species, as I hinted earlier, are very sedentary, in the sense that individual members have a low tendency to move elsewhere during their lives; consequently, there is an appreciable coefficient of inbreeding within their local populations, and the number of relations that any individual possesses among his contemporaries and neighbours is likely to be high. This Hamilton describes as a 'viscous' population, and within it the tendency of selection to promote mutually beneficial activities between the members will clearly be strong. For the same reason, the dilution that an individual's genes suffer in the gene pool during subsequent generations is less than it would be in a population of lower viscosity because the gene pool is smaller. There seems little doubt that in these conditions the basic social adaptations could easily have originated.

A good example of a viscous population is found in the Red grouse, a bird that is able to fly faster than 1 km/min when it has to; yet over 90 per cent of all grouse die within 5 km of where they were born (Jenkins, Watson and Miller, 1963). Good evidence of the immense evolutionary importance generally attached to retaining one's citizenship in a local group is revealed in the amazing powers of homing developed by migratory animals, be they birds, mammals, sea turtles or fish: powers that enable them to return with great precision to breed in the place where they were born, or at least to breed repeatedly in the place where they have become established.

Viscosity of populations varies very much from species to species. The rook is another bird that has been the subject of a long-term study at Aberdeen, and it has been found that there is a wide scatter of the young birds during their adolescence, sometimes to distances of hundreds of kilometres. Rooks are particularly abundant in our region, and to find even wing-tagged young that have gone away and become established in distant rookeries is intrinsically difficult. To prove they are actually breeding there is harder still, but a few have been found dead in the breeding season in rookeries as far as 25 or even 40 km from their birthplace. This suggests that their effective interbreeding populations are considerably larger than those of the Red grouse; we even have single rookeries containing more than a thousand nests, and possibly the difference is as much as one or two orders of magnitude. Nevertheless, the rook has a social organisation, and conventional patterns of behaviour which, though quite different in detail, do not appear to be any less restrictive or compelling than those of the Red grouse. Not all rooks succeed in maturing at the same age, and they have to compete to win a nest site in a rookery. They are inhibited or prevented by other rooks from nesting in complete isolation, or on any site not acceptable to their local community, even inside the rookery.

We have to ask the question, therefore, whether it is possible that social restraints on individual freedom, evolved through kin selection during a highly viscous population regime, can be retained by a species which afterwards undergoes a reduction in viscosity, giving it a freer gene flow and larger interbreeding units. Can social adaptations achieved at one stage in the evolutionary history of a species survive in a subsequent stage, when the effects of kin selection are no longer large enough to affect an individual's fitness, beyond perhaps caring for its young?

Before I attempt to answer this question it is worth looking at the state of social adaptations as they exist in our own species, since at least it is a subject on which we have abundant evidence at first hand. We must be very cautious about applying conclusions derived from man to other mammals, if only for the reason that man now lives in societies in which there is no universal social restriction on population growth. Social pressures no longer cause marked differences between individual people in their Darwinian fitness; on the contrary it can happen that sections of the population under particular stress, for instance because of their low earning capacity, reproduce faster than those placed higher on the social scale. What I would postulate as two of the primary adaptive functions of our social system have been lost entirely by civilised man. They are, or were, both directed to safeguarding the survival of future generations, first by controlling population density so as to prevent us from over-taxing our habitat, and second by favouring the reproduction of successful parents, thus actively suppressing the spread of burdensome genes. But in countless other ways, needless to say, our way of life depends on maintaining an efficient social system, above all for the protection of life and property and the preservation of the rule of law.

The tendency in the past thousand years has been towards a lowering of population viscosity, especially in the western world. In more primitive human societies, as in the rural backwaters of this country not so long ago, there were communities with a far higher viscosity than we generally have in Britain now — a viscosity not vastly different perhaps from that of the Red grouse. Even today our population is far from being panmictic, that is, breeding at random. If one can rely on the samples represented by the 1974 telephone directory, the surname Noble, reckoned per thousand of the population, is 13 times as common in Fraserburgh as it is in Aberdeen, only 40 miles away.

For all that, the coefficient of inbreeding in our society cannot be regarded as high. Nevertheless, where the general atmosphere of morality and social conduct is sufficiently good, our consciences can still easily be trained, and make the majority of us into upright citizens. Even among the antisocial minority it is safe to conclude that there are some people who would have made good citizens if they had had a more fortunate upbringing and a better chance. There are no strong grounds yet for thinking that conscience itself, considered as an innate biological adaptation, has been significantly impaired. It is true that crime statistics indicate a downward trend in the average standard of social responsibility at the present time; but this still seems likely to result more from a change in the social milieu, a weakening of the spiritual and temporal reinforcement available to the individual, than from genetic deterioration.

If our spouses are included, it is normal for us to feel more disposed to put ourselves out for our own kith and kin than for non-familial acquaintances. To a certain extent this inclination may be acquired by experience, as part of the bond of affection we can obviously forge with any person (or even a pet animal) near and dear to us. But there seems to be a deeper-seated, intuitive basis for what we call family pride, for the feeling that regardless of intimacy or personal affection 'blood is thicker than water', and for the apparently subconscious recognition that the obligations of kinship diminish as blood relationships get more remote. If so, it suggests that man too has genes promoting inclusive fitness.

Interest in family ties does not prevent us from feeling strong loyalties towards other members of the particular community that we belong to. Within it we join social organisations, perform public and private services, subscribe to good causes and, in time of emergency risk our own safety for the common good. Most of us want to be thought good citizens and merit the approval of others. The motivation behind this, whether conscious or unconscious, is not wholly altruistic. Our Darwinian fitness as individuals depends materially on our continued acceptability as members of society: the penalties of rejection are severe. Nevertheless, acts of generosity and self-sacrifice are performed every day, anonymously and to persons unknown, in circumstances where the doer can have no other reward than the inward joy of doing good. This reward, abstract and immaterial as it is, is enough reinforcement only, I suspect, because we are genetically programmed to feel it so.

Biologists will not quarrel with the statement that temperamental differences between individuals are in some part genetically determined also, and that some people are inherently more inclined to do good deeds than others. Those who are generous in this respect, when they are deciding what action or effort is appropriate to the circumstances of the moment, weigh the cost of their action against the benefit they expect it to confer on the receivers, and think not at all of the gratitude or reward they themselves might receive. Indeed, unworthier mortals entertaining baser thoughts will in the end be disillusioned, for as the German proverb says, "ingratitude is the world's reward". I do not think there is much room for doubt that what Haldane called "genes for absolute altruism" do exist in man.

Other species of vertebrates are not much less dependent for their survival than we are on the maintenance of an adequate social system. Without the homeostatic restraints it imposes on population size and thereby on the rate of food consumption, many species would starve themselves out and even destroy their habitats, to their own lasting detriment and ultimate extinction.

Turning to antisocial behaviour, most serious human crime has, from the earliest history until very recent times, been officially punishable by death. Thus society itself had long assumed the role of selector, taking direct action to rid itself of antisocial traits. This leads one to ask, in the light of recent revelations of the intensity of social selection in animals such as the Red grouse, whether the same results could somehow be achieved by them, so that their societies too could purge themselves of individuals that transgressed permitted norms. But there is no evidence, except in human societies, of the expulsion and elimination of individuals except on grounds of being subordinate in status.

There remains the possibility that altruistic adaptations could evolve and be maintained by a process of selection between local self-perpetuating stocks or demes – the hypothetical process of group selection, which I intuitively favoured when writing *Animal Dispersion*. It has, however, been examined by some of the ablest evolutionary theorists in recent years, and their verdicts have not been encouraging. MacArthur and Wilson (1967) have shown that the number of species of fauna and flora found on islands tends in time to reach an equilibrium, at which the rate of new colonisation just balances the rate of extinction of species already on the island. A balance is evidently reached sooner on smaller islands than on larger ones, and sooner on remoter islands than on those nearer sources of colonisation.

This gives us a conception of deme extinction as a normal statistical event; rates can be roughly estimated and there is no reason why the process should not occur also under mainland conditions. The great theoretical objection to group selection is that, unless the rate of deme extinction were improbably high, a 'selfish' mutant that gave advantage to its bearers would increase and be carried to other demes long before the eventual climax or crash that wiped the first deme out. It is the long time lag that prevents group selection from reversing individual selection. Wilson (1973; 1975), reviewing the mathematical models that have been suggested to account for the evolution of altruistic genes, concludes that it would be an improbable event for them to evolve by pure interdemic selection, and improbable above all in stable populations like those of the long-lived vertebrates.

Clearly there are concepts and principles still to be elucidated here: the phenomena themselves are real enough. The fact is that every adaptation for population homeostasis shares the essential property of altruism, in being able to elicit sacrifices from some individuals for the benefit of others – others who in this case have been internally or socially selected to perpetuate the stock. Such adaptations are able to curtail the ruthless pursuit of self-advantage at all times, and most conspicuously when prudential ceilings of population density are being approached. Moreover, it has been shown that sometimes (and in the Red grouse virtually every year) only a minority of individuals survive the homeostatic thinning-out, and the majority are sacrificed. The adaptations operate through receptor inputs to the individual's neuro-endocrine system, controlling his physiology and behaviour to achieve whatever degree of sacrifice is required. The sacrifices are accepted automatically, in effect willingly, and therefore with a minimum cost to the established individuals who carry the responsibility of perpetuating the group. The evolution of none of these mechanisms can be explained through Darwinian selection alone.

This chapter was originally delivered as a presidential address to Section D, British Association for the Advancement of Science, at Stirling in September 1974, followed by the contributions of Dr Adam Watson and Dr C. Hilary Fry on which chapters 3 and 12 of this volume are based.

References

Aleksiuk, M. (1968). Scent-mound communication, territoriality and population regulation in beaver (*Castor canadensis*). *J. Mammal.*, **49**, 759 - 762.

Bradt, G. W. (1938). A study of beaver colonies in Michigan. *J. Mammal.*, **19**, 139 - 162.

Darwin, C. (1859). *The Origin of Species.* John Murray, London. (6th edn., with additions and corrections, 1872).

Fisher, R. A. (1930). *The Genetical Theory of Natural Selection.* London. (2nd revised edn., Dover Publications, New York, 1958.)

Geist, V. (1971). *Mountain Sheep.* University of Chicago Press, Chicago and London.

Haldane, J. B. S. (1932). *The Causes of Evolution.* Longmans, Green and Co., London, New York and Toronto.

Hamilton, W. D. (1964). The genetical evolution of social behavior, I & II. *J. theor. Biol.*, **7**, 1 - 16, 17 - 51. (Reprinted with an addendum in Williams G. C. (1971). *Group Selection.* Lieber - Atherton, Chicago, pp. 23 - 43, 44 - 89.)

Jenkins, D., Watson, A. and Miller, G. R. (1963). Population studies on red grouse, *Lagopus lagopus scoticus* (Lath.) in north-east Scotland. *J. Anim. Ecol.*, **32**, 317 - 376.

Lack, D. (1968). *Ecological Adaptations for Breeding in Birds*. Methuen, London.

Lorenz, K. (1966). *On Aggression.* Harcourt-Brace, London. (Translated from *Das Sogennanten Böse.* Vienna, 1963.)

MacArthur, R. H. and Wilson, E. O. (1967). *Theory of Island Biogeography*. Princeton University Press, Princeton, New Jersey.

Nicholson, A. J. (1955). Compensatory reactions of populations to stresses and their evolutionary significance. *Aust. J. Zool.*, **2**, 1-8.

Schaller, G. B. (1972). *The Serengeti Lion.* University of Chicago Press, Chicago and London.

Smith, J. Maynard (1964). Group selection and kin selection. *Nature, Lond.*, **200**, 623 - 626. (Reprinted in Williams, G. C. (1971). *Group Selection.* Lieber - Atherton, Chicago, pp. 119 - 122.)

Snow, D. W. (1962). A field study of the black and white manakin. *Zoologica, New York*, **47**, 65 - 104.

Snow, D. W. (1963). The evolution of manakin displays. *Proc. XIII Int. Orn. Congr.*, 553 - 561.

Williams, G. C. (1966). *Adaptation and Natural Selection.* Princeton University Press, Princeton, New Jersey.

Wilson, E. O. (1973). Group selection and its significance for ecology. *BioScience*, **23**, 631 - 638.

Wilson, E. O. (1975). *Sociobiology: The New Synthesis.* Harvard University Press, Cambridge, Massachusetts.

Wynne-Edwards, V. C. (1962). *Animal Dispersion in Relation to Social Behaviour.* Oliver and Boyd, Edinburgh and London.

Wynne-Edwards, V. C. (1970). Feedback from food resources to population regulation, in *Animal Populations in Relation to their Food Resources.* (ed. A. Watson), Blackwell Scientific Publications, Oxford and Edinburgh, pp. 413–427.

Silliman, R. P. (1968). Interaction of food-level and exploitation in experimental fish populations. *Fishery Bull. Fish Wildl. Serv. U.S.A.,* **66,** 425–439.

Silliman, R. P. and Gutsell, J. S. (1958). Experimental exploitation of fish populations. *Fishery Bull. Fish Wildl. Serv. U.S.A.,* **58,** 214–252.

3 Population limitation and the adaptive value of territorial behaviour in Scottish Red grouse, Lagopus l. scoticus

Dr A. Watson*

Recent research on Red grouse in Scotland has concentrated on mechanisms which limit populations (Watson and Moss, 1972). This chapter concerns the adaptive value of territorial behaviour in population limitation, an aspect of evolutionary significance largely developed by David Lack, and of lasting interest to him.

Lack's view, summarised in two of his books (1954; 1966) was that birds breed as fast as they can within limits set by the available food supply, other environmental factors and their own inherent capacities. Density-dependent mortality is expressed mainly in a contest for food, resulting in the displacement of less dominant individuals and their death mainly by starvation, or less often by predation and disease. In his interpretation, territorial behaviour spaced out those individuals which survived the contest, but did not limit their numbers in the first place. His statement (1954, p. 260) that, if territorial behaviour limited population density, ' . . . the size of the territories should be nearly constant in each species', was an assumption not supported by good evidence. His further conclusion that territorial behaviour could not limit numbers in species in which the size of the territory varied greatly, thus involved a circular argument. Accepting Huxley's (1934) analogy of territories as rubber disks — the more they are compressed, the more they resist further compression (Lack, 1966, p. 78) — Lack argued that territorial behaviour is important for population control only if the limit of compression is often reached, and only ' . . . if it corresponds to the minimum quantity of food needed by the pair to raise their young' (1954, p. 260), for which he found no evidence.

From his 17-year study of Great tits (*Parus major* — a species which fluctuates greatly in density from year to year) he concluded that territorial behaviour could have limited numbers only once — in the year of highest density — and was therefore clearly unimportant as a factor restricting population in most years (Lack,

*Adam Watson graduated in zoology at the University of Aberdeen. His Ph.D. study was on ptarmigan, at McGill University and the University of Aberdeen. Later he received a D.Sc. at Aberdeen, and became a Fellow of the Royal Society of Edinburgh in 1971. He is at present Senior Principal Scientific Officer at the Institute of Terrestrial Ecology, Blackhall, Banchory, Scotland, working mainly on population regulation and social behaviour of Red grouse.

1966). His conclusion depended on the unnecessary and preconceived assumption about peak density when he interpreted the rubber disk model, rather than good primary evidence. Lack took this same assumption further (1969) when he considered stability of breeding numbers over several years to be the criterion of population control by territorial behaviour, even in situations where behaviour had not been studied, and in spite of the fact of population stability in non-territorial species. As Watson and Moss (1970) have noted, long-term stability is not one of the conditions required to demonstrate population control by territorial behaviour. When discussing the model mentioned above, Lack (1966; p. 78) postulated that territorial behaviour fluctuates in relation to food — especially the food supply for the young — and accepted that varying territory size might be considered to limit breeding numbers if the variation could be shown to be adaptive. However, his main conclusion was that territorial behaviour evolved to benefit the individual by maximising its chances of producing young, any function in limiting numbers being secondary.

Wynne-Edwards (1962) has argued that territory size has evolved to limit populations, with other functions secondary. His view is that animals compete not for food itself, but for a convention — territory — allowing the successful an undisputed share of food. In reducing numbers, territorial behaviour prevents animals from over-exploiting food resources, and thus damaging future stocks. Altruistic behaviour by individuals for the benefit of the group as a whole, and lower-than-maximal rates of reproduction which restrict numbers, are evolved by group selection which overrides individual selection. Watson and Moss (1970) found no critical evidence from wild populations which either supported or opposed the Wynne-Edwards hypothesis, and continuing lack of such evidence indicates the difficulty of research in this field. In current theory, selection among close kin (Hamilton, 1964) is likely to occur more frequently than group selection, and many interesting models have been proposed, but data from wild populations remain scarce.

Our research on Red grouse has emphasised experiments on population limitation in relation to territorial behaviour and food (Watson and Moss, 1972). Red grouse feed almost entirely on green heather tips. Only a small percentage of the total amount is eaten over the year (Savory, 1974) but heather is a poor-quality food and grouse are selective, taking only the better shoots, especially before and during egg-laying (Moss, 1972). The cock and hen birds look after their chicks until they are fully grown in August. Between August and October, families break up and old cocks begin to defend their old territories against other cocks, and to court the hens. Between late September and November, young cocks contest for a territory and young hens for pairing; the pattern of territories and pairings therefore changes. Some of the old birds are evicted or leave, usually to die non-violently soon after on undefended ground nearby, where food and cover are scarce.

This reshuffle determines which cocks occupy territories, which hens pair up with them, and how many birds will form the breeding stock next spring. Grouse usually rear more young than are needed to replace adult losses. Whether shooting occurs or not, many cocks do not establish territories, and many hens do not pair with territorial cocks. In the absence of shooting, on average more than half the total population fails. These non-territorial birds form a recognisable social class with different behaviour. They often move on to the territories of other birds and occasionally challenge the owners, but almost invariably are driven off to undefended ground where food is scarce. On the few occasions when territorial birds die, non-territorial ones usually replace them quickly, and areas of new habitat are soon colonised by non-territorial birds. However, most do not get a territory and many will die during the winter due to predation and accidents. Those that are left in spring are in poor condition and die non-violently, often on undefended ground. Virtually none remains alive when the hens start incubating in May (Watson, 1967), whereas territorial birds survive the winter well, very few of them falling to predators. Thus, although the immediate cause of death is usually predation, the initial cause is socially-induced within the population itself. This point was illustrated by an experiment in which I shot some or all of the population on different areas (Watson and Jenkins, 1968). After I had shot territory owners, non-territorial birds took over those territories and then survived well and bred. Clearly, the number of territories taken in autumn was the immediate factor limiting breeding stocks.

As grouse now live densely in a man-made habitat different from the presumed open woodland where they evolved, Lack (1966) found it hard to see the true function of territorial behaviour in regulating numbers at densities 'far higher than those in which the behaviour in question was evolved'. In fact, densities are not higher than in Willow grouse *Lagopus lagopus* in primaeval habitats in Siberia, Norway and Newfoundland (see, for instance, Myrberget, 1972). Certainly, most habitat for Red grouse is maintained by burning and grazing, with gamekeepers controlling the predators. However, the high Scottish moors, such as the Cairnwell (Moss, Watson and Parr, 1975), can support densities similar to those on the lower moors, without burning, with little grazing and hardly any management of predators. Unpublished studies at the Cairnwell show that population limitation by territorial behaviour occurs there too.

A major problem is understanding how and why numbers in an area fluctuate over the years. During one intensively-studied fluctuation, young cock grouse which had been reared in summers when breeding stocks were declining from spring to spring took bigger territories in their first autumn than older cocks (Watson and Miller, 1971). Young cocks reared in summers when breeding stocks were increasing from spring to spring maintained smaller territories in their first autumn and were less aggressive than older cocks. Although varying in the same direction, the territory size of older cocks changed less from one autumn to another, and contributed less to the fluctuation than the change in territoriality of each year's young cocks at the annual reshuffle of territories in the autumn.

Changes in the sex ratio also contribute to fluctuations. In years of increase, breeding populations tend to consist of equal numbers of cocks and hens, but in years of decline the cocks often outnumber hens, frequently with a ratio of 2:1, and sometimes even 3:1. This greatly reduces breeding production. It is not due to too few hens being reared, as plenty of spare hens are still available in February and even March, but most of them do not pair up with territorial cocks. Many unmated cocks have tiny territories perhaps too small to support a hen as well, and a different division could hypothetically produce more broods. However, by taking territories, these extra unmated cocks ensure their survival until the next year.

Grouse rear bigger broods when breeding stocks are increasing from one spring to another, mainly because fewer newly-hatched chicks die in such summers than in others. Much of this mortality is already determined by egg quality (Jenkins, Watson and Picozzi, 1965), which in turn is influenced by the hens' food before incubation (Moss, Watson and Parr, 1975). Artificially fertilising the heather improves the food supply so that grouse can support bigger broods and take smaller territories. Our recent model (Watson and Moss, 1972) has two relationships between better food and subsequent lesser aggressive behaviour and smaller territories of young grouse. One procedes indirectly, *via* egg quality linked to maternal food before egg-laying, the other more directly to food in late summer–early autumn on ground where the young were reared.

Food also explains much of the difference in mean numbers between areas (Watson and Moss, 1972). Heathery ground supports higher densities and smaller territories than more grassy ground. Moors over rich rocks such as limestone, where the vegetation is more nutritious, on average support bigger breeding stocks and broods than moors over poor rocks such as granite.

Animals do not need to own a territory either as an address where the pair can find one another, or for copulation without interference, or for dispersal to reduce disease and nest predation; these activities can occur with other, non-territorial kinds of spacing. In Red grouse, provision of food for the young can be ruled out, as most broods leave the territories on the day after hatching. However, non-territorial cocks and unpaired hens die before they can breed. A territory is a pre-requisite for a cock to attract, court and pair up with a hen. A cock must have a territory for a hen to be attracted, to court and pair up with him, and subsequently breed. It provides an almost exclusive food supply in late winter and early spring, when food is scarcest and hens need good food to nourish their eggs. Limiting the number of territories results in successful grouse getting a greater, exclusive food supply than if all were to settle.

Wynne-Edwards' (1962) view that population limitation avoids over-exploitation of food probably does not apply in the case of the Red grouse, as their own grazing has only a very minor effect on their food supplies. Individual selection is sufficient to explain the observations; group selection and kin selection are not necessary. However, it is much easier to obtain data on individuals than on kin, and on kin than on group. Thus, obvious adaptive value to the individual does not exclude selection for kin and group, unless these are studied.

Local breeding populations may well occur in grouse. The timing of social events, such as the family breaking up or the reshuffle of territories, often varies so greatly that grouse in one place may have completed an important social event which has not even begun less than a kilometre away. Although most grouse in east Scotland occupy continuous moorland with no geographical isolation in the sense of 'islands' of heather surrounded by uninhabitable woods or fields, this variation in timing might isolate local breeding populations from one another. Henderson (1976) found that populations of grouse only 3.5 km apart are genetically different.

One of our present aims is to study local breeding populations. We mark every chick over an area of about 200 ha and later find which get territories. So far, during years of low numbers, increase and peak, very few young cocks have taken territories more than a kilometre from where they hatched. As son attacks father or uncle in autumn, often displacing them, male ownership of the same small area keeps within one family for at least several generations. This strong homing, along with variation in the timing of social events, suggests that local populations may be composed of related kin. In the new study we are also trying to find which birds contribute most to future generations. When discussing adaptive advantage, Lack (1954; 1966) often mentioned numbers of young reared, but the number recruited and surviving to breed is important for selection, not the number reared.

Another problem concerns the many cocks failing to get territories and hens failing to pair, which are thus mostly doomed. As non-territorial cocks take territories if given a little extra androgen (Watson, 1970), some people have wondered why more cocks do not produce this themselves. Possibly this case might involve altruistic behaviour by cocks sacrificing themselves for kin or group. However, it might be best for their individual selection if they stay in the home ranges where they were reared, without making any serious challenges. Possibly they might then have a better chance — however small — of replacing one of the few territory owners which die, than if they exhaust themselves by continual challenges in their home ranges, or move to unfamiliar ground elsewhere. Apparent altruistic behaviour sometimes occurs in summer, when an occasional hen with chicks leaves her young to distract men and dogs disturbing a neighbouring hen's brood. In view of the previous paragraph about local populations, altruistic behaviour by individuals for kin other than their own young seems a reasonable possibility.

In any one year, some cocks occupy bigger territories than others. Watson and Miller (1971) found that cocks with the biggest territories had two hens; cocks with the smallest territories had no hen and fared worse over the winter. Cocks with big territories sang more often, had more encounters with neighbouring cocks, and more often than not won these encounters. Is a cock more aggresive because he has a bigger territory to defend? Or does he take a bigger territory because he is more aggressive in the first place? The second possibility seems likely from experiments where hormone pellets were implanted (Watson, 1970). With extra androgen, thin non-territorial birds became more aggressive and then took territories, survived well and improved in condition, while territorial cocks became even more aggressive and later took bigger territories.

A large variation in territory size was Lack's (1966) main reason for not accepting that territorial behaviour limits grouse numbers, mainly because there was then no evidence for the territories being adaptively related to food. He suggested (1966, p. 205) that, if territorial behaviour limits breeding numbers, the size of territories should be functionally related to the state of the heather. There is now evidence that variation in the amount and quality of heather can be sufficient to account for some of the variation in territory size. For instance, Miller and Watson (unpublished) found that territories held by cocks with one hen had about twice the amount of green heather shoots as the smaller territories of unmated cocks.

As for fluctuations between years, what needs explaining in terms of adaptive value is why young cocks took larger territories than old cocks during a decline in numbers (Watson and Miller, 1971) associated with a deterioration in food (Jenkins, Watson and Miller, 1963; Watson, Miller and Green, 1966), and smaller territories than older birds during an increase in numbers associated with an improvement in food. Yet only a minute percentage of green heather shoots is eaten by grouse.

The explanation stems from the fact that grouse feed selectively (Moss, 1972). The main point to emerge from recent studies (Moss, 1975) is that the better the quality of food selected by a grouse from what is available, the fitter the bird should be in competing with neighbours. A bird's digestive tract alters to suit a high-quality diet, and it then probably needs a big territory to maintain that diet. With a smaller territory it would have less scope for selection and a lower plane of nutrition, and so would be less fit. Thus, it should be adaptive for an individual to take a territory big enough to give scope for the selective feeding to which it has become physiologically suited, but not too big to defend. The territory size of old cocks, which take up their old territories again in August – September, may provide an approximate reference point for young birds. Moss (personal communication) has contributed the following suggestion. When the quality of the food is deteriorating, it is probably adaptive for the young cock to take a bigger territory than the older bird. He would probably fail to get a territory if he tried for a very big one, but a very small one would lower his plane of nutrition and his fitness. When food is improving, the young cock can afford to take a smaller territory than the old cock, without paying any nutritional penalty, and may be more likely to get a territory by doing so.

Thus, population fluctuations between years, as well as limitation within years and difference between areas, can be simply explained in terms of food, territory and adaptiveness to individuals.

Research on grouse in eastern Scotland illustrates five general hypotheses explaining why dense populations fall quickly to lower densities. The first emphasises the role of predators (Pearson, 1966), the second shows that the animals starve (Lack, 1966), and the third that populations fluctuate irregularly, depending on changes in weather, the type of animal involved, and the heterogeneity of other features in the environment, not necessarily depending on density (Andrewartha and Birch, 1954). The fourth hypothesis is that physiological stress from the increased hostility and fighting at high density kills many and reduces reproduction (Christian, Lloyd and Davis, 1965), and the fifth is that more aggressive animals of a different genetic type compete better at high densities, oust the more tolerant individuals, and take more space, independent of any lowering of the quality of food (Chitty, 1967). Since 1970, a very unusual run of mild winters has occurred in eastern Scotland, with little scorching of heather, little overgrazing by cattle or sheep, early growth in spring with nutritious new shoots, high quality of eggs and big broods reared. Very high numbers and small territory sizes have occurred for five springs from 1971 to 1975. Thus the necessity for the mechanisms in hypotheses 1, 4 and 5 to start operating immediately high densities are reached can be refuted, but this cannot yet be said of hypothesis 3. Hypothesis 2 can be rejected as unnecessary; a big decline has now begun in 1975 – 76, with no evidence of starvation of territorial birds. Whether these mechanisms, or our recent model (Watson and Moss, 1972), will be sufficient or necessary to explain events to the end of the decline, remains to be seen. I hope the model runs into trouble, as this is usually exciting and a sign of better progress in understanding. It would also be a sign that the rather dull-looking grouse in its uniform-looking Scottish moorland home has more adaptive tricks up its evolutionary sleeve than one might have thought.

Acknowledgements

I thank Dr Robert Moss for his comments on the manuscript, and particularly for an unpublished suggestion about nutrition. This paper has criticised some of David Lack's writings, but the thinking behind it owes much to his questioning of the grouse research. Some of the best scientific arguments I can remember were with him on the top of the Cairngorms. I wish to acknowledge his deep interest in my work and the stimulus I received from his writings and discussions.

References

Andrewartha, H. G. and Birch, L. C. (1954). *The Distribution and Abundance of Animals.* University of Chicago Press, Chicago, 782 pp.

Chitty, D. (1967). The natural selection of self-regulatory behaviour in animal populations. *Proc. ecol. Soc. Aust.,* **2**, 51 – 78.

Christian, J. J., Lloyd, J. A. and Davis, D. E. (1965). The role of endocrines in the self-regulation of mammalian populations. *Rec. Prog. Hormone Res.,* **21**, 501 – 571.

Hamilton, W. D. (1964). The genetical evolution of social behavior, I and II. *J. theor. Biol.,* **7**, 1 – 16, 17 – 51.

Henderson, B. A. (1976). The genetics of natural populations of red grouse (*L. lagopus scoticus* Lath.) at different densities. Ph. D. thesis, University of Aberdeen.

Huxley, J. (1934). A natural experiment on the territorial instinct. *Br. Birds,* **27**, 270 – 277.

Jenkins, D., Watson, A. and Miller, G. R. (1963). Population studies on red grouse, *Lagopus lagopus scoticus,* (Lath.) in north-east Scotland. *J. Anim. Ecol.,* **32**, 317 – 376.

Jenkins, D., Watson, A. and Picozzi, N. (1965). Red grouse chick survival in captivity and in the wild. *Trans. Congr. int. Un. Game Biol.,* **6**, 63 – 70.

Lack, D. (1954). *The Natural Regulation of Animal Numbers.* Clarendon Press, Oxford, 343 pp.

Lack D. (1966). *Population Studies of Birds.* Clarendon Press, Oxford, p. 341.

Lack, D. (1969). Population changes in the land birds of a small island. *J. Anim. Ecol.,* **38**, 211 – 218.

Moss, R. (1972). Food selection by red grouse (*Lagopus lagopus scoticus* (Lath.)) in relation to chemical composition. *J. Anim. Ecol.,* **41**, 411 – 428.

Moss, R. (1975). Different roles of nutrition in domestic and wild game birds and other animals. *Proc. Nutr. Soc.,* **34**, 95 – 100.

Moss, R., Watson, A. and Parr, R. (1975). Maternal nutrition and breeding success in red grouse (*Lagopus lagopus scoticus*). *J. Anim. Ecol.,* **44**, 233 – 244.

Myrberget, S. (1972). Fluctuations in a north Norwegian population of willow grouse. *Proc. int. orn. Congr.,* **15**, 107 – 120.

Pearson, O. P. (1966). The prey of carnivores during one cycle of mouse abundance. *J. Anim. Ecol.,* **35,** 217 – 233.

Savory, C. J. (1974). The feeding ecology of red grouse in N.E. Scotland. Ph.D. thesis, University of Aberdeen.

Watson, A. (1967). Social status and population regulation in the red grouse (*Lagopus lagopus scoticus*). *Proc. R. Soc. Pop. Study Grp,* **2,** 22 – 30.

Watson, A. (1970). Territorial and reproductive behaviour of red grouse. *J. Reprod. Fert., Suppl.,* **11,** 3 – 14.

Watson, A. and Jenkins, D. (1968). Experiments on population control by territorial behaviour in red grouse. *J. Anim. Ecol.,* **37,** 595 – 614.

Watson, A. and Miller, G. R. (1971). Territory size and aggression in a fluctuating red grouse population. *J. Anim. Ecol.,* **40,** 367 – 383.

Watson, A., Miller, G. R. and Green, F. H. W. (1966). Winter browning of heather (*Calluna vulgaris*) and other moorland plants. *Trans. bot. Soc. Edinb.,* **40,** 195 – 203.

Watson, A. and Moss, R. (1970). Dominance, spacing behaviour and aggression in relation to population limitation in vertebrates, in *Animal Populations in relation to their Food Resources* (ed. A. Watson). Blackwell Scientific Publications, Oxford and Edinburgh, pp. 167 – 220.

Watson, A. and Moss, R. (1972). A current model of population dynamics in red grouse. *Proc. int. orn. Congr.,* **15,** 134 – 149.

Wynne-Edwards, V. C. (1962). *Animal Dispersion in Relation to Social Behaviour.* Oliver and Boyd, Edinburgh and London, 653 pp.

4 Natural selection and the regulation of density in cyclic and non-cyclic populations

Professor D. Chitty*

What determines population density? Animal ecologists debated this question in the postwar years with as much fervour as bishops at an Ecumenical Council, and with as little thought of settling their differences experimentally. As one of the heretics, I was in schism with the orthodox doctrine of density dependence to which David Lack adhered (Nicholson, 1933; Smith, 1935; Lack 1954a). One of Lack's contributions was putting this doctrine into an evolutionary context; it is therefore ironical that certain ideas he disapproved of (Chitty, 1952; Lack 1954b) should, after a long metamorphosis, have led to my outdoing him in attempting to tie together natural selection and the regulation of numbers (Chitty, 1967a; 1970). However gross the errors of this still-untested view, they are at least in a direction Lack approved of.

A passionate devotion to one's own point of view is probably essential to a scientist's need to ignore, explain away, or otherwise cope with difficulties that would stop normal people from even getting started. On the other hand, without colleagues who are critical, but not too unkind and discouraging, most of us would go on believing the wrong ideas we generally start off with. Some go on doing so anyway, especially population ecologists, who have no easy way of settling disputes by appeal to prediction. For example, it is now more than 25 years since Lack and I began disagreeing, yet no evidence so far collected is good enough to have banished either of our views. Therefore, since I have no new revelations to pass along, I shall restrict the present chapter to describing these differences of opinion, and how they arose.

Lack's views about the balance of numbers may be summarised in the following two excerpts. The first is from Lack (1954a, p. 276): 'The factors most likely to cause a higher death rate at higher population densities, and so to produce this balance, are food shortage, predation or disease, as these can act more severely

*Professor Dennis Chitty graduated in honours biology at the University of Toronto in 1935 and joined the Bureau of Animal Population, Oxford, as assistant to Mr Charles Elton. Elton's long-term studies of cycles in voles *Microtus agrestis* introduced Chitty to a problem which he worked on until 1939, and again from 1946 until 1961; during the Second World War, like other members of the Bureau, he was part of a team which worked on the control of rats, mice and rabbits. He is now a Professor of Zoology at the University of British Columbia, where, after a break of 11 years, he once again went back to research, still hoping to find out what regulates the numbers of *Microtus*. He is still hoping.

when numbers are higher. Food shortage appears to be the chief natural factor limiting the number of many birds, of various carnivorous and herbivorous mammals . . .' The second is from Lack (1966, p. 291): ' . . . the absence of field evidence does not, and will not, make the advocates of density-dependent regulation change their minds. This is . . . because, given certain assumptions about the persistence of natural populations, the existence of density-dependent regulation becomes a logical necessity.'

To those working on cyclic populations, however, it had become clear by 1939 that the regulation of numbers of voles *Microtus agrestis,* in Great Britain (Elton, 1942; Chitty, 1952) or snowshoe hares *Lepus americanus* in North America (Green, Larson and Bell, 1939) was not due to changes in the severity of the usual mortality factors. Christian (1950) also reviewed the evidence against existing explanations, and substituted his stress hypothesis. Thus the conventional wisdom, applied to two intensively studied species, had been found wanting.

Faced with new and awkward evidence, a scientific community can (*a*) ignore or dismiss it, (*b*) accept it, but deny its relevance to existing theories, (*c*) reinterpret it to leave existing theories intact, or (*d*) use it to modify or reject existing theories. In the apparent clash between general population theory and the evidence about cycles, Lack opted for (*c*) − reinterpreting the awkward evidence (Lack, 1954*a*; *b*), I opted for (*d*) − modifying existing theories (Chitty, 1952; 1954; 1955), neither of us opted for (*b*) − accepting different solutions for cyclic and non-cyclic populations. So, in another tribute to human ingenuity, each used the same data to support an opposite point of view; our respective analyses of the report of the Committee of Inquiry on Grouse Disease (1911) persuaded Lack that regulation in Red grouse *Lagopus scoticus* was due to starvation and parasitism (Lack, 1954*a*), and persuaded me that it was due to changes in physiological condition (Chitty, 1954). My conclusion from this review was as follows: 'We should now perhaps regard infectious disease as only one of a complex group of biotic and physical factors any one of which may be the actual cause of death in physiologically deranged animals. The individual contribution of these numerous factors to the death rate may therefore be entirely dependent on the accidents of local conditions. The problem then becomes of more general interest, i.e. to determine what change in the population makes its members more vulnerable to normal hazards.'

This interpretation was consistent with the current evidence for voles, snow-shoe hares and grouse, but clashed with an unstated assumption vital to the whole doctrine of density dependence: the assumption that quality remains the same regardless of changes in population density. If this assumption were false, as I claimed, the action of weather was unlikely to be independent of density; thus partitioning the environment into density-dependent and density-independent factors was unjustified. The same claim was also made by Andrewartha and Birch (1954), who held that populations are as a rule unregulated, a proposition I did not go along with.

That one should not neglect qualitative differences between populations had already been shown by Leslie and Ranson (1940), who suggested that a simple change in age structure might increase the death rate, and so bring about a decline in the number of voles. Although this suggestion did not work out (Chitty, 1952), other changes in properties did seem to account for the differences between peak and declining populations. The changes were apparently congenital, yet occurred too fast to seem compatible with selection. So, with selection unlikely, maternal stress was the best alternative. The advantage of this hypothesis was that it was readily testable; its disadvantage was that, year after year, it kept giving encouraging leads, all but one of which (Dawson, 1956) turned out to be will-o'-the-wisps. For example, stressed voles whose lactation is affected produce underweight young (Chitty 1955); yet wild young are no lighter than usual in the winter before a decline (Chitty and Chitty, 1962). Finally, after 'shock disease' in snowshoe hares also turned out to have been an artifact (Chitty, 1959), the way was prepared for asking a new kind of question, namely, what is the selective advantage of a change in quality? As Lack (1954*b*) had already pointed out, the hypothesis that the change was purely pathological was hard to reconcile with natural selection.

At this turning point (Chitty, 1958) I had the benefit of my wife's evidence that adrenal weight — and by inference stress — was no greater in declining or regulated populations that in expanding or unregulated populations (Chitty, 1961). The maternal-stress hypothesis had thus failed its crucial tests both in the field and in the laboratory; but thanks to a transfusion given by Christian and Davis (1964) it is taking a long time to die.

I am also indebted to Janet Newson for her work on splenic hypertrophy (Dawson 1956; Newson and Chitty, 1962), which in various unexpected ways set the stage for the discovery, reviewed by Krebs and Myers (1974), that large and abrupt changes in selection pressure are indeed associated with changes in numbers of *Microtus* species. Although the observed changes in gene frequency do not tell us whether selection is a necessary condition for regulation, or merely one of its effects, they at least confirm the revised prediction that genetic changes are associated with the regulation of numbers.

The rival model that gradually emerged from these events includes these assumptions: in harsh habitats, or any being recolonised, selection favours individuals with the highest intrinsic rates of increase; but in good habitats selection with the highest intrinsic rates of increase; but in good habitats selection favours those whose behaviour forces their neighbours to emigrate, or reduces their chances of surviving and reproducing (Chitty, 1967*a*; 1970). According to this view, territorial behaviour in birds has the same function as mutual intolerance in voles and spacing-out behaviour in other organisms; and selection for these forms of behaviour has the incidental effect of preventing unlimited increase in numbers.

As this model had worn conventional selectionist dress since 1958, it escaped the righteous indignation heaped on the more dangerously heretical ideas of those denying regulation (Andrewartha and Birch, 1954) or affirming group selection (Wynne-Edwards, 1962) — see Lack (1966). At this time, however, Lack was not prepared to change his views about the function of territorial behaviour (Lack, 1966, Chitty, 1967*b*); only later (Lack, 1969) did he acknowledge that in two species of birds the upper limits to numbers were set by this means — a small but promising advance!

Evidence of selection being associated with the regulation of typically stationary populations has not yet shown up, which is by no means surprising, as such populations deny the observer any chance of seeing what an unregulated population looks like. He is thus unable to take the fundamental step of comparing the phenomenon with its control and so eliminate irrelevant variables. Belief in the interrelationships of natural selection and regulation has nevertheless become more popular since MacArthur and Wilson (1967) introduced the idea of *r*- and *K*-selection, which has much in common with the model presented above.

The difficulty one ecologist finds in trying to change another ecologist's mind emphasises the need for experimental methods. Contrary to general belief, the essence of experimenting is in testing an hypothesis; it does not necessarily involve taking a problem into the laboratory or 'messing-about with nature' (Medawar, 1957). And the essence of testing is in arranging for prediction to be confronted by observation, which satisfies the criterion that explanations in science must have falsifiable consequences (Popper, 1959). A purely descriptive approach, which is what most population studies consist of, merely provides correlations *a posteriori* — a deficiency that may explain why most people are able to support their own hypotheses, and why the resulting controversies seem interminable.

The controversy described here need never have started, of course. Why, then, did neither Lack nor I take option (*b*) mentioned above of accepting the evidence for cyclic populations without expecting it to fit non-cyclic populations as well? I cannot speak for Lack (who may have changed his mind later), but assume his reasons were the same as mine, namely, that the more varied the facts you try to explain, the better your chances of correcting yourself, or — more rarely — of unifying your discipline (Lack, 1954*a*, chapter 1). The chance of being wrong is thus an occupational hazard; but as Darwin put it: 'False views, if supported by some evidence, do little harm, for every one takes a salutary pleasure in proving their falseness; and when this is done one path towards error is closed and the road to truth is often at the same time opened' (Darwin, 1874). This form of salutary pleasure is one that David Lack and I seldom denied ourselves.

To summarise the main points of our controversy: in Lack's view (1954*a*), animal populations are regulated by density-dependent biotic factors, that is those acting more severely at high than at low densities. In my own view (Chitty, 1952; 1954; 1955), the action of any factor, including weather, depends not only on the severity of the factor itself, but also on the animals' susceptibility, which is likely to vary with population growth. The distinction between density-dependent and density-independent factors is thus unjustified. Both Lack (1954*a*) and I (Chitty, 1955) assumed that the process of regulation is the same in cyclic and non-cyclic species, an assumption that meant reinterpreting the evidence about cycles or modifying current ideas about density dependence. Lack took the first approach, I took the second. The resulting controversy has not yet been resolved, largely because ecologists have trouble testing their views experimentally. The two sets of ideas have nevertheless converged, thanks to evidence that qualitative changes in vole populations are at least partly genetic, and that population density in birds is at least sometimes determined by territorial behaviour.

References

Andrewartha, H. G. and Birch, L. C. (1954). The Distribution and Abundance of Animals. University of Chicago Press, Chicago.

Chitty, D. (1952). Mortality among voles (*Microtus agrestis*) at Lake Vyrnwy, Montgomeryshire in 1936 – 9. *Phil. Trans. R. Soc. B.,* **236**, 505 – 552.

Chitty, D. (1954). Tuberculosis among wild voles: with a discussion of other pathological conditions among certain mammals and birds. *Ecology,* **35**, 227 – 237.

Chitty, D. (1955). Adverse effects of population density upon the viability of later generations. *The Numbers of Man and Animals,* (ed. J. B. Cragg and N. W. Pirie). Oliver and Boyd, London, pp. 57 – 67.

Chitty, D. (1958). Self-regulation of numbers through changes in viability. *Cold Spring Harb. Symp. quant. Biol.,* **22**, 277 – 280.

Chitty, D. (1959). A note on shock disease. *Ecology,* **40**, 728 – 731.

Chitty, D. (1967*a*). The natural selection of self-regulatory behaviour in animal populations. *Proc. ecol. Soc. Aust.,* **2**, 51 – 78.

Chitty, D. (1967*b*). What regulates bird populations? *Ecology,* **48**, 698 – 701.

Chitty, D. (1970). Variation and population density. *Symp. Zool. Soc. Lond.,* **26**, 327 – 333.

Chitty, H. (1961). Variations in the weight of the adrenal glands of the field vole, *Microtus agrestis. J. Endocr.,* **22**, 387 – 393.

Chitty, H. and Chitty, D. (1962). Body weight in relation to population phase in *Microtus agrestis. Symp. Theriologicum, Brno,* 1960, Czechoslovak Academy of Sciences 77 – 86.

Christian, N. J. (1950). The adreno-pituitary system and population cycles in mammals. *J. Mammal.,* **31**, 247 – 259.

Christian, J. J. and Davis, D. E. (1964). Endocrines, behavior, and population. *Science,* **146**, 1550 – 1560.

Committee of Inquiry on Grouse Disease. (1911). The grouse in health and in disease: being the final report of the Committee of Inquiry on Grouse Disease. Smith, Elder and Co., London, 2 volumes.

Darwin, C. (1874). *The Descent of Man, and Selection in Relation to Sex* (2nd. edn). J. Murray, London, 1890.

Dawson, J. (1956). Splenic hypertrophy in voles. *Nature, Lond.,* **178**, 1183 – 1184.

Elton, C. (1942). *Voles, Mice and Lemmings: Problems in Population Dynamics.* Oxford University Press, Oxford.

Green, R. G., Larson, C. L. and Bell, J. F. (1939). Shock disease as the cause of the periodical decimation of the snowshoe hare. *Am. J. Hyg.,* **30B**, 83 – 102.

Krebs, J. C. and Myers, J. H. (1974). Population cycles in small mammals. *Adv. Ecol. Res.,* **8**, 267 – 399.

Lack, D. (1954*a*). *The Natural Regulation of Animal Numbers.* Clarendon Press, Oxford.

Lack, D. (1954*b*). Cyclic mortality. *J. Wildl. Mgmt.,* **18** 25 – 37.

Lack, D. (1966). *Population Studies of Birds.* Clarendon Press, Oxford.

Lack, D. (1969). Population changes in the land birds of a small island. *J. Anim. Ecol.,* **38**, 211 – 218.

Leslie, P. H. and Ranson, R. M. (1940). The mortality, fertility and rate of natural increase of the vole (*Microtus agrestis*) as observed in the laboratory. *J. Anim. Ecol.,* **9**, 27 – 52.

MacArthur, R. H. and Wilson, E. O. (1967). *Theory of Island Biogeography.* Princeton University Press, Princeton, New Jersey.

Medawar, P. B. (1957). *The Uniqueness of the Individual.* Methuen, London.

Newson, J. and Chitty, D. (1962). Haemoglobin levels, growth and survival In two *Microtus* populations. *Ecology,* 43, 733 - 738.

Nicholson, A. J. (1933). The balance of animal populations. *J. Anim. Ecol.,* 2, 132 - 178.

Popper, K. R. (1959). *The Logic of Scientific Discovery.* Hutchinson, London.

Smith, H. S. (1935). The rôle of biotic factors in the determination of population densities. *J. Econ. Ent.,* 28, 873 - 898.

Wynne-Edwards, V. C. (1962). *Animal Dispersion in Relation to Social Behaviour.* Oliver and Boyd, Edinburgh.

5 Factors affecting population density in the wild rabbit, Oryctolagus cuniculus (L.), and their relevance to small mammals

Dr J. A. Gibb*

"The whole field of population regulation in nature, and theories about how it works, has got into a rather peculiar state where a number of strongly held views exist that are at first sight incompatible with each other" (Elton 1966, p. 380).

Introduction

Because many species of animals live in diverse situations, it is unwise to generalise about what determines their population density. Some ecologists argue and others deny that the restricted fluctuations in numbers observed in nature can be explained only by some form of density-dependent regulation (Nicholson, 1954; *cf.* Andrewartha and Birch, 1954; see also Ehrlich and Birch, 1967). Some argue that animal numbers are usually limited by intraspecific competition for food (Lack, 1954*a*); others that once animals possess a piece of ground 'they can do the actual food-getting in perfect peace and freedom, entirely without interference from rivals' (Wynne-Edwards, 1962, p. 12). Hairston, Smith and Slobodkin (1960) argued that because most green matter falls to the ground uneaten, the numbers of herbivores generally are not limited by food shortage; though in fact food may be limiting even when only a small fraction of the stock is eaten (Huffaker, 1966). Many believe that 'predation is centred upon over-produced young . . . upon what is identifiable as the more biologically expendable parts of the population' (Errington, 1963, p. 184); yet Pearson (1971) suggests that carnivore predation alone is 'responsible for the amplitude and timing of the microtine cycle'. Ehrlich and Birch (1967) expect regulatory factors to vary among populations of the same species and through time; while Chitty (1958), Watson and Moss (1970), and Krebs and Myers (1974) seek one 'universal explanation' for what prevents unlimited increase in small mammals. If there is any justification for these beliefs, they cannot be quite so incompatible as at first sight appears.

*Dr John A. Gibb was born in Puddletown, Dorset, and educated at Sherborne School and St Edmund Hall, Oxford, where he read law. He served in the Royal Artillery, 1940–46, and was on the staff of the Edward Grey Institute from 1946 to 1957, receiving a doctorate for his research on titmice. He was awarded the Bernard Tucker Medal in 1956 as organiser of the British Trust for Ornithology Nest Record Scheme. In 1957 he emigrated to New Zealand, where he developed interests in the population ecology of mammals, and was appointed Director of the Ecology Division, New Zealand Department of Scientific and Industrial Research, in 1965.

Rabbits can be seen at times to be limited by food shortage or disease (for example, myxomatosis), or by 'predation' by man as occurs in much of New Zealand today. The influence of carnivore predation remains obscure but is probably underestimated. Surveying rabbit populations in Australia, Ridpath (unpublished paper to the Australian Vermin Control Conference, Canberra, August 1973) obtained no concensus of opinion: most studies have been concerned either with the rabbits or with their predators, not with both.

Comparison of the factors affecting the numbers of rabbits with those affecting other small mammals is of interest because of the similarities between them, and because for decades the cyclic microtines have focused the conflicting opinions of ecologists (see Elton, 1942).

A ten-year study of a confined population of wild rabbits in New Zealand, described in detail by Gibb, Ward and Ward (in press), sheds some light on the roles of food shortage, predation and behaviour in determining their numbers. Here I discuss some of the salient points without repeating the supporting data.

Study Area and Methods

This study was undertaken from 1958 to 1967 on sheep pasture about 115 km east of Wellington, North Island. The enclosure (figure 5.1) of 8.5 ha, later reduced to 4.3 ha, was on a limestone fault scarp rising to 533 m above sea level. The area had supported a dense population of rabbits until they were

Figure 5.1 View of the study area looking north, January 1960.
The main warren was on the hill to the right, and a hide may be seen across the valley to the left.

reduced by poisoning in the late 1950s. There was almost no above-ground cover and no standing water. The mean annual rainfall of 1180 mm was well-distributed through the year and temperatures were moderate (mean air temerature 5.8°C in July, 15.8°C in January).

The enclosure was surrounded by a rabbit-proof fence, which was no obstacle to feral house cats (*Felis catus*) and ferrets (*Mustela putorius furo*), the main carnivores. In late 1963, the fence was heightened and electrified and an attempt was made to exclude the carnivores for the rest of the study.

No further attempt was made to influence the numbers of rabbits: no supplementary food was provided and hunting was forbidden. Sheep were grazed in the area when the rabbits left surplus feed, but for long periods at high density the rabbits alone grazed the pasture much closer than was acceptable for sheep.

The study area was visited at least monthly for 4–5 days at a time. At each visit the rabbits above ground were counted from hides at about hourly intervals from dawn to dusk on one day, and irregularly at night using binoculars and a spotlight. They were also mapped and their activities recorded under six headings (feeding, resting or sitting, toilet, social activities, on the move and digging). Since most rabbits are active above ground at dusk, the size of the population was estimated from the numbers counted then. Variable numbers of rabbits were trapped alive for examination and marking and were then released.

Results

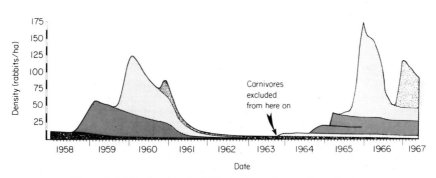

Figure 5.2 The density of rabbits in the study area from January 1958 to May 1967, together with the approximate age structure by cohorts. Feral cats and mustelids had free access in and out of the enclosure until late 1963, but were mostly excluded from then on.

Figure 5.2 shows two peaks in rabbit numbers — in the summers of 1959 – 60 and 1965 – 66; the first when carnivores had unrestricted access and the second after their exclusion. During the first two breeding seasons, numbers built up to a peak in the summer of 1959 – 60, when further increase was prevented by food shortage. The pasture was badly depleted with much bare ground, and this persisted through

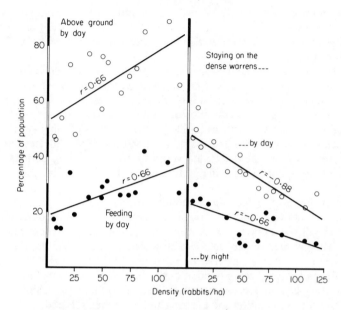

Figure 5.3 The steadily increasing amount of time that the rabbits spent above ground and feeding by day as their density increased; and the decreasing amount of time that they spent on the dense warrens (as opposed to more distant feeding grounds) by day and by night.

most of the following winter. Adult rabbits did not die of starvation then and young born early in the 1959 season put on weight and survived quite well; but young born later in the season grew much more slowly, were permanently undersized and survived poorly.

The reality of food shortage was more clearly evident from the rabbits' daily activities than from their weights or survival. Rabbits are mainly nocturnal and spend most of the day in warrens underground. Their working 'day' commences in late afternoon and most are active by dusk. They spend nearly all night above ground and almost as many are still up at dawn as at dusk. The numbers above ground then fall to a minimum in mid-morning, the main rest period. In contrast, when food was short during periods of high population density the rabbits spent much more of the day above ground and feeding, and much less time on the warrens because they had to forage further afield (figure 5.3); at night they resorted to distant communal feeding grounds that were otherwise little used, though not much more than 100 m away.

As the better-sited warrens became crowded there were fewer burrows per rabbit and more rabbits left to colonise less favourable sites (figures 5.4 and 5.5). The overcrowding of burrows at high population density may have impeded escape from predators.

There was ample rainfall in the spring and early summer of 1960, and the pasture as well as the rabbits remained in good condition. The growth of the

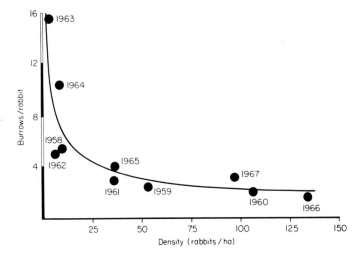

Figure 5.4 The rabbits extended the warrens as their density increased, but there were fewer burrows per rabbit as the warrens became more crowded.

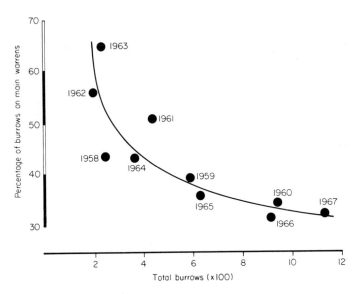

Figure 5.5 The study area contained one main warren and numerous subsidiary ones. The main warren contained a smaller proportion of all the burrows as their total numbers increased; in other words, more rabbits colonised less favourable ground.

young was not retarded as it had been in the previous season, yet they survived very poorly after leaving the nest and the total population was only slightly higher at the end of the breeding season in December 1960 than it had been at the start of it. Most of the young were being killed by predators.

The cats and ferrets took both young and full-grown rabbits, the cats hunting above ground and the ferrets in the burrows. The harrier *Circus approximans* was common, but concentrated on young rabbits of up to about 500 g and carrion. No cats and only one or two ferrets were seen in 1958. There were still only few in 1959, but many more were seen in 1960 and more again in 1961. By 1961 four or five cats and about six ferrets were more or less resident in the study area. Few rabbits were actually seen being killed, but fresh remains were common during the decline. The maximum rate of the rabbits' decline (roughly 100 animals per month in the autumn of 1961) was consistent with the food requirements and feeding habits of the carnivores (see Gibb, Ward and Ward, in press).

The long decline in rabbit numbers following the 1959 - 60 peak was interrupted only briefly in 1960 - 61 and continued through 1961, 1962 and most of 1963. Only one or two young rabbits were seen in the breeding seasons of 1961 - 62 and 1962 - 63; yet adult females palpated were heavily pregnant and nests were found. Cats were seen looking for nests and although they did not dig them out they evidently caught all the young soon after they left the nest. Unlike the cats, ferrets ate nestlings as well as older rabbits.

The population of rabbits peaked at about 1000 in early 1960. Three years later only 16 animals were left, of which at least 12 were males. It was one thing to demonstrate that predators ate many young and adult rabbits, but quite another to suggest that they so reduced the population that this prevented successful breeding in 1961, 1962 and most of 1963. We therefore removed the carnivores in late 1963, so as to compare the rabbits' survival without them with that of the preceding years. The results were spectacular.

It was not until late October 1963 that the fence was heightened and electrified and the carnivores removed. The population of rabbits was then down to 14, and one of three surviving females was actually seen being killed by a cat on November 7. Still no young had been seen and the 1963 season was following the same pattern as in 1961 and 1962. Then, two weeks after switching on the fence, a number of young appeared above ground and survived – for the first time since 1960. The population then doubled in what little remained of the 1963 - 64 season.

With carnivores still mostly excluded, rabbits of all ages survived exceptionally well. The population again doubled in 1964 - 65, and in the autumn of 1965 the rabbits bred again and redoubled their numbers. This was the only substantial autumn breeding experienced in the 10 years. The population stood at about 140 at the start of the 1965 - 66 breeding season, which began early and continued until density peaked at about 172 rabbits/ha in December. Starvation inevitably followed.

It was again mainly the young rabbits that died in the summer of 1965 - 66. Mortality was heavy until late winter despite a brief reprieve with the growth of new pasture in the autumn. The crisis came in winter when heavy grazing had depleted the pasture. First the younger rabbits and then the old ones lost weight and died of starvation. The body weights of some full-grown animals fell by

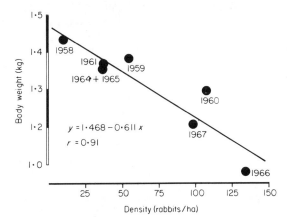

Figure 5.6 The mean body weights of first-year rabbits in autumn decreased as population density increased.

almost 50 per cent before they died, though others survived a 45 per cent loss; and males fared much worse than females. The relationship between the autumn body weights of first-year rabbits and population density is illustrated in figure 5.6.

Population density fell from about 150/ha in January 1966 to only about 50/ha in September. Conditions then improved rapidly as the pasture began to grow in the spring. The body weights of the rabbits were almost normal by November 1966, when the first young of the new season appeared above ground, very late; but numbers never regained their former level. The study ended in May 1967.

Discussion

Four fundamental questions may be asked of this study:

(1) Was the population regulated in a density-dependent way?
(2) Did the rabbits' social behaviour buffer them from the constraints of food shortage, predation and disease?
(3) What were the actual constraints on population density? In particular, what limited peak density and what prolonged the decline following the peak?
(4) How relevant was all this to similar problems among small mammals?

Before dealing with these questions, we must consider very briefly the possible effects of confinement on the population.

Effects of Confinement

The study area was not uniformly suitable for rabbits; parts were occupied only when the rabbits were numerous, and even then only sparsely despite the general food shortage. Fluctuations in population density were accompanied by marked changes in dispersion as must occur in unconfined populations. Although, obviously, some rabbits would have dispersed more widely had there been no fence, most young in fact settled within 100 m of their birthplace; thereafter they rarely travelled more than 100 m in any direction or shifted their home ranges by more than 20 m. The study area was thus much larger than that normally covered by one rabbit in its lifetime, and we suspect that most free-ranging rabbits are equally sedentary.

Although the population processes described here did not occur in a closed system, the changes in the numbers of rabbits and predators were dictated primarily by forces arising within the study area. It is important to appreciate that, due to the normal operations of the local Rabbit Board, the rabbits were always much scarcer outside than inside the study area.

Krebs (1971) described the fence-effect on *Microtus* populations, which increased unusually fast to densities three or four times those of unfenced populations, and over-grazed their habitat, probably because they could not disperse. Although changes in the numbers of our rabbits may have been exaggerated by their confinement, it is as well to remember that some unconfined populations of rabbits fluctuate markedly and over-graze their pasture. Within sensible limits, findings from enclosed populations need not be treated as 'complete artifacts' (*cf*. Chitty, 1967).

Density dependence

Ten years was too short a period from which to judge the reality of density-dependence; none the less, numerous facets of the rabbits' biology changed in ways conducive to density-dependent regulation (for details see Gibb, Ward and Ward, in press):

(1) During the first 5 years, before carnivores were removed, breeding began later and finished earlier the higher the initial density of rabbits; the percentage increase in numbers was thus reduced at high density.

(2) Mortality between breeding seasons increased with density.

(3) With increasing density the rabbits spent longer above ground and feeding, much less time resting, and more time dispersed from the warrens (figure 5.3); they also tended to ignore predators (and man) under the urgent pressure of feeding.

(4) With increasing density the numbers of burrows available per rabbit were reduced (figure 5.4); this may have facilitated predation.

(5) The proportion of the population occupying the main warren decreased with increasing density, as surplus rabbits colonised less favourable ground (figure 5.5).

(6) With increasing density, the amount and quality of pasture deteriorated.

(7) Also the size of the rabbits' territories contracted, and the rabbits were forced to use extra-territorial, communal feeding grounds remote from the safety of the warrens.

(8) With increasing density, first-year rabbits carried heavier loads of helminth parasites (Bull, 1964); and their body weights (figure 5.6) and survival were reduced.

(9) Until removed in 1963, the carnivores responded numerically and functionally to the increased density of rabbits; numbers frequenting the area increased, as did the time they spent in it and the amount of rabbit in their diet.

(10) Many such trends took place evenly over the whole span of rabbit densities (figures 5.3 and 5.6), not just or predominantly at peak density (figures 5.4 and 5.5). Thus density-dependent restrictions may begin to become effective at a very low density. (The term "carrying capacity" inappropriately suggests the imposition on numbers of a relatively 'hard' ceiling or threshold level, below which all is well but which cannot be exceeded.)

Role of social behaviour

A series of situations may be recognised in which the social behaviour of animals becomes increasingly important in limiting density, as follows:

(*a*) Numbers are determined directly by available resources or by the incidence of predation or disease, with reproduction and mortality distributed evenly through the population. Resources may become depleted to the detriment of the long-term survival of the species. Examples of such a regime may be confined to the simplest of organisms.

(*b*) As in (*a*), but with social behaviour distributing reproduction and mortality unevenly through the population without materially affecting overall rates. Such populations may still deplete their resources.

(*c*) Numbers determined primarily by available resources or by the incidence of density-dependent predation or disease but with social behaviour keeping numbers just below the maximum supportable by available resources. Dominant animals may benefit from appropriating more than their 'fair' share of resources at the expense of subordinates. Numbers are thus mildly buffered from environmental extremes by the animal's social behaviour. This regime may provide improved long-term stability and be appropriate for animals depending on quite predictable resources.

(*d*) Further development of (*c*) might enable an animal's social behaviour to keep numbers well below the resource limit for most of the time, yet with density still ultimately limited by available resources, predation or disease. In this case an elite minority may corner a major share of the resources, leaving others to die. This regime may be the product of natural selection if the extra energy expended by the elite in cornering most of the resources occasionally pays off in enhanced survival.

(*e*) The final extreme envisages a population of animals whose social behaviour keeps numbers permanently below the resource limit in order to avoid shortages of any kind and to ensure the long-term survival of the species even at the expense of short-term gains. The existence of such a regime in nature is difficult to accept, for there is no known way in which selection may work for the long-term survival of the species unless it also improves the short-term survival of individuals or their families.

We place the rabbits we studied between points (*b*) and (*c*) in this series. Their well-developed social behaviour certainly distributed reproduction and mortality unevenly through the population, but did not exempt them from food shortage

and predation. In effect, social behaviour seemed to determine which rabbits
survived and bred, rather than how many. If dispersal is much more important in
unconfined populations than we observed, then the rabbit may approach point
(*c*). Reduced reproduction and heavy mortality are forced on rabbits at high
density, not sought or accepted for the sake of long-term survival. 'What I have
seen of Nature's way . . . is the ruthless way, little resembling any mysteriously
benign process of falling birth rates' (Errington, 1957).

Placing the rabbit between points (*b*) and (*c*) is appropriate for an animal
exploiting an uncertain food supply. The rabbit contrasts with the Brown hare
Lepus europaeus, in this respect. In New Zealand, as in Australia and Europe,
rabbits and hares commonly live sympatrically in similar though recognisably
distinct habitats. They are very different animals ecologically. Hares browse shrubs
and tall herbaceous vegetation as well as the short pasture favoured by rabbits. The
food supply of hares is rather predictable for most of the year and may reasonably be measured by the standing crop of vegetation, whereas that of rabbits is uncertain even from week to week and is better measured by the growth rate of the
vegetation. Correspondingly, the rabbit is an opportunist breeder while the hare
keeps rigidly to a set breeding season (Flux, 1965). Rabbits are more gregarious
than hares, they reach much higher densities and their numbers fluctuate more
directly with their food supply. Hares avoid heavy predation by being alert and
by the generally low density permitted by their social behaviour; while rabbits
'accept' occasional heavy predation, relying on rapid reproductive rates and
synchronised breeding to compensate for it and to swamp the predators. Thus
we may expect to place the hare between points (*c*) and (*d*) in the series.

Food shortage, predation and disease

Food shortage and predation were both important in determining the numbers
of our rabbits, disease apparently much less so – there being no myxomatosis in
New Zealand. Coccidiosis may have killed many young rabbits at high density, but
not enough to prevent later starvation among the survivors. Bull (1964) found
heavier loads of helminth parasites in first-year rabbits at high density than at
low density in our population. Heavy parasite loads may aggravate the effects of
food shortage by increasing food requirements. In a very real sense food shortage
invites predation and disease at high density. We saw no overt signs of disease, but
it was not necessarily unimportant. In Australia, where myxomatosis is well
established, food shortage, predation and disease have all been implicated in the
declines in rabbit numbers (Myers, 1971), and latent myxomatosis may become
lethal in stressed animals.

The effects of food shortage at peak density in 1960 and 1966 were very
obvious in our study. The pasture became badly depleted; the growth, body
weights and survival of young were severely affected; and with few if any
carnivores present in 1966, many older as well as first-year rabbits died of
starvation. The carnivores may have largely anticipated the death of rabbits
from starvation in 1960.

Food shortage drastically altered the behaviour of rabbits in ways which
probably rendered them more liable to predation. Few carnivores were seen in
1958 or 1959, partly because there were not many rabbits to attract them but
perhaps also because many carnivores had been killed in the immediately preceding years by poisons laid for rabbits; carnivores are susceptible to secondary

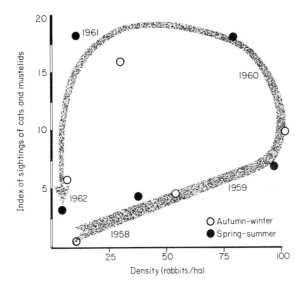

Figure 5.7   As the density of rabbits increased through 1958 and 1959, so did the number of sightings of feral cats and ferrets; but the number of carnivore sightings continued to increase throughout 1960–61 while the rabbits crashed. Many fewer carnivores were seen by 1962. This circular predator–prey relationship resembled the theoretical model of Lotka and Volterra. (See Gibb, Ward and Ward, in press for details of the carnivore index.)

poisoning from eating rabbits killed with sodium monofluoroacetate (compound '1080'). It was not until 1961, long after the rabbits' peak, that cats and ferrets were most often seen. The delayed response of carnivores to increased rabbit density produced a Lotka–Volterra type of oscillation (figure 5.7). Their response was both numerical and functional: the number of different cats and ferrets frequenting the study area increased, as did the time they spent inside instead of outside the enclosure and the amount of rabbit in their diet. Once carnivores gained the ascendancy, they reduced the rabbits far below the level at which disease and/or resources were limiting – a situation which persisted for a long time. Indeed, with only two female rabbits left in late 1963, the population was near extinction when we removed the carnivores.

We do not believe that the population would have continued to oscillate if left undisturbed, though it might have fluctuated more than surrounding populations now strictly controlled by man. With rabbit populations at generally low densities, Pest Destruction Boards favour night shooting instead of poisoning for routine control, and this does not kill the carnivores. We believe (Gibb, Ward and Ward, 1969) that many formerly dense populations of rabbits are now naturally controlled at low density by carnivores for most of the time. None the less control by carnivores is uncertain and may be disrupted by the chance deaths of one or two local cats. However, this is largely speculation and there is much need for research.

Relevance to population studies of other small mammals

Extensive research has been devoted to populations of small mammals, especially the cyclic microtines (Elton, 1942). Much of it has concentrated on what prevents unlimited increase (see Krebs and Myers, 1974). Rather less research has been done on what accounts for the prolonged decline after peak density, which Chitty (1967) called their 'most puzzling feature', and less still on the causes of synchronous breeding among widely separated populations. Our study of rabbits has particular bearing on the first two of these problems.

Some authors (for example, Lack, 1954*b*; Pitelka, 1964; Schultz, 1964) have been impressed with the effects of voles and lemmings on the vegetation at peak density, others (for example, Maher, 1970) with the impact of predation on them, whereas Chitty (in many papers), Krebs and Myers (1974) and others have searched far and wide to explain observed population changes.

Like Ehrlich and Birch (1967), we expect limiting factors to vary in different situations. Yet, in trying to explain what prevents unlimited increase, we sympathise with Krebs and Myers's (1974) aversion for multiple-factor hypotheses and with their search for 'a universal explanation' (see also Chitty, 1958; Watson and Moss, 1970). Krebs's single-factor outlook is clear: 'If you believe that food shortage is the important resource involved, you must view these genetic experiments as meaningless . . . if you believe that predation is the driving force, you would view food shortage experiments as meaningless' (Krebs, 1971, p. 254).

Chitty, Watson and Moss, and Krebs and Myers all rejected starvation, predation and disease as factors universally limiting density because no one of them was reported in every case. But if even quite mild food shortage affects the animals' behaviour or condition in such a way as to invite intensified predation or disease, then there is no need to look for one common mortality factor. For at high density it is as if there is a race between food shortage, predation and disease, to reduce the population. In this race food shortage comes hand in hand with predation and disease, so not surprisingly sometimes one, sometimes another, seems almost by chance to win. In effect we adopt one hypothesis allowing a choice of mortality factors.

The long decline of our rabbits following peak density resembled that 'most puzzling feature' of cyclic microtines. Lack (1954*b*) suggested that the delayed recovery of the vegetation might explain the long decline; but the decline of the rabbits continued long after the vegetation recovered. A long decline is to be · expected, however, in prey populations subjected to a buildup of predators that lags behind and eventually overshoots them.

The rabbits' decline from predation by cats and ferrets recalls that of Pearson's voles (1964; 1966; 1971) eaten by house cats in California. In both cases alternative foods allowed some carnivores to survive and continue killing their main prey long after the latter were so reduced that they alone could not support the carnivores. A choice of food for predators is often thought to buffer main prey species from excessive predation; but in Pearson's case and ours it exaggerated and prolonged their decline.

Both these studies, and Fitzgerald's (in press) on weasel predation on voles in California with its implications also for lemmings (MacLean, Fitzgerald and Pitelka, 1974), all emphasise predation as a dominant force in the population dynamics of small herbivores. In addition, we have shown how food shortage drastically affected the rabbits' habits, predisposing them to predation at high density; and we have suggested that social behaviour determined which rabbits survived rather than how many. Other animals may behave similarly. We found nothing specially puzzling in the long decline of the rabbits following peak density, no need to look for one single limiting factor common to all situations, and no need to invoke self-limitation of numbers by behavioural mechanisms.

Acknowledgements

It is a pleasure to acknowledge the help of C.P. and G.D. Ward, who collaborated in the original study, and all those mentioned in our earlier paper; but especially my colleagues Drs B.M. Fitzgerald and J.E.C. Flux for numerous discussions about predation and other things to do with this paper, and Henrik Moller of Auckland University, for constructive comments on the manuscript. I am also grateful to Dr M. G. Ridpath, of the CSIRO Division of Wildlife Research, for letting me quote his paper to the Australian Vermin Control Conference in 1973.

References

Andrewartha, H. G. and Birch, L. C. (1954). *'The Distribution and Abundance of Animals'.* University of Chicago Press, Chicago, 782 pp.

Bull, P. C. (1964). Ecology of helminth parasites of the wild rabbit, *Oryctolagus cuniculus* (L.) in New Zealand. *N. Z. Dep. Sci. industr. Res. Bull.* **158**; 147.

Chitty, D. (1958). Self-regulation of numbers through changes in viability, in *'Population Studies: Animal Ecology and Demography'. Cold Spring Harb. Symp. quant. Biol.* **22**; 277 - 280.

Chitty, D. (1967). The natural selection of self-regulatory behaviour in animal populations. *Proc. ecol. Soc. Aust.* **2**, 51 - 78.

Ehrlich, P. R. and Birch, L. C. (1967). The 'balance of nature' and 'population control'. *Am. Nat.,* **101**, 97 - 107.

Elton, C. S. (1942). *'Voles, Mice and Lemmings'.* Clarendon Press, Oxford, 496 pp.

Elton, C. S. (1966). *'The Pattern of Animal Communities'.* Methuen, London, 432 pp.

Errington, P. L. (1957). *'Of Men and Marshes'.* Macmillan, New York, 150 pp.

Errington, P. L. (1963). The phenomenon of predation. *Am. Scient.* **51**, 180-192.

Fitzgerald, B. M. (in press). Weasel predation on a cyclic population of the montane vole *(Microtus montanus)* in California. *J. Anim. Ecol.*

Flux, J. E. C. (1965). Timing of the breeding season in the hare, *Lepus europaeus* Pallas, and rabbit, *Oryctolagus cuniculus* (L.). *Mammalia,* **29**, 557 - 562.

Gibb, J. A., Ward, G. D. and Ward, C. P. (1969). An experiment in the control of a sparse population of wild rabbits (*Oryctolagus cuniculus* (L.)) in New Zealand. *N. Z. Jl Sci.,* **12**, 509 - 534.

Gibb, J. A., Ward, C. P. and Ward, G. D. (in press). The Kourarau enclosure, 1958 - 1967. Effects of food shortage, predation and behaviour on a population of wild rabbits, *Oryctolagus cuniculus* (L.). *N. Z. Dep. Sci. industr. Res. Bull.*

Hairston, H. G., Smith, F. E. and Slobodkin, L. B. (1960). Community structure, population control, and competition. *Am. Nat.,* **94**, 421 - 425.

Huffaker, C. B. (1966). Competition for food by a phytophagous mite: the roles of dispersion and superimposed density-independent mortality. *Hilgardia* **37**, 533 - 567.

Krebs, C. J. (1971). Genetic and behavioural studies on fluctuating vole populations, in *'Dynamics of Populations'.* (ed. P. J. den Boer and G. R. Gradwell) *Proc. Adv. Study Inst. Dynamics numbers Popul. (Oosterbeek, 1970)* pp. 243 - 256.

Krebs, C. J. and Myers, J. H. (1974). Population cycles in small mammals. *Adv. Ecol. Res.,* **8**, 267 - 399.

Lack, D. (1954a). *'The Natural Regulation of Animal Numbers'.* Clarendon Press, Oxford, 280 pp.

Lack, D. (1954b). Cyclic mortality. *J. Wildl. Mgmt,* **18**, 25 - 37.

MacLean, S. F., Fitzgerald, B. M. and Pitelka, F. A. (1974). Population cycles in arctic lemmings: winter reproduction and predation by weasels. *Arctic Alpine Res.,* **6**, 1 - 12.

Maher, W. J. (1970). The pomarine jaeger as a brown lemming predator in northern Alaska. *Wilson Bull.,* **82**, 130 - 157.

Myers, K. (1971). The rabbit in Australia, in *'Dynamics of Populations'.* (ed. P. J. den Boer and G. R. Gradwell) *Proc. Adv. Study Inst. Dynamics numbers Popul. (Oosterbeek, 1970)* pp 478 - 506.

Nicholson, A. J. (1954). An outline of the dynamics of animal populations. *Aust. J. Zool,* **2**, 9 - 65.

Pearson, O. P. (1964). Carnivore-mouse predation: an example of its intensity and bioenergetics. *J. Mammal.,* **45**, 177 - 188.

Pearson, O. P. (1966). The prey of carnivores during one cycle of mouse abundance. *J. Anim. Ecol.,* **35**, 217 - 233.

Pearson, O. P. (1971). Additional measurements of the impact of carnivores on California voles (*Microtus californicus*). *J. Mammal.,* **52**, 41 - 49.

Pitelka, F. A. (1964). The nutrient-recovery hypothesis for arctic microtine cycles. I. Introduction, in *'Grazing in Terrestrial and Marine Environments'.* (ed. D. J. Crisp) *Symp. Br. ecol. Soc.* **4**, 55 - 56.

Schultz, A. M. (1964). The nutrient-recovery hypothesis for arctic microtine cycles. 2. Ecosystem variables in relation to arctic microtine cycles, in *'Grazing in Terrestrial and Marine Environments'.* (ed. D. J. Crisp) *Symp. Br. ecol. Soc.,* **4**, 57 - 68.

Watson, A. and Moss, R. (1970). Dominance, spacing behaviour and aggression in relation to population limitation in vertebrates. in *'Animal Populations in Relation to Their Food Resources'.* (ed. A. Watson) *Symp. Br. ecol. Soc.,* **10**, 167 - 220.

Wynne-Edwards, V. C. (1962). *'Animal Dispersion in Relation to Social Behaviour'.* Oliver and Boyd, Edinburgh, 653 pp.

6 Song and territory in the Great tit Parus major

Dr J. R. Krebs*

Introduction

It is common knowledge that bird song has one or both of two functions: territorial advertisement and mate attraction (Thorpe, 1961; Armstrong, 1973). However, beyond the simple correlation between the seasonal peak of singing and territorial or reproductive activity, there is remarkably little direct evidence for either proposed function of song. Here I am concerned with the role of song in territorial behaviour in the Great tit (*Parus major*). The spring peak of singing coincides with the establishment of a territory, which occurs after pairing in the Great tit (Hinde, 1952; Krebs, 1971), perhaps indicating that mate attraction is less important than territorial advertisement in this particular species. The question I attempt to answer is 'Does song act as a signal to keep intruders out of a territory?'. I also discuss how the organisation of song might relate to territorial exclusion.

Many studies have shown that territorial males will attack, approach or sing at a loudspeaker 'singing' in their territories, but I know of only two studies that have considered the question of whether singing by a resident bird acts as a 'Keep out' signal to deter intruders. The advantage to a resident of having such a signal is obvious: as long as the intruder retreats on hearing the signal, a bird can defend its territory without having to continuously patrol the boundaries. Peek (1972), in an excellent series of field experiments, found that if he muted male Red-winged blackbirds *Agelaius phoeniceus* during early spring, by severing the syringeal muscles, they lost their territories. This experiment, which has been questioned recently by Smith (1976), does not specifically isolate the effect of *song* as a 'Keep out' signal, but does show clearly that vocalisations as a whole have an essential role in successful territorial defence. Göransson, Högstedt, Karlsson, Källander and Ulfstrand (1974) report an experiment on Thrush nightingales *Luscinia luscinia* in which territorial occupation of parts of a wood was simulated by playing song continuously over loudspeakers. This is the most direct approach to testing the territorial function of song, and it is the one I adopt in the present study. The results of Göransson *et al.* seemed to show some effect of song in deterring intruders, but the results are reported in such little detail that no firm conclusions can be drawn.

*After completing a degree in zoology at Oxford John Krebs joined the Animal Behaviour Research Group for postgraduate work on the role of territory in the population ecology of Great tits in Wytham Wood. He then spent a year as Demonstrator in the Edward Grey Institute and three years as an Assistant Professor at the University of British Columbia, Vancouver, Canada, where he worked mainly on herons. He is now back at Oxford, in the Edward Grey Institute studying bird song and foraging behaviour.

Loudspeaker Replacement Experiments

In previous work, I found that when territorial pairs were removed from an optimal habitat (mixed deciduous woodland) in early spring, they were rapidly replaced by new birds coming in from surrounding farmland, a suboptimal habitat (Krebs, 1971). The rapid reoccupation of territories showed that competition for good territories was severe. In the present experiment, I used these results as the basis for an experiment to test the role of song in the maintenance of a territory. In early spring, I removed all the territory holders (eight mated pairs) from a piece of mixed woodland. Three of them were replaced with loudspeakers reproducing the songs of the particular birds I had removed, three were replaced with control loudspeakers making an irrelevant noise of similar volume and time pattern to the songs and the remaining territories were left empty. If song acts to deter intruders, one would predict that the two types of control territories should be reoccupied by invaders more rapidly than the experimental song territories. Any difference between the two treatments is a reflection of the effectiveness of song alone as a territorial maintenance signal.

Methods

The work was done in Higgin's Copse, an outlying area of mixed deciduous woodland of about 6 ha, on Wytham Estate near Oxford (figure 6.1). The wood is surrounded by agricultural land containing hedges and trees which are occupied by Great tits*, but is known to be suboptimal in terms of reproduction (Krebs,

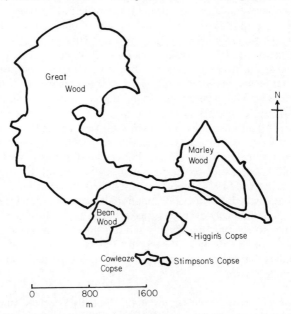

Figure 6.1 Wytham Woods, showing Higgin's Copse

*In the eighteenth century a field at one edge of Higgin's Copse was appropriately named Great Titts (sic) Close.

1971). The nearest large pieces of woodland are Bean Wood and Marley Wood, both about 600 m distant, which have vegetation similar to Higgin's Copse (described in Krebs, 1971 and Perrins, 1965, respectively).

In January 1975, I caught (at feeding stations) and colour ringed all the resident birds. Probably as a result of the mild winter, all the birds had paired and settled on territories by mid-January. Territory boundaries were plotted by accumulating a large number of observations of singing, feeding, disputes, and so on, and subsequently plotting them on a large-scale map using 50 m grid posts as reference points. In the Great tit, the area utilised for feeding is similar in size to the defended area, except for occasional forays out of the usual territory into other parts of the wood, so that the territorial boundaries shown in the figures were obtained by encircling all observations of a particular pair except for a few outlying points. At the end of January, there were eight territories in the wood (figure 6.2). Having plotted the boundaries, I recorded the songs of each resident male, using a Uher 4000 IC tape recorder and Beyer M101NC (with a Grampian 24 inch parabola) or

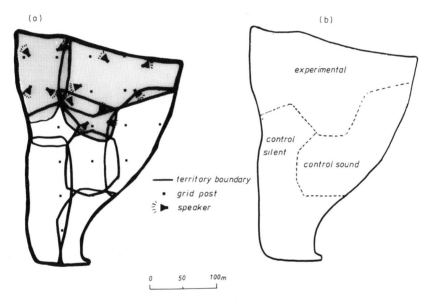

Figure 6.2 (a) Territory boundaries before the first removal experiment (February 1975). Experimental territories are stippled, and approximate positions of loudspeakers in these territories are indicated. (b) The areas designated as control silent, control sound, and experimental

Sennheiser MKH 815 microphone. Each male Great tit has a repertoire of two to seven song types, and I recorded the complete repertoire of each bird. On each of 5 days before the start of the experiment I carried out standard patrols to establish the baseline levels of activity in the three parts of the wood designated as control speaker (two territories), control silent (three territories) and experimental (three territories) (figure 6.2(b)). A standard patrol consisted of a 1 h circuit of the wood, stopping at each of 12 places for 5 min and moving rapidly between stops. The circuit was so arranged that all parts of the wood were given equal attention and the circuits, four each day, were carried out at all hours between dawn and dusk. These pre-experimental observations showed that the difference in numbers of observations of birds in the future control and experimental areas was correlated with the size of the areas (table 6.1). From this I concluded that all comparisons

Table 6.1 Comparison of size of experimental and control areas and relative frequency of observations of birds in February 1975, before the start of the experiment.

	Ratio of areas	Ratio of number of observations
Control silent	0.61	0.63
Control speaker	0.71	0.69
Experimental	1.00	1.00

of levels of activity between areas should be corrected for the relative size of the areas. Once corrected in this way, the pre-experimental levels of singing, boundary disputes, and total observations were similar in the three areas (figure 6.3). This suggests that there were no important differences between the three areas regarding their attractiveness to intruders.

On the morning of February 18, 1975 I removed all the territorial residents, eight males and eight females, by mist netting. The broadcasts began at 1400,

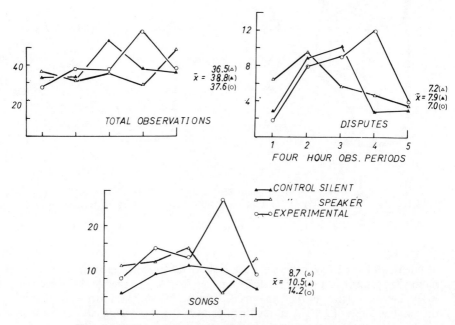

Figure 6.3 Corrected levels of activity per 4 h in 20 h of observation on five days preceding the experiment I.

continued throughout February 19, and were switched off at 2400 on February 20, when the wood was fully reoccupied. The loudspeakers were arranged as follows. Each experimental territory had a tape recorder (Uher 4000 IC) with a continuous loop tape playing at 19 cm/s containing the repertoire of the removed owner. The tape recorder was linked through a multiway switch and a Sinclair Z40 15 W amplifier to four Midax 750 metal horns, each of which was 50 m from the tape recorder (in the centre of the territory) and positioned near the boundary (figure 6.2(a)) 2 – 3 m off the ground. The tapes did not play con-tinuously, but each recorder was switched on for 8 min (one complete playing of the loop) each hour from dawn till dusk. This corresponds roughly to the normal level of singing of a territorial bird at that time of year. During the 8 min of play-back, each of the four loudspeakers was active, in turn, for 2 min. The three experimental birds had repertoire sizes of three, four, and five song types, and the loops contained an equal amount of each song type in the repertoire (2 min of each song for the four-song bird, 2.67 min of each for the three-song bird, and so on). The songs were broadcast at roughly normal volume, and each territory was 'active' in sequence.

The two control territories in which sound was broadcast were equipped with the same loudspeaker arrangement as the experimental territories, and were operated in a similar manner. The control sound was a repetitive two-note phrase played on a tin whistle. The control silent territories were simply left empty.

After starting the broadcasts, I monitored the patterns of reoccupation of the wood by continuously retracing the 1 hour circuit of the wood used in the pre-experimental period, during all hours of daylight. This arrangement ensured that

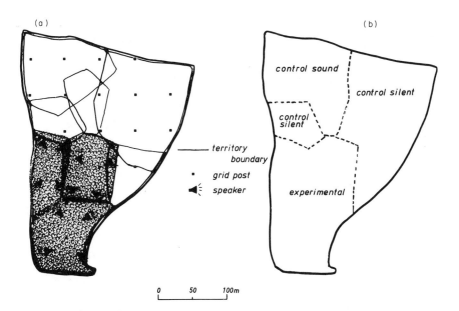

Figure 6.4 (*a*) Positions of experimental territories in the second removal experiment (March 1975). (*b*) The three treatment areas

each part of the wood was visited regularly, so that the detailed pattern of reoccupation could be observed hour by hour. During this continuous observation, I recorded all sightings or auditory locations of birds and the type of activity involved. In calculating rates of activity in the three areas, I used corrected values as mentioned earlier. I was able to plot the territory boundaries of the new settlers more or less hour by hour, since some of the new males were already colour ringed, and the others could be identified by their songs.

After the first experiment, the wood was eventually fully reoccupied with eight pairs, so that it was possible to do a second experiment three weeks later. The design was similar: a pre-experimental observation period; removal of all the birds on March 10; broadcasts started at 0930 on March 10 and continuing until 1230 on March 12, when the wood was fully reoccupied. Again, three territories were designated as experimental, three as control silent, and two as control speaker. The positions of the three areas in the wood were interchanged (figure 6.4).

Results

Experiment 1

Figure 6.5 shows the pattern of reoccupation of the wood after removal. The critical conclusion is that, although both areas were occupied fully after about 2 days of loudspeaker playback, the control areas were both occupied much more rapidly than the experimental, with no difference between the control areas. The criterion for territorial occupation by an invader was to observe a male regularly singing loudly, and/or chasing other birds, as opposed to skulking quietly in the undergrowth. The control areas (which are treated as one in the subsequent discussion) were both completely occupied within 8 h of daylight of the start of the experiment, whereas the experimental area was not fully reoccupied until about 20 h of daylight (2 days) after the start. At the time of year of the experiment there were about 9.5 h of daylight per day. A further difference between experimental and control areas in the reoccupation pattern was that the experimental area tended to be occupied by gradual encroachment from the edges (by birds H, C, and F in figure 6.5) whereas the control area was occupied simultaneously by several individuals.

Figure 6.6 gives a more quantitative picture of the pattern of resettlement. It shows the cumulative number of observations, singing bouts and territorial boundary disputes. All three measures show a steeper rise in the control areas. There were already an appreciable number of singing observations in the control area within 6 h of starting the experiment, which indicates how rapidly new birds established themselves in the empty areas. Figure 6.7 shows these same data expressed as rates of activity per 4 h.

Although one might suppose that the eventual occupation of the experimental area followed after the new birds had simply habituated to the loudspeakers, my observations show that this was not the whole story. At the beginning of the experiment, when an intruder started to sing near one of the loudspeakers, and the loudspeaker replied, the intruder would immediately stop singing and retreat quietly (the loudspeakers were operating on a fixed time schedule and so did not

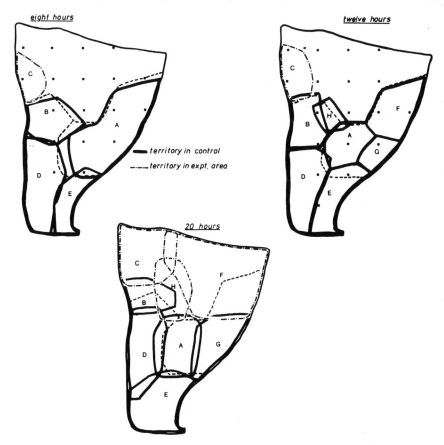

Figure 6.5 The position of reoccupation of the control and experimental areas after the first removal. The letters A – H indicate the order in which the birds settled, and the times are daylight hours after the start of the experiment (about 9.5 h of daylight per day). The broken lines indicate the boundaries of the three areas in figure 6.2 (*b*).

always reply). During the later part of the experiment, the behaviour of intruders to the loudspeakers gradually changed: instead of retreating, they countersang loudly against the speakers, at first from the control area, and later from inside the experimental territories. At this point the intruders were showing the same sort of reaction to the loudspeakers as would be shown by a territorial resident; only later still did the new residents show signs of totally ignoring the speakers. The fact that intruders initially sang in response to the loudspeakers only from within the control areas tends of course to increase the number of song observations in these areas shown in figure 6.6. As figure 6.5 shows, eventually eight pairs replaced the eight that had been removed.

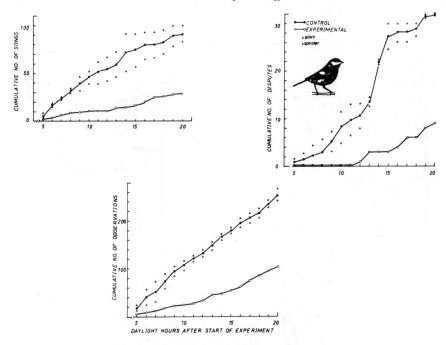

Figure 6.6   Cumulative number of observations, song bouts and
territorial disputes in the control and experimental areas after the
first experiment. Note that observations did not start until 5 h
after the loudspeakers were turned on

Experiment II

Figure 6.8 shows that as in experiment I, the two control areas were occupied
before the experimental area, the former being occupied in 10 h, the latter after
30 h of daylight. In this experiment only four pairs replaced the eight removed,
with one hedgerow bird (figure 6.8 D) occupying a small corner of the wood. Ten
days after the experiment, a further pair established themselves, so that in May
1975, six pairs attempted to breed (figure 6.8). As in experiment I, the experimental area was occupied by gradual encroachment (figure 6.8, pairs C and E).
Figures 6.9 and 6.10 show the cumulative activity and rates of activity in the
experimental and control areas, showing results similar to those of experiment I.

Origin of new birds

In previous work (Krebs, 1971) I concluded from removal experiments that
some birds (mainly young ones) are excluded from mixed woodland by territorial
behaviour and that these individuals all breed in suboptimal habitats. The results

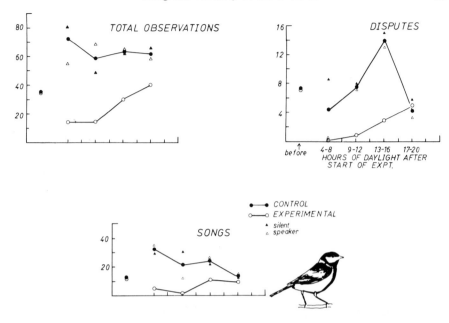

Figure 6.7 The same data expressed as rates of activity per 4 h

of this study apparently do not confirm the second conclusion, but support the first. In experiment I, at least some of the eight replacement pairs were known to have come from hedgerow territories and, in contrast to earlier results, these hedgerow territories were rapidly refilled by birds of unknown origin, but presumably non-territorial. In experiment II, only one replacement pair was known to have come from hedgerows, but some of the others had been seen irregularly as 'floaters' in the wood. As far as I could tell, they did not belong to any territories outside the wood, so I conclude that they must have been non-territorial surplus birds, although it is possible that they would eventually have established territories elsewhere before the onset of egg laying in April. They characteristically foraged quietly and immediately withdrew if chased by a resident. The fact that they

Table 6.2   Age of terriotrial males in Higgin's Copse at different stages of the season

	Adult	First year	Unknown
Before first removal experiment (February 1975)	8	0	0
Before second removal experiment (March 1975)	3	5	0
After second removal experiment (April 1975)	1	4	1

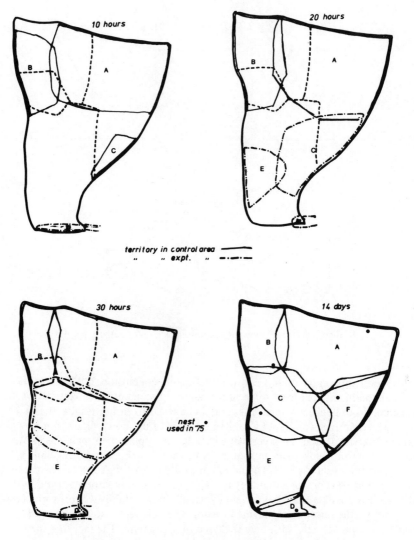

Figure 6.8 Pattern of reoccupation after the second removal (details
as for figure 6.5)

were paired by the time they established territories shows either that pairs can be
formed within a few hours by the surplus birds of both sexes, or that they were
paired previously. Table 6.2 shows that the replacement males tended to be
first-year birds.

Release of territorial birds

After the second experiment, I released five of the territorial pairs that had
been removed on March 10. One pair was released by accident 4 miles from the

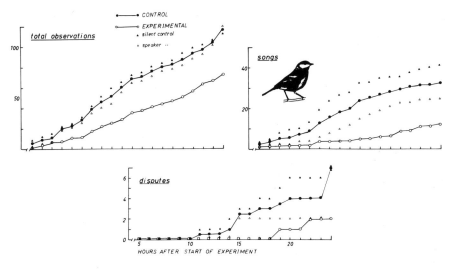

Figure 6.9 Cumulative activity scores after second removal

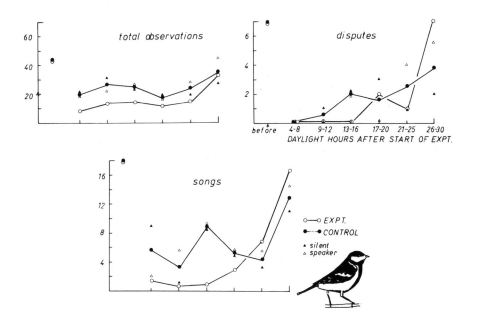

Figure 6.10 Data in figure 6.9 expressed as rates of activity per 4 h

wood, in Oxford on March 19, and on March 21 the other four pairs were released in their old territories. None of these birds succeeded in re-establishing themselves in the wood, although two pairs persisted in trying to oust the new residents for some hours, and a further pair for several days. Two of the ten birds (one male and one female) were subsequently found in different parts of the surrounding hedgerow, about 200 m from the wood. The results contrast with those of Harris (1970). He found that two out of fourteen pairs of Oystercatchers *Haemotopus ostralegus* successfully ousted new residents when they were released after three weeks in captivity, and that several other pairs re-established themselves later. This raises the interesting question of how long a resident has to be off its territory for a reversal of dominance to occur.

Breeding success of replacement birds

Since the final occupants of the wood were, so to speak, the reserve replacements, it would have been interesting to see how they fared at breeding. Unfortunately I could not make a reliable comparison between the breeding success of these birds and those in other parts of Wytham Wood, because several nests were destroyed by human predators early in the breeding season.

Discussion

The loudspeaker replacement experiments show that song has a role in territory maintenance of the Great tit, or to put it another way, that potential settlers use song to assess the occupancy of an area. Not surprisingly the speakers have only a short-term deterrent effect, both because they are a suboptimal song stimulus (for example, they do not respond immediately to an intruder) and because song is only one of many methods of territorial advertisement. Göransson *et al.* (1974) claimed to have successfully deterred intruding Thrush nightingales over several weeks, although they give no detailed data. In view of my results it is surprising that they observed such a long-lasting effect. Whitney and Krebs (1975) carried out an experiment of similar design on Pacific tree frogs, and were able to show that continuous playback of male mating calls in an enclosure discouraged other males from establishing calling territories.

The reoccupation of empty territories by new pairs after the removal experiments confirms the conclusion that the breeding density in optimal habitats in the Great tit is limited by territorial behaviour (Krebs, 1971). Even after two successive removal experiments the breeding density was not much lower than the starting density of territorial pairs before the removals, and although some of the replacement birds came from suboptimal habitats it seems as though there were also some pairs which would have failed to establish territories at all — a true non-breeding surplus.

A male Great tit, in common with most other territorial song birds, has more than one form of its characteristic song. Each male has a repertoire (usually) of between two and seven song types, some of which may be shared by other males nearby (Gompertz, 1961 and personal observation). All the song types in a repertoire are used at the same time of year and in the same contexts: a male usually sings one song type for a minute or two before switching to another song type and so on. It appears that the repertoire is redundant. The problem of why birds have song repertoires has been discussed by many authors (see Armstrong, 1973; Bertram, 1970; Hartshorne, 1973) and I have suggested a way in which repertoires might increase the effectiveness of song as a territory maintenance signal (Krebs, 1976, 1977). Whatever the mechanism, repertoires must increase fitness if they are the product of natural selection, and I will briefly describe two attempts to correlate repertoire size with fitness.

One possibility is that repertoires are sexually selected. Great tits are mono-gamous and have a roughly equal sex ratio, so for a male to benefit as a result of being sexually selected by females, he must obtain a female that is ready to breed early and/or is capable of laying a large clutch (Fisher, 1958). Howard (1974) found that in a sample of ten Mockingbirds *Mimus polyglottus*, males with larger repertoires mated earlier than those with smaller repertoires, but unfor-tunately age was not measured and could have been a confounding variable. As mentioned earlier, the Great tit pairs before the spring peak of singing, which makes it unlikely that repertoires are to do with sexual selection, but neverthe-less I tested the idea as follows. In May 1975 I measured the song repertoire sizes of 31 males in Marley Wood, where the breeding biology of the Great tit has been studied for 28 yr, so that the effects of age on breeding date and clutch size are well known. The repertoires were recorded by stimulating birds to sing with playback near the nest. I visited each territory until no new songs were recorded during a visit. The prediction is that males with larger repertoires should breed earlier and/or have larger clutches and broods. The results are shown in table 6.3: there are no statistically significant trends in breeding date, clutch size

Table 6.3 Repertoire size, laying date and breeding success, Marley Wood 1975 (means ± s.e.)

	Repertoire size				
	1	2	3	4	5
N	1	13	13	2	2
Date of egg laying*	30.0	30.7 ± 0.65	30.9 ± 1.13	33.5 ± 3.5 _____ 35.0	36.5 ± 3.5
Clutch size	7.0	7.9 ± 0.27	7.8 ± 0.36	8.0 ± 0.0 _____ 7.25	6.5 ± 0.5
No. of young fledged	2	4.31 ± 1.0	5.77 ± 0.81	3.0 ± 3.0 _____ 4.75	6.5 ± 0.71
Percentage adult ♂	100	54.5	60	50	
Percentage adult ♀	100	54.5	80	33	

(*April 1)

or brood size correlated with repertoire size. If anything, females mated to males with larger repertoires breed slightly later than those mated to males with small repertoires, but clutch size and number of young fledged do not show any consis-tent trends. Female age is known to influence both clutch size and breeding date in Marley Wood, first year females lay about 2 days later and their clutches are 0.2 eggs smaller than those of older females (Moss, 1972). However, as table 6.3 shows, the age distribution of females does not significantly vary with repertoire size of the males, so these correction factors do not alter the results.*

The second measure of fitness I examined was territory size. Within Marley Wood, territory size is negatively correlated with predation risk during nesting (Krebs, 1971), so that securing a large territory would be advantageous. Terri-tory size in Great tits is influenced by habitat quality (Krebs, 1971), male age (Dhondt, 1971) and pressure from neighbours (Krebs, 1971), and these factors

*The results of a similar study in 1976 confirm the conclusions of table 6.3.

have to be considered before dealing with repertoire size. Within Higgin's Copse
the density of trees is fairly uniform, although species composition varies slightly
from one part to another (some areas are predominantly ash, some elm, others
hazel and maple), so it is unlikely that habitat quality has any major influence on
variations in territory size. Table 6.2 shows that in February and March 1975
there were eight pairs in the wood (the first eight were removed in mid-February
and the second eight were replacements). The first eight males were all adults,
and in the second group the mean territory size for first-year birds did not differ
from that for adults (0.68 ha against 0.8 ha), so that in this particular study there
is no measurable effect of age on territory size. The effect of neighbour pressure
is shown in figure 6.11: birds at the edge of the wood have some boundaries
of their territories not shared with neighbours, and these birds tend to have
larger territories. This effect is of course quite likely to be an artifact of my small
isolated study area, but nevertheless it has to be taken into account. Figure 6.12
shows the deviations in territory size from the regression of figure 6.11 plotted as
a function of repertoire size. After correcting for the effect of neighbour pressure,
there is a significant positive correlation between repertoire size and territory
size. Since I observed that repertoire size did not change during territory establishment I conclude that larger repertoires enable males to obtain larger territories.
The question of the underlying mechanism and the problem of why birds do not
evolve larger and larger repertoires are discussed elsewhere (Krebs, 1977).

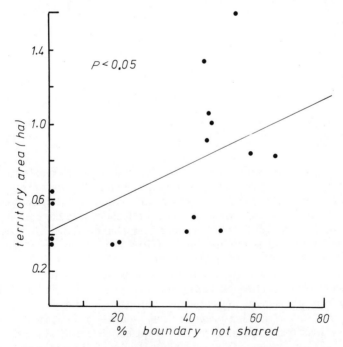

Figure 6.11 The relationship between territory size and proportion
of boundary that is not shared with a neighbour (see text for
explanation).

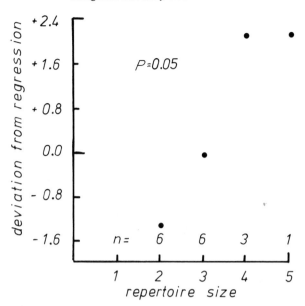

Figure 6.12 The relationship between deviations in territory size from the regression in figure 6.11 and repertoire size.

Acknowledgements

I thank Shelly Cohen for her help with the field work. The research was financed by the Science Research Council. Mike Webber and Alex Kacelnik kindly criticised the manuscript.

References

Armstrong, E. A. (1973) *A Study of Bird Song*. Dover, London.

Bertram, B. (1970). The vocal behaviour of the Indian Hill Mynah *Gracula religiosa*. *Anim. Behav. Monogr.* **3**, 81 – 192.

Dhondt, A. A. (1971). Some factors influencing territory in the Great Tit, *Parus major* L. *Le Gerfaut,* **61**, 125 – 135.

Fisher, R. A. (1958). *The Genetical Theory of Natural Selection*. Dover, London.

Gompertz, T. (1961). The vocabulary of the Great Tit. *Br. Birds,* **54**, 359 – 394, 409 – 418.

Göransson, G., Högstedt, G., Karlsson, J., Källander, H. and Ulfstrand, S. (1974). Sångensroll för revirhållandet hos näktergal *Luscinia luscinia* några experiment med playbackteknik. *Vår Fågelvärld,* **33**, 201 – 209.

Harris, M. P. (1970). Territory limiting the size of the breeding population of the Oystercatcher (*Haematopus ostralegus*) – a removal experiment. *J. Anim. Ecol.,* **39**, 707 – 713.

Hartshorne, C. (1973). *Born to Sing: an Interpretation and World Survey of Bird Song*. Indiana University Press, Bloomington, Indiana.

Hinde, R. A. (1952). The behaviour of the Great Tit (*Parus major*) and some other related species. *Behaviour Suppl.*, 2, 1 - 201.

Howard, R. D. (1974). The influence of sexual selection and interspecific competition on mockingbird song (*Mimus polyglottos*). *Evolution,* 28, 428 - 438.

Krebs, J. R. (1971). Territory and breeding density in the Great Tit, *Parus major* L. *Ecology*, 52, 2 - 22.

Krebs, J. R. (1976). Habituation and song repertoires in the great tit. *Behav. Ecol. Sociobiol.,* 1, 215 - 227.

Krebs, J. R. (1977). The significance of song repertoires: the Beau Geste hypothesis. *Anim. Behav.,* 25, 475 - 478.

Moss, D. (1972). A statistical analysis of clutch size in the Great Tit, *Parus major* L. MSc Thesis, Oxford.

Peek, F. W. (1972). An experimental study of the territorial function of vocal and visual display in the male Red-winged Blackbird (*Agelaius phoenicius*). *Anim. Behav.,* 20, 112 - 118.

Perrins, C. M. (1965). Population fluctuations and clutch size in the Great tit, *Parus major* L. *J. Anim. Ecol.,* 34, 601 - 647.

Smith, D. G. (1976). An experimental analysis of the function of red-winged blackbird song. *Behaviour,* 56, 136 - 156.

Thorpe, W. H. (1961). *Bird Song.* Cambridge University Press, London.

Whitney, C. L. and Krebs, J. R. (1975). Spacing and calling in the Pacific Tree Frog, *Hyla regilla. Can. J. Zool.,* 53, 1519 - 1527.

Section 2 Feeding adaptations and ecological segregation

The four chapters in this section refer to feeding adaptations of birds and mammals; three relate to ways in which animals partition the feeding habitat between them and the fourth concerns the breeding implications of courtship feeding. First, Dr M. P. Harris, a former research assistant of Lack's, examines the seabirds of the Galapagos Islands to see how 18 of the 19 breeding species divide up the available food supply between them. In chapter 8 Dr Bryan Nelson studies the different patterns of feeding adaptation to be found within the single avian order Pelecaniformes, relating feeding patterns to breeding strategies and showing a number of points of evolutionary convergence among the families. Birds as a group have rather simple methods of catching, manipulating and consuming food. In consequence the size of food item which a species can process efficiently is rigorously determined, and different species often show clear ecological separation in the size of food items preferred. In chapter 9 Dr Rosenzweig points out, based on field studies of heteromyid rodents in the deserts of southwestern USA, that rodents are not subject to the same limitation, and do not divide the feeding habitat among themselves in the same way as Darwin's finches; their speciation is clearly based on different mechanisms. In chapter 10 Dr Nisbet takes up an early speculation by David Lack concerning the possible role of courtship feeding as a determinant of clutch size in birds. His study of Common terns in the eastern USA suggests strongly that females rely on food supplied by their partners to reach full breeding potential; this confirms his earlier view that the number and size of eggs in a clutch depends partly at least on the foraging efficiency of the mate.

7 Comparative ecology of seabirds in the Galapagos Archipelago

Dr M. P. Harris*

Introduction

If the numbers of birds are regulated by availability of food, as the evidence suggests, species which live together in the same area must have evolved means of reducing interspecific competition for food. They might feed in different places, at different times, in different ways, or on different prey (Huxley, 1942); any two species may differ in more than one of these ways. How these separations are brought about may be difficult to determine because they may take effect only when food is short, the species overlapping in almost all respects when food is plentiful.

Most studies on seabirds have been made during the breeding season, when they are easiest to observe. In some tropical areas, for example the Galapagos Islands, many species either breed throughout the year or have peaks of nesting at different times each year. Those with non-annual breeding probably breed whenever the female can get enough food to produce an egg (Harris, 1969a). The output of young per breeding cycle is always lower in these areas than in temperate regions. These facts suggest that there is no predictable 'best' time for breeding, and super-abundances of food, if they occur, are as likely at one season as another. Competition is thus likely to be intense for much or all of the year and inter-specific differences more obvious.

This study attempts to show how the various seabirds nesting in the Galapagos Islands partition food resources, a task set me by David Lack when he was re-sponsible for starting my studies on these birds. In many ways the situation is similar to that described by Ashmole and Ashmole (1967) for Christmas Island (Central Pacific) but there are important differences. Whereas they had detailed information on the food of only 8 of the 17 species of seabirds nesting on a relatively small coral atoll, I have data for 18 of the 19 species which breed in Galapagos (table 7.1). Also, whereas birds on Christmas Island had to nest on

*Dr M. P. Harris was a member of the Edward Grey Institute from 1963 to 1972. During this time he started the Institute's long-term seabird studies on Skokholm Island and also spent four years at the Charles Darwin Research Station in the Galapagos Islands studying many of the resident seabirds. He is now at the Institute of Terrestrial Ecology working on factors influencing Puffin populations.

Table 7.1 Feeding ecology of Galapagos seabirds

Species	Origin	Main feeding methods	Observed distribution at sea	Average incubation stint (h)	Feeds per day	No. of colonies	Food brought to the young
Galapagos penguin *Spheniscus mendiculus*	H	Pursuit–diving	Inshore	c 48	?	Many	Medium-sized, mid-water and one reef fish
Waved albatross *Diomedea irrorata*	?	Surface–seizing	Pelagic	112–528	0.3–0.5	1	Squid (mean weight 150 g) flying and other fish, crustacea
Audubon's shearwater *Puffinus lherminieri*	T	Surface–seizing, pursuit–diving	Inshore	50–140	0.7	29+	Planktonic larval fish (to 7 mm), minute crustacea, few minute cephalopoda
Hawaiian petrel *Pterodroma phaeopygia*	T	?	Pelagic	240–312	0.5	5	Medium squid and fish
Madeiran storm petrel *Oceanodroma castro*	T	Dipping	Pelagic	144	0.7	15	Fish 30–50 mm long, very small cephalopoda
Galapagos storm petrel *Oceanodroma tethys*	H	Dipping	Pelagic/between islands	120	c 0.5	3	Fish (mean length 28 mm) few cephalopoda and crustacea
Elliot's storm petrel *Oceanites gracilis*	H	Pattering	Inshore	?	?	?	Very small fish and crustacea
Red-billed tropicbird *Phaethon aethereus*	T	Deep–plunging	Pelagic	>144	c 1.0	30	Flying and other fish, squid to 200 mm
Brown pelican *Pelecanus occidentalis*	T	Surface–plunging	Inshore	?	?	Many	Fish from 40 to 350 mm
Blue-footed booby *Sula nebouxii*	T	Deep–plunging	Inshore	18–25	1.8	35+	Medium fish, maximum length 250 mm
Masked booby *Sula dactylatra*	T	Deep–plunging	Between islands	27	1.4	23	Medium fish, sometimes squid

Species	Origin	Main feeding methods	Observed distribution at sea	Average incubation stint (h)	Feeds per day	No. of colonies	Food brought to the young
Red-footed booby *Sula sula*	T	Deep – plunging	Pelagic	60	0.9	5	Medium fish, sometimes squid
Flightless cormorant *Nannopterum harrisi*	?	Pursuit – diving; bottom feeding	Inshore	c7	c1 – 2	112+	Bottom living fish, eels, octopuses
Great frigatebird *Fregata minor*	T	Dipping; cleptoparasitism	Pelagic	240	0.5 – 0.8	12	Flying and other fish, sometimes squid
Magnificent frigatebird *Fregata magnificens*	T	Dipping; cleptoparasitism	Inshore/between islands	70	2.2	12	Fish, including reef species, offal
Swallow-tailed gull *Creagrus furcatus*	?	Surface – plunging	Between islands	<12	1 – 2	55+	Flying and other fish, squid to 200 mm (rarely larger)
Lava gull *Larus fuliginosus*	?	Scavenging	Coastal	?	?	Solitary nester	Mostly scavenges, few small fish
Brown noddy *Anous stolidus*	T	Dipping	Inshore	?	?	Many	Small fish, usually 15 – 20 mm long, rarely 50 mm
Sooty tern *Sterna fuscata*	T	Dipping	Pelagic	?	?	1	No data, elsewhere fish and squid up to 80 mm

(1) Feeding methods are as defined by Ashmole (1971) – *dipping* is picking food from the surface of the sea without landing and without using the feet; *pattering* is similar except that the feet are used; *surface – plunging* and *deep – plunging* are where the bird dives into the water from a height and is either incompletely or completely immersed; birds swimming on the surface can either seize food directly (*surface – seizing*) or dive and chase it (*pursuit – diving*). Cleptoparasitism is the chasing of other species to make them regurgitate or drop any food they are carrying.

(2) *Inshore* refers to birds feeding within a mile of the coast, *inter-island* or *offshore* to species feeding among the islands without being inshore, *pelagic* to birds feeding well away from the islands.

(3) Small fish and squid are up to 50 mm long, medium 50 – 150 mm, large above 150 mm.
(4) Probable origin – T indicates tropical or subtropical, H cool water *via* Humboldt Current, ? of doubtful source.
(5) Details from Nelson (1968, 1970), Boersma, de Vries and Gordillo (personal communications), and my published and unpublished data.

the single island in uniform sea conditions, or not breed at all, those in Galapagos can choose a site near the food source, and in either a tropical marine (sea surface temperature of 25 °C) or a temperate regime (15 – 20°C). This is because different parts of the archipelago are influenced by first, the cold and productive waters of the Humboldt Current, which predominate to the south and east of the islands, secondly, the tropical unproductive water to the north and west, and thirdly, the Cromwell Current which runs eastwards below the westward flowing surface waters and upwells at the westernmost islands.

Associated with the variable marine environment, birds from a range of habitats breed in Galapagos; twelve species are typical of tropical and/or sub-tropical avifaunas, three are associated with the Humboldt Current and four are of doubtful affinities. At unpredictable intervals, traditionally about every 7 years the Humboldt Current 'fails' and warm waters flow south through Galapagos seas to Peru. This results in a high mortality among seabirds in Peru although not Galapagos seabirds, perhaps because Galapagos species are better adapted to a varying environment, or have access to a wider range of food outside the local current system.

Ten of the fifteen Galapagos species studied in detail suffered periodic breeding failures resulting from decrease in size and frequency of feeds brought to the young, lengthening of incubation stints, deaths of young, non-laying by potential breeders and desertions of colonies. Most breeding failures were probably due to failures in the food supply (Nelson, 1969; Harris, 1969b), although massive egg losses by the Waved albatross (scientific names of Galapagos species are given in table 7.1) and Galapagos (= Wedge-rumped) storm petrel are unexplained (Harris, 1973).

Although they appear to be breeding as quickly as possible (Lack, 1954), seabirds in general, and tropical species in particular (table 7.2), recruit their young into the nesting population at a low rate and have a low associated mortality. Goodman (1974) has shown that in the Red-footed booby of Galapagos the result is a damping of population fluctuations. Only rarely will the population be far below the 'carrying capacity' of the environment. This perhaps explains why few migrant seabirds winter in the area for there would seldom be room for extra birds. Only the Red-necked (= Northern) phalarope *Lobipes lobatus* occurs in numbers and this takes smaller food than any resident seabird. Twenty-seven other species have been recorded at various times, so the paucity of immigrant species is not due to their inability to get there.

Comparative Feeding Ecology: Food Preferences

On Galapagos each species had its own individual food spectrum, as was also found for eight species on Christmas Island by Ashmole and Ashmole. There was, however, a major difference in that in the Central Pacific squid formed a major part of the diet of all species, and only two species of terns took less squid than fish. In Galapagos, squid were recorded from only 10 of the 18 species: medium sized species were important for Waved albatrosses, Swallow-tailed gulls and Red-billed tropicbirds, and very small species for Storm petrels. Hawaiian (= Dark-rumped) petrels ate squid of intermediate size, as did Red-footed boobies and Great frigatebirds on Tower Island in 1964 (Nelson, personal communication).

Table 7.2 Breeding and survival in Galapagos seabirds

Species	Intervals between breeding of individuals (yr)	Clutch size	Nesting success young per pair per breeding cycle	Age of first breeding (yr)	Annual adult survival (%)
Galapagos penguin	< 1	2	low	?	High
Waved albatross	1	1	0.16 – 0.20	4	95
Audubon's shearwater	< 1	1	0.26	?	92 – 95
Hawaiian petrel	1	1	0.06	?	?
Madeiran storm petrel	1	1	0.30	?	?
Galapagos storm petrel	1	1	< 0.23	?	?
Red-billed tropicbird	≤ 1	1	0.32 – 0.55	5	78
Brown pelican	< 1	2 – 3	? *c* 1.0	?	?
Blue-footed booby	< 1	2 – 3	0.45	3	?
Masked booby	1	2	0.63	3	83
Red-footed booby	⩾ 1	1	0.08	?	?
Great frigatebird	1 or 2	1	*c*0.15	?	?
Magnificent frigatebird	⩾ 1	1	*c*0.50	?	?
Flightless cormorant	< 1	3	0.40	2	90
Swallow-tailed gull	< 1	1	0.33	4	97
Brown noddy	⩾ 1	1	0.40	?	High

(1) Annual survival based on recaptures of ringed birds returning to colonies in later years.
(2) Source — Nelson (1968, 1970), Snow (1965), Snow and Snow (1967), my published and unpublished data.

Squid occurred in 28 of 38 samples of food of the Brown noddy from Christmas Island but in none of 10 from Galapagos. Brown noddies also took squid at Aldabra in the Indian Ocean (Diamond, 1971) and one was found in 13 samples from Ascension in the Atlantic Ocean (Dorward and Ashmole, 1963). I did not record squid from the stomachs of tuna taken at Galapagos whereas almost a half of the tuna sampled near Christmas Island had fed on squid. Squid are presumably less abundant in the eastern than in the central Pacific.

All species in Galapagos ate fish, the size being correlated with the size of the bird except that the Lava gull caught small fish in tide pools and the Brown pelican filtered very small fish from the water, although the latter also ate fish up to 350 mm long. I made few specific identifications of prey, but the easily recognised flying fish of the family Exocoetidae were regurgitated regularly only by Great frigatebirds, Red-footed boobies and Red-billed tropicbirds, all of which fed well away from land (see below), and occasionally by Swallow-tailed gulls, Masked boobies and Waved albatrosses. Fish regurgitated by Magnificent frigatebirds tended to be species that lived in shallow water whereas those thrown up by Great frigatebirds were pelagic. Whether this difference reflected preferences for different foods or different feeding areas is not clear. Both species of frigatebirds chase other seabirds to pirate their last feeds. Great frigatebirds nested near seabird colonies of mixed species whereas Magnificent frigatebirds nested alone or with the Blue-footed booby which feeds inshore. Food differences between these congeners could be due to the different seabirds which they cleptoparasitised. The smallest fish (often larval forms) were taken by Brown noddies and Audubon's shearwaters, but even when feeding together the former took slightly larger prey. Storm petrels caught larger fish and not larval forms.

Small crustacea were important in the diet of Audubon's shearwaters and Galapagos storm petrels but not of Madeiran (= Band-rumped) storm petrels; larger crustacea were an important food of Waved albatrosses. Swallow-tailed gulls fed on pelagic crabs when these were available, these birds feeding by day only when crabs were present.

Magnificent frigatebirds and Lava gulls regularly scavenged inshore, perhaps in competition; Hailman (1963) has suggested that the all-dark coloration of this gull evolved as a cryptic defence against the stronger frigatebird.

The diets varied with both time and place even within Galapagos. In 1963-64 Snow and Snow (1967) noted that the commonest foods of Swallow-tailed gulls on Plaza Island were fish (recorded in 37 cases) and squid (recorded in 12). My observations at the same colony in 1965-67 were similar — 96 sampled had eaten fish and 36 had eaten squid. During both studies squid sometimes constituted a high proportion of the diet, but these occurrences could not be linked with any obvious oceanographic changes. Observations in 1962 (Hailman, 1964) and 1966-67 at another colony 100 km north suggested that Swallow-tailed gulls there fed almost entirely on squid. Nelson (personal communication) also found that squid were a common food of Red-footed boobies and Great frigatebirds in 1964, though my own casual observations extending over four years did not confirm this. During 1965-67 flying fish were frequently seen at sea and many species brought them to their young, but after 1969 they were rarely seen.

Feeding Methods

Red-footed boobies caught flying fish in the air; Great frigatebirds picked them from the surface, Swallow-tailed gulls caught them just below the surface, Masked boobies caught them some distance below and Waved albatrosses took them while swimming on the surface. Similarly, Brown noddies hovered to take the smallest fish, storm petrels made a quick pounce and Audubon's shearwaters pursued them underwater. Deep-diving birds have more prey available than those which fish only at the surface, but are less manoeuvrable; Masked boobies have little chance of changing target once their dive has started, while lighter Blue-footed boobies can easily stop short or take their prey by dipping. Seabirds have morphological adaptations associated with their different feeding methods (Ashmole and Ashmole, 1967).

Many Christmas Island seabirds depended largely on predatory fish to chase the small fish and squid towards the surface where they were easier to catch.

The only species dependent on such fish in Galapagos was the Brown noddy. Flocks of shearwaters, pelicans and Brown noddies were often seen with large schools of filter-feeding fish, but these were often cases of mutual attraction to a common food source. Small fish were not being chased to the surface. It is possible that the birds which feed far from the island (including the Sooty tern, which I never saw feeding) attend shoals of tuna in offshore waters.

Time of Feeding

Apparently flying fish and squid are more abundant nearer the surface, and hence more accessible to birds, by night than by day. The most specialised nocturnal seabird is the Swallow-tailed gull which usually feeds only at night. Hailman (1964) realised that nocturnal feeding would allow this gull to occupy an otherwise empty feeding niche, but thought that nocturnal habits had evolved to enable both birds of the pair to protect the nest from diurnal frigatebirds. Certainly any gull with food which leaves the shelter of the cliffs by day is attacked by frigatebirds, but non-breeding Swallow-tailed gulls off the coast of Peru feed nocturnally even where there are no frigatebirds. Presumably the exploitation of an untapped food source was also a major factor in the evolution of this habit. Swallow-tailed gulls had a very similar diet to Red-billed tropicbirds but the species fed in different areas and only the latter caught its prey by diving. Galapagos storm petrels obtained much of their food at night. They are unique in being the only storm petrel species to fly around their colonies by day. At night their place over land was taken by Madeiran storm petrels which were entirely nocturnal, even where Galapagos storm petrels were absent, but fed by day. Owls killed both species of storm petrels, so their time of flying did not protect either species completely against predation. Different times of feeding may, however, have allowed the two species to take different foods. The reason why Audubon's shearwater was diurnal in Galapagos is obscure; elsewhere it comes to land only at night. Few other seabirds are known to feed regularly at night, though several are active at dusk and may bring much of the food to their young after dark, for example boobies, frigatebirds (T.de Vries, personal communication).

Feeding Places and Breeding Ecology

In Galapagos waters most species had well defined feeding areas (table 7.1). Species which were rarely, if ever, encountered feeding were inferred to be foraging well away from land. As well as direct observations, the measuring of incubation stints and feeding frequencies gave some indication as to whether the birds were feeding nearby or far away; the longer these took, the further away the bird was likely to be feeding. These measures cannot separate birds which feed inshore well away from the colonies from those which feed pelagically, and they also assume that birds return immediately after finding food. Incubation stints of Audubon's shearwaters were shorter (2.9 days) when food was plentiful (as judged by size and frequency of foods brought to the young) than when food was scarce (6.3 days), indicating that birds returned as quickly as possible after feeding (Harris 1969*b*). Adults rearing

young on an erratic food supply are likely to feed the chicks as frequently
as possible so that the young can build up reserves to carry them through
hard times. Again there may be difficulties of interpretation as birds may
feed their young several times on one load of food or, conversely, may return
empty having lost the food to frigatebirds near the colony. There was a
significant correlation ($r = 0.8, P < 0.01$) between length of incubation
stint and the frequency of feeding young within a species, which suggests
that each measurement gives a fair idea of how far the birds are going.

Eight species fed within a mile or two of the shore, the closest feeder was
the Flightless cormorant which needed shallow water and so did not venture
more than 100 m from land. Seven species were pelagic and four intermediate.
Among the latter, the Swallow-tailed gull fed its young a few hours after
leaving the colony at dusk, so must have been foraging either between the
islands or, less probably, close inshore. On Christmas Island Sooty terns and
Brown noddies had very similar diets, but the former foraged well away from
land, the latter at an intermediate distance. Inshore waters of Christmas
Island were the domain of two other species of terns which are lacking from
Galapagos. In Galapagos waters Brown noddies occupied both inshore and
offshore zones, though most fed close to land.

Inshore feeders breed in relatively small colonies situated fairly close
together, whereas offshore and pelagic birds have much larger colonies
situated further apart (Lack, 1967). The Flightless cormorant rarely moved
more than 2 km from its natal colony and pairs nested as close to the feeding
grounds as possible. More than a few nests together occurred only where
landing places were restricted in an otherwise suitable habitat. The scavenging
Lava gull nested solitarily, unlike most other species of gull, and pairs were well
dispersed. Other species which fed close to land, for example Audubon's shearwater, Swallow-tailed gull, Brown noddy and Brown pelican, also had dispersed
colonies (figure 7.1*a* and *c*). Species which were pelagic had few large colonies,
for example Waved albatrosses (12 000 pairs on one island), Galapagos storm
petrels (figure 7.1*c*; three colonies, including one of 200 000 pairs), and Hawaiian
petrels (figure 7.1*c*). The Red-billed tropicbird is atypical; despite pelagic feeding
habits, the few thousand pairs are split between small colonies spread rather
uniformly through the area (figure 7.1*b*). There seem to be plenty of nest sites in
all colonies, yet intraspecific competition for the best sites results in many nesting
failures (Snow, 1965). Similar losses due to intraspecific competition were noted
in Galapagos storm petrels. The reason why birds do not expand their colonies to
prevent these losses is not obvious to me.

The relationship between colony dispersion and feeding place is well shown
in the Sulidae (discussed in Nelson, 1970). The inshore feeding Blue-footed
booby breeds in small colonies throughout the main part of the archipelago
(figure 7.1*f*) and its absence from the northern islands is linked with the lack of
shallow water needed for feeding. Masked boobies feed between the islands and
have fewer but larger colonies (figure 7.1*e*); the distant feeding Red-footed booby
is restricted to three large and two small colonies (figure 7.1*d*). Colonies of Red-
footed boobies remain in place much longer than those of Blue-footed boobies
which not infrequently move: Goodman (1974) estimated that Red-footed booby
colonies on Tower Island had lasted more than 6000 years. A similar colony

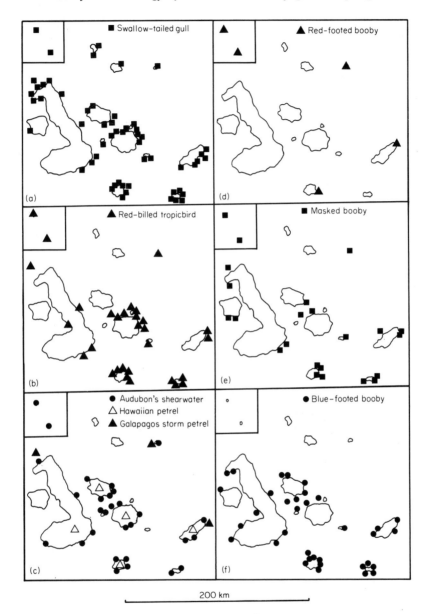

Figure 7.1 The distribution of breeding colonies of some Galapagos seabirds. The two inserted islands are situated 100 km north-west of the next most northern island. ●, positions of colonies of inshore feeders; ■, offshore species, △, pelagic species. Some positions are offset slightly for clarity.

dispersion is shown by the two species of frigatebirds. The offshore feeding Great frigatebird has colonies dispersed around the periphery of the archipelago, whereas most of the less common Magnificent frigatebirds, an inshore feeder, breed in colonies among the central islands. Isolated pairs also occur in most colonies of Great frigatebirds.

Discussion

It is difficult to decide which of the various mechanisms discussed above are most important in reducing interspecific competition. Cody (1973) concluded that there was so much overlap in the diets and diving depths of six species of Alcidae that only different feeding areas could bring about ecological separation — assuming that this was needed — but Bédard (1976) has demonstrated that Cody's data are unconvincing. In the North Sea there was considerable overlap in the sizes and species of fish taken by different species of seabirds, and Pearson (1968) thought that ecological separation during the breeding season was brought about mainly by birds fishing at different distances from the colony and at different depths. Ashmole and Ashmole (1967) considered that, between species which were not closely related, competition was reduced mainly by differences in feeding methods; in closely related species differences in body size and morphological characteristics were critically important. On Ascension Island there was a large overlap in the species eaten by the nine seabirds whose diet was studied but the larger species ate the largest prey (Stonehouse, 1962). Although all the birds fed out of sight of land there were some differences in foraging range, for example between Masked booby and Brown booby *Sula leucogaster*, and feeding methods (Dorward, 1962). On Aldabra Island, Diamond (1971) found that the Red-tailed tropicbirds *Phaethon rubricauda* and Yellow-billed tropicbirds *Phaethon lepturus* ate squid and flying fish, the latter taking smaller prey and more squid than the former. The Great frigatebird and the Lesser frigatebird *Fregata ariel*, co-existing on Aldabra, had diets which were very similar overall, the only statistically significant differences being seasonal changes in the proportions of fish and squid. Surprisingly, the larger of two species did not take larger prey, though it brought larger feeds to the young. Differences in the time of day the young were fed suggested different feeding areas.

Although there are considerable interspecific overlaps in each of the ways in which Galapagos seabirds might reduce competition for food, the main differences were clear, and largely maintained even when food seemed to be plentiful. Even allowing for differences between year and location, each species had a distinctive diet, as regards both type and size of prey taken. The interspecific differences were more marked than those found in other areas. Whether each species deliberately chose different prey, or whether the observed differences in diet were due entirely to different feeding techniques (requiring morphological specialisations), to different feeding zones, to different feeding times, or to a combination of these factors, is an open question. It is instructive and intellectually stimulating to look for the most obvious ways by which closely related species differ in how they obtain their food. These differences vary from group to group. For instance in Galapagos boobies and frigatebirds the distance between the breeding colony and feeding place appears to be the most important factor, whereas one species of storm petrel feeds by night while its congener feeds by day.

However, comparisons between species in different families are probably of little value. Doubtless all the isolating mechanisms have been influenced by selection and little is gained by attempting to rank the adaptations, whether physical or behavioural, in order of importance. It is unlikely that any species would ignore potential prey if it could catch it, so ecological separation is probably due to a combination of all the factors.

Acknowledgements

Drs D. Boersma, J. B. Nelson and T. de Vries and Sr. J. Gordillo supplied unpublished information. I thank Drs D. Jenkins and I. Newton for helpful criticism of the manuscript.

References

Ashmole, N. P. (1971). Seabird ecology and the marine environment, in *Avian Biology*, Vol II (ed. D. S. Farner and J. R. King), pp. 224 – 286.

Ashmole, N. P. and Ashmole, M. J. (1967). Comparative feeding ecology of seabirds of a tropical oceanic island. *Peabody Mus. Nat. Hist. Bull*, **24**.

Bédard, J. (1976). Coexistence, coevolution and convergent evolution in seabird communities: a comment. *Ecology*, **57**, 177 – 184.

Cody, M. L. (1973). Coexistence, coevolution and convergent evolution in seabird communities. *Ecology*, **54**, 31 – 44.

Diamond, A. W. (1971). The ecology of sea-birds of Aldabra. *Phil. Trans. R. Soc. Lond. B.*, **260**, 559 – 569.

Dorward, D. F. (1962). Comparative biology of the White Booby and the Brown Booby *Sula* spp. at Ascension. *Ibis*, **103b**, 174 – 220.

Dorward, D. F. and Ashmole, N. P. (1963). Notes on the Brown Noddy *Anous stolidus* on Ascension Island. *Ibis*, **103b**, 447 – 457.

Goodman, D. (1974). Natural selection and a cost ceiling on reproductive effort. *Am. Nat.*, **108**, 247 – 268.

Hailman, J. P. (1963). Why is the Galapagos Lava Gull the colour of lava? *Condor*, **65**, 528.

Hailman, J. P. (1964). The Galapagos Swallow-tailed Gull is nocturnal. *Wilson, Bull.*, **76**, 347 – 354.

Harris, M. P. (1969*a*). Breeding seasons of sea-birds in the Galapagos Islands. *J. Zool., Lond.*, **159**, 145 – 156.

Harris, M. P. (1969*b*). Food as a factor controlling the breeding of *Puffinus lherminieri*. *Ibis*, **111**, 139 – 156.

Harris, M. P. (1973). The biology of the Waved Albatross *Diomedea irrorata* of Hood Island, Galapagos. *Ibis*, **115**, 483 – 510.

Huxley, J. (1942). *Evolution. The Modern Synthesis*. Harper and Bros., London, 645 pp.

Lack, D. (1954). *The Natural Regulation of Animal Numbers*. Oxford University Press, Oxford, 343 pp.

Lack, D. (1967). Inter-relationships in breeding adaptations as shown by marine birds, in *Proc. XIV Int. Ornith. Congr.* (ed. D. W. Snow). Blackwell Scientific Publications 405 pp.

Nelson, J. B. (1968). *Galapagos, Island of Birds*. Longmans, London, 338 pp.

Evolutionary ecology

Nelson, J. B. (1969). The breeding ecology of the Red-footed Booby in the Galapagos. *J. Anim. Ecol.,* **38**, 181 – 198.

Nelson, J. B. (1970). The relationship between behaviour and ecology in the Sulidae with reference to other seabirds. *Oceanogr. Mar. Biol. Ann. Rev.,* **8**, 501 – 574.

Pearson, T. H. (1968). The feeding biology of sea-bird species breeding on the Farne Islands, Northumberland. *J. Anim. Ecol.,* **37**, 521 – 552.

Snow, D. W. (1965). The breeding of the Red-billed Tropic Bird in the Galapagos Islands. *Condor,* **67**, 210 – 214.

Snow, D. W. and Snow, B. K. (1967). The breeding cycle of the Swallow-tailed Gull *Creagrus furcatus. Ibis,* **109**, 14 – 24.

Stonehouse, B. (1962). Ascension Island and the British Ornithologists' Union Centenary Expedition 1957 – 59. *Ibis,* **103b**, 107 – 123.

8 Some relationships between food and breeding in the marine Pelecaniformes

Dr J. B. Nelson*

Introduction

David Lack stimulated me to look at seabird ecology from an evolutionary view-point and to become particularly intrigued by the phenomenon of 'holistic' adaptation, which he chose as the theme of his Presidential address to the 1966 International Ornithological Congress. As soon as the adaptive significance of one trait, morphological, ecological or behavioural begins to emerge, it becomes apparent that it is part of a complex adaptive web. Perhaps the single most influential selection pressure is food. Lack built massively on this foundation, basing his theories about reproductive rates and the control of populations in higher animals on the concept of direct, density-dependent competition for food (Lack, 1954; 1966; 1968). In this chapter I trace some of the ways in which, in the Pelecaniformes, feeding habits have determined breeding strategies. The latter include the timing of breeding (the seasonal timing, the degree of synchron-isation and the frequency), the lengths of the components of the breeding cycle, brood size and the age of first breeding. Are they all adapted towards the *maximum* production of young? In pursuing this theme I will stress social (behavioural) stimulation of breeding, which can be very important but which Lack seemed, to me at least, to underestimate. Although he removed some of the ambiguities in Darling's early work on social factors he did not undertake research into them, or exploit the potential for detailed etho-ecological work on social stimuli and breed-ing success, of the Tinbergen – Lack axis, even though the results would have been highly germane to his great interest in recruitment.

Food and Feeding Habits

The wide range of feeding habits encompassed by the uniformly fish-eating pelecaniforms — plunge diving in boobies and tropicbirds, splash-diving and scooping in pelicans, surface-snatching in frigatebirds, swim-diving and under-

*Bryan Nelson (born 1932) followed a first in Zoology (St Andrews) with a D. Phil. at Oxford where (1959 – 60) he worked under David Lack. Following three years on the Bass Rock studying gannets (1961 – 63) he and his wife went off to uninhabited islands in the Galapagos (1964), extending his work on the Sulidae. In 1967 he worked on Christmas Island (Indian Ocean) and in 1968 – 69 in the Jordanian Desert, at Azraq. Since 1969 he has lectured in the Zoology Department at the University of Aberdeen. Currently, he is involved with work on Aldabra's frigatebirds.

water pursuit in shags and cormorants – may be pondered from many angles. Seabirds cannot nest dispersed among their food; they must congregate for breeding on land. There they are subject to the nature of the terrain, including its climate, to the limitations of their morphology evolved primarily for feeding, to competition from other species and to the properties of the surrounding seas in terms of the numbers and availability of their prey. Just as the pelecaniforms are but one major seabird group, exploiting in overlapping fashion a broad food band, so the families, genera, species and even sexes within the order further subdivide the food resources. A major way of doing so is to become adapted not only to particular species and size-ranges of prey, but to their distribution and density. The latter is of great significance in that it determines the foraging and feeding pattern of seabirds, to which morphology is adapted, as are energetics, breeding strategy and social factors in breeding. So, in each species, an adaptive web has evolved in which the main strand is food.

Feeding Habits, Brood Size and Breeding Success

Within the Pelecaniformes, all five species of frigatebirds and all three species of tropicbirds invariably lay a single egg. Two of nine sulids (the Blue-footed booby *Sula nebouxii* and the Peruvian booby *S. variegata*) produce broods of more than a single chick though two more (the Brown booby *S. leucogaster* and the Masked booby *S. dactylatra*) usually lay clutches of two eggs, leaving obligative siblingmurder to reduce the brood. All six species of pelicans lay clutches of two or three eggs and may rear two or more chicks per brood. The twenty-nine species of cormorants and shags typically lay about four eggs and may rear them all, and the same applies to the two darters.

Brood size is correlated with the frequency with which the young are fed, those foraging furthest having the smallest broods. Species rearing a single chick are all pelagic: these include the tropicbirds, two of the three species of bluewater boobies (Red-footed booby *S. sula* and the Masked booby and Abbott's booby *S. abbotti*). Frigatebirds, which also rear a single chick, are often pelagic. All of these are characterised by low fledging success due to food-related deaths, long incubation periods, (40–42 days in tropicbirds, 55 days in frigatebirds, compared with about 30 days in the shag) slow growth, and prolonged post-fledging parental care. Probably none of the frigatebirds rears more than 15 per cent of the eggs laid (Nelson, 1975); Red-footed and Brown boobies are, in some places, equally unsuccessful, whereas Yellow-billed tropicbirds *Phaethon lepturus* bred successfully on Ascension, 30 per cent of the fledglings surviving (Stonehouse, 1962). Adaptations such as relatively large-yolked eggs and energy conserving habits reduce further the amount of food that the parents are required to provide per unit of time, and relative to adult body weight. Adaptations of chicks include special resting postures to aid heat loss in Red-footed boobies and frigatebirds, and temperature control by gular-fluttering and by the evaporation of excreta deliberately deposited on the webs in, at least, some sulids. Not unexpectedly, wherever conditions are eased, such adaptations are less marked. Thus, in two populations of the same species, those feeding in less impoverished blue waters may produce smaller eggs, have a higher breeding success and faster growth. Still further, incidentally, such 'locality differences' then link with associated behavioural ones, in so far as the latter serve, for example, different functions of the pair bond. Thus quite different pair and social relationships may have arisen in the Brown boobies of Ascension island compared with those of Christmas Island in the Indian Ocean. The effects of long foraging stints on pair and social behaviour can be traced much further, notably in the frigatebirds where it under-lies highly unusual territorial and pair behaviour (see below).

This consistent picture is apparently spoiled by the three species of gannets (Atlantic, *Sula [bassana] bassana*; Cape, *S. [b] capensis* and Australasian, *S. [b] serrator*). Despite abundant food (especially in the Atlantic), all invariably lay one small egg, although they grow rapidly and have high fledging success. The gannets form a series from Australasian through Cape to Atlantic, the first probably being the nearest to the ancestral sulid and the last the most recently diverged and highly specialised. The single egg of the latter may be seen as a means of producing a fat-laden chick in the shortest time, in a breeding cycle curtailed by seasonal (climatic) factors, thus exploiting seasonally abundant food to the fullest extent. The associated and notable abandonment, by the gannets, of the otherwise universal sulid practice of feeding the young after fledging is consistent with this interpretation. In the primarily uniparous Sulidae, species rearing more than one chick do so under special feeding circumstances. This is clearly the case with the Peruvian booby, which feeds on the anchovies of the Humboldt Current and is highly productive except in 'niño' years, and with the Blue-footed booby, whose breeding distribution is on the fringe of productive areas. Where (as in the Galapagos) it rears more young per brood than other species of booby on the same island, its sexual division of labour is implicated (see below).

All the pelicans lay two or three eggs. In general, their foraging habits are intermediate between those of the sulids and phalacrocoracids (cormorants). Although typically they do not forage very far from the breeding area they can feed at a considerable distance from it. The Great white pelican *Pelecanus onocrotalus*, for example, may feed up to 150 km away. Moreover, an unusually abundant supply of fish is a necessary adjunct for breeding in this species (Brown and Urban, 1969). Not only does the local foraging habit and abundant food generally enable pelicans to rear larger broods than pelagic families but it maintains much more rapid growth than in tropicbirds, boobies and frigatebirds, even though a young pelican may weigh between 10 and 30 times as much as those.

Typically, breeding success is significantly lower in pelicans than in most cormorants. The Great white pelican often loses half its young after hatching; colonies of the Australian pelican *P. conspicillatus* often fail completely and Brown pelicans *P. occidentalis*, are not notoriously prone to desert their young. In some cases desertion is almost certainly due to sudden deterioration in the food supply, thus the Great white pelican occasionally congregates to breed opportunistically only to find (as at Lake Natron and Lake Shala in East Africa) that the fish die. Similar losses affect the Australian pelican. Sometimes, however, failure is due to maladaptive social behaviour on the part of parents and young and to the effects of the sun and of man.

The cormorants and shags, the most widespread and numerous of the Pelecaniformes, all rear broods of two to six chicks. Moreover, even in the large species such as the common cormorant *Phalacrocorax carbo* growth takes less than half as long as in the gannet. Breeding success is high; in the Shag, *P. aristotelis*, for example, it averaged 61 per cent over 4 years on Lundy and the figures for chicks fledged from those hatched reached 95 per cent in the larger broods (Snow, 1960). This picture is broadly typical of the family and depends on the much larger amount of food, per adult, brought by phalacrocoracids compared with sulids. This is possible only because they feed very near to their breeding sites. In fact, breeding shags on the Farnes spent only 4 - 8 per cent of daylight hours on fishing activities, and foraging trips lasted, on average, only 0.77 hours (Pearson, 1968). Where feeding conditions are exceptionally favourable, as in the Humboldt, Benguela and Californian Current regions, this is reflected not so much in brood size, which is already near the maximum, but in the huge size and dense concentration of the local population.

Colony Size and Density

Throughout the Pelicaniformes, species which congregate in the biggest colonies also concentrate the most densely, though in all cases, density is much more constant than colony size. Both features are related to food.

Absolute colony size depends principally on the amount and distribution of suitable breeding habitat relative to the population and to the foraging habits of the species concerned. Colony density depends on the amount of physical space relative to demand and on the social function of density. Clearly, several widely different combinations of factors can determine both size and density.

In general within the order certain features can be classified by colony size:

(1) Extremely large (100 000 pairs or more) and dense colonies occur only where food is exceptionally plentiful and nesting sites somewhat limited. Such colonies may be composed either of inshore or of relatively offshore foragers, or often a mixture of species, as in the vast colonies of Guanay cormorants *Phalacrocorax bougainvillea*, Brown pelicans *Pelecanus occidentalis thagus* and Peruvian boobies, and the equivalent mixed colonies on the South African and Californian Guano islands.

(2) Large (20 000 pairs or more) and dense colonies may occur relatively independently of physical shortage of sites, but they occur only in areas where food is rich; colonies of the Atlantic gannet, which is a somewhat offshore feeder, are examples. The Great white pelican in the African Rift Valley forms colonies containing between 20 000 and 40 000 pairs, but these depend on unusually abundant food in highly productive lakes; here, nesting areas, which must be inaccessible to predators, are in short supply.

(3) Large (but not necessarily dense) colonies may occur in impoverished tropical blue water where islands are scarce. There are, for example, probably more than 100 000 pairs of Red-footed boobies on Tower Island in the Galapagos and there used to be perhaps 15 000 pairs of Masked boobies on Malpelo and 'immense numbers' on Clipperton. All these islands are in the empty wastes of the eastern tropical Pacific, and their large colonies are composed exclusively of species foraging a long way offshore.

(4) Small colonies are typical of inshore or local feeders in either relatively rich or impoverished waters; for example, most cormorants and shags, Brown and Pink-backed pelicans *P. rufescens*. Colonies of offshore feeders may be small if sites are limited, as in the hole nesting tropicbirds, none of which nests

in colonies of much more than 1000 – 2000 pairs.

Density is not merely a function of space; food and social factors are important also. Shortage of space forces Guanay cormorants to congregate, but, being inshore feeders, they are able to thrive at such densities only because they live among unimaginably prolific prey. They could not do so in the Galapagos because they could not fly far enough, or fish appropriately enough, to catch sufficient flying fish. Often, as in many cormorants and pelicans, density is partly determined by the space limitations of a small island, even though the total numbers of birds is small. Density is usually typical of a species, and its role is often a social one, for example in the local synchronisation of laying and in pair formation. This is the main reason for its consistency where space is not a limiting factor, but the precise nature of the social role must be determined for each species.

Timing of Breeding and the Composition of the Breeding Cycle

Breeding strategy has evolved under intense selection pressure. In a strictly uniform environment the timing of breeding would be neutral, but no environment is uniform, and even if it were, the composition of the cycle, that is, the absolute and relative lengths of its parts, would offer plenty of material for natural selection. Energy budgeting is important. The times at which females produce eggs and males endure the stress of site establishment and pair formation, at which adults moult and young are fed, and at which they reach certain critical points in their progress towards proficient independence, must all be adaptive. Food, though by no means the only selective agent, is probably by far the most critical.

The following pelecaniform breeding strategies are closely related to food and feeding habits.

(i) Strictly annual, highly seasonal breeders

Pelecaniforms whose mean laying date is about the same each year, all have a predictably seasonal climatic and food regime, such as that found in the subarctic/boreal seas of the Northern Hemisphere. Probably this generalisation applies strictly only to the pelecaniforms of the Northern Palaearctic. For example, the mean laying dates of shags on Lundy, in the three breeding seasons 1954 – 56 inclusive, differed by only 6 days (Snow, 1960). Over 5 years, the spread of laying covered 3 months but was only 5 weeks in 1958; the breeding season seems to be timed to coincide with the availability of the shag's main food, the sand-eels *Ammodytes* spp.

Altlantic gannets of the Canadian population return late in April to their icy ledges. Nevertheless, by mid-May over 80 per cent of the population on Bonaventure have laid (Poulin, 1968). The compressed pre-laying period enables them to achieve a mean laying date little later than that of British gannets, which return in January or February. In both populations, the annual mean laying date is constant to within about a week. The southwards migration of the newly-fledged juveniles before late September helps them to avoid stormy weather during the critical period when, completely without help from their parents, and highly vulnerable to starvation, they must become proficient plunge divers. Although post-fledging mortality as a result of the weather is almost certainly the main discriminant against late fledgers, the timing of fledging with regard to climate is not the only factor determining the laying date. First, early fledging, linked as it is in the gannet to rapid growth and large mass (huge fat deposits), can be achieved only when food is abundant. Thus the breeding distribution of the gannet coincides with plentiful herring and mackerel and its chicks grow when these fish are most available. The gannet's weight (the greatest in the family) and strong bill enable it to penetrate deeply when it dives and cope with large and muscular prey, while its sexual isomorphism means that both sexes have this ability. Rapid growth of large young is thus doubly facilitated. Sexual isomorphism may depend in part on the lack of any competitor in the gannet's feeding niche, and hence lack of pressure to diversify its feeding technique, but is equally consistent with shared defence of the nesting site. Secondly, the pre-laying period, which is of great importance in the gannet's social system of re-establishing the site and reforming the bond, is a time of considerable stress for the male, who spends more than two-thirds of the daylight hours guarding and displaying on his site. During this period, he loses weight. Presumably, this is why the whole cycle cannot occur even earlier in the year. The formation of the egg is probably of little importance with respect to energy conservation. In this system, social (behavioural) factors which make the laying date earlier are likely to be very important. It is in this way that a socially adequate site acquires its undoubted adaptive value and why it is the object of such fierce conflict. This argument did not convince Lack, who preferred to attribute the obtrusive phenomenon of site competition in the gannet to shortage of physically suitable sites which are free from predation. These factors are, however, certainly an inadequate explanation.

(ii) Strictly annual, loosely seasonal breeders

Most seabirds breed annually, but where seasonal pressures are relatively imprecise the mean laying date of a population may vary in different years by several weeks or even months. This may arise either by individual females laying at about the same time each year but with a wide spread in the population, or by each female varying the lengths of the intervals between successive layings but always laying once in each calendar year. Rarely do we know which it is. In theory, the former seems more likely, but variable between-cycle periods certainly do occur in some sulids.

   This category (though often it grades into a seasonal breeding) embraces most of the Pelecaniformes, particularly those in the higher latitudes between the tropics, and north and south of these, where seasons differ enough in day length, temperature, wind systems and (for marine species) oceanographic features (currents, convergences, upwellings) to affect food and hence the timing of breeding, but to do so only broadly. It applies to the Cape and Australasian gannets (though these tend strongly towards the category (i) end of the spectrum), the Peruvian booby, most populations of all three pantropical boobies, probably all the pelicans (though "loose" becomes very loose, almost continuous in some areas, for example among the Great white pelicans of East Africa), some two-thirds of the cormorants and some populations of (probably) all three tropicbirds. Others, for example Red-billed tropicbirds in the Galapagos and both Red-billed and Yellow-billed tropicbirds on Ascension, breed continuously through the year.

Unlike the Atlantic gannet, the Australasian, laying in the southern spring, varies by several weeks from year to year. Its spread of laying is also greater than that of the Atlantic gannet; thus a smaller proportion will lay on the same date from year to year regardless of annual variation in mean laying date. This may mitigate the effects on chicks of the severe and unpredictable, though brief, shortages of food which, on Horu Horu (New Zealand) are periodically catastrophic (Stein, personal communication). In addition severe climatic conditions occasionally wipe out 90 per cent of chicks, mostly of a particular age. These two hazards may be linked, and both must work against too high a degree of synchronisation. Perhaps related to this are subtle differences in the social nature of the colony, when compared with those of Atlantic gannets. Australasian gannets attend their sites less in the pre-laying period (perhaps food is less dependable) and other signs indicate slightly weaker attachment to the site; thus fighting, site display and site fidelity are all reduced. This could imply that in the Australasian gannet, the site has less social significance. The reduced synchronisation which does occur is consistent with this. Finally, growth is slower. Since extreme compression of the breeding cycle within fixed seasonal bounds is not important here, slow growth matters less than it would in the case of the Atlantic gannet. In any case, it may be enforced by less abundant food; Australasian gannets failed to rear twins when given an extra chick (Robertson, personal communication), whereas Atlantic gannets easily did so. In fact the young fledge with proportionately smaller reserves of fat than those of the juvenile Atlantic gannet. This may be linked with interesting differences in their fledging technique. Australasian gannets practise flying before fledging (which Atlantic gannets cannot do) and may wander to the edge of the colony and return to be fed. Some, perhaps most, can fly properly — for example, taking flight from the waters — as soon as they are fledged, which no Atlantic juveniles can do. Australasian juveniles grow their primary feathers to adult length before fledging; Atlantic juveniles finish growing their flight feathers while swimming away from the colony. Thus the nature of the initial stages of the migration of the juveliles might differ in the two species of gannets. The Australasian is nearer to the booby pattern of feeding, growth and fledging procedure, whereas the Atlantic is highly specialised to exploit a particular food and feeding niche.

At the other extreme of category (ii) lies the Galapagos Masked booby, every island population of which lays in many months of the year. Except when the groups are small, synchronisation is poor, growth is slow and breeding success low. These are all classical features of the poor-food, far-foraging bird in a relatively aseasonal area. The timing of breeding is clearly not crucial and yet the species, whenever it appears, continues to show only annual breeding; this is not true of the tropicbirds, which in some areas (as on Ascension) depart from the annual cycle.

(iii) Less-than-annual, seasonal breeders

Besides the five frigatebirds, so far as we know, there is only one species, Abbott's booby (*Sula abbotti*) in the order that consistently lays once in two years (excluding replacement eggs). All six lay seasonally, although in most frigatebird populations synchrony is loose. Only some Galapagos Red-footed boobies are known to lay at intervals of more than one year but less than two.

Frigatebirds are excellent examples of species in which food and feeding technique has distorted the entire breeding regime and social system far from the ordinal norm (see Nelson, 1975). They feed by snatching flying fish and squid from near the surface or in the air, and by robbing other seabirds; they have the lowest wing loading of any bird and weak, only partially-webbed feet, and they lack waterproof plumage. Typically, they inhabit warm, blue-water areas and breed on oceanic islands. The combination of relatively scarce food and a specialised feeding technique enforces an extraordinarily long breeding cycle. The chick requires 5 to 6 months to fledge (more than twice as long as sulids inhabiting the same areas) and is then fed as a free-flying juvenile for as long again, sometimes much longer. The egg, which is larger than that of any other pelecaniform, has an incubation period almost twice as long as that of a pelican weighing five or six times as much. In the Galapagos, incubation stints, implying equally long foraging stints, are several times as long as those of pelagic boobies on the same island and more than 100 times as long as those of the Galapagos flightless cormorant *Nannopterum harrisi*. During these foraging flights a frigatebird may have to regain up to a fifth of its weight, lost during its previous incubation stint. Presumably, patchy and sparse food is best exploited by staying away as long as necessary to find enough good feeding spots, even if this entails foraging for thousands of miles over 10 – 15 days. It is the same for an adult foraging for its chick, hence the latter's slow growth. The juveniles long dependence on adults after fledging, and the extraordinarily protracted period before breeding for the first time (see below) further testify to the difficulty of becoming proficient in this feeding niche. The long cycle precludes annual breeding and this in turn makes it impracticable for frigatebirds to remain faithful to site and mate (see Nelson, 1975, for details). Consequently, the whole nature of territorial behaviour and pair relations differs from that found in the rest of the order. Despite this, laying remains loosely seasonal, though, in relatively aseasonal areas the nature of the proximate trigger remains unknown.

(iv) More than annual, non-seasonal breeders

Perhaps oddly, the only pelecaniforms which are known to have shortened their cycle to consistently less than 12 months are two species of sulids in populations inhabiting notably impoverished areas (the Blue-footed booby of the Galapagos and the Brown booby of Ascension) and the Yellow-billed tropicbird. Several phalacrocoracids of the Southern Hemisphere have two distinct widely-separated breeding peaks, which are thought to involve different pairs. The same probably applies to some pelicans with a practically continuous breeding season. It might seem surprising that species in impoverished waters, like the boobies and tropic-birds mentioned above, should attempt more than the normal amount of breeding in a year. But this may increase the success of individuals, measured over a lifetime, by maximising the number of chances of hitting upon favourable periods. This strategy requires special underpinning, which in the Galapagos Blue-footed booby appears to be a shortened cycle based on sexual dimorphism and an associated division of labour; inshore fishing by the small male minimises foraging time and reduces the risk of starvation in small chicks. In the Ascension

Brown booby, Simmons (1967) has shown that the social system has been remarkably modified to allow continuous occupation of the site and continuous readiness to start a new cycle, following failure at any point. This, similarly, maximises the number of cycles and compensates for the many failures imposed by the unpredictability of the food supply.

Breeding strategies thus appear to be ultimately adapted towards maximum production of young and the variety of means used to this end fit well with the known dictates of the species' feeding situation. The proximate triggers in seasonal latitudes are presumably rhythms based on photo-period and temperature periodicity, and the pituitary and gonads, while in aseasonal latitudes, for some species, food has been directly implicated as a proximate factor (Nelson, 1969).

Deferred Maturity

The length of the pre-breeding periods known or strongly suspected for various pelecaniforms are: frigatebirds, 7 – 11 yr; Atlantic gannet 4 – 6 yr; Australasian gannet 3 – 5 yr; Cape gannet 2 – 4 yr; Masked, Red-footed, Brown, Blue-footed and Peruvian boobies 2 – 4 yr; pelicans 2 – 3 yr; all the cormorants and shags 2 – 3 yr.

The age at which a bird breeds for the first time is an important determinant of its recruitment rate. In general, body size is not decisive. Fulmars, for example, do not breed until they are at least 7 years old whereas cormorants do so at 3 years of age. The particular demands of the species' social structure may be important. Site acquisition and pair formation are not uniformly difficult. Gannets, for instance, usually hold a site for a year before breeding on it, whereas shags and cormorants do not. The role of food in affecting deferred maturity is unclear but may relate mainly to the time required for learning the species' feeding lore, rather than technique. Technique as such surely cannot require that some pelecaniforms should remain among the ranks of the pre-breeders for as long as they do. Probably, Atlantic gannets use different fishing areas at different times of the year and in different weather conditions. The marked difference between colonies in the timing of laying is probably due to local food factors; it is not correlated with temperature or severe weather along a north – south or east – west axis. The best use of fishing time must be important, and may well depend on detailed experience of localities and weather conditions, which takes time to acquire. The same is unlikely to be true for pantropical boobies or tropicbirds, whose food, more widely dispersed, is not similarly dependent on locale and weather. It is certainly untrue for Peruvian guano birds (boobies, pelicans and cormorants) whose food is everywhere around them. It does not apply with equal force to British shags and cormorants, which have much more restricted foraging areas, or to the Cape gannet, which has the more uniformly distributed food supply of the Benguela Current nearby. Correspondingly, all these have shorter deferred maturities than the Atlantic gannet. The fulmar supports the general hypothesis since it is known to utilise precisely located feeding areas a long way from its breeding grounds and, like the gannet, has a complex social system. Frigatebirds, with the longest deferred maturity of any pelecaniform have, on other evidence, the greatest difficulty in catching food.

Using the headings in this chapter it should be possible to construct, for any pelecaniform, a generalised profile of breeding strategy in relation to food.

Discussion

In sketching these relationships I have omitted much and in most cases, for brevity, have not attempted to support general statements with the available facts or even references. The aim has been to show in general, rather than review terms, how food shapes breeding ecology and behaviour. From these generalisations about the timing of breeding, brood size, breeding success, deferred maturity and social behaviour, it seems that recruitment rates in pelecaniforms are maximal. This seems particularly clear in the sulidae where, with the exception of the Atlantic gannet, there is convincing evidence that every species is highly adapted for rearing as many young as it can. Brood size and breeding strategy can be convincingly related to the nature of each species' food and feeding technique, and the other basic aspects of breeding are significantly varied to cope with the different conditions encountered over its range. In fact the range and detail of the adaptations which result in maximising breeding are impressive. Atlantic gannets are in one respect anomalous in that they can rear twins, but never lay two eggs. This could be due to time lag in evolving to suit changed conditions. In no other way, apparently, do they withhold breeding as part of a social mechanism to maintain population homeostasis. On the other hand, there are many indications that they are in no sense involved in keen, density-dependent, direct competition for food. This must surely apply, also, for example, to tropical sulids and to the Peruvian boobies of the anchovy-rich Humboldt Current. In the former, food may often be disastrously scarce, but it is extremely unlikely that the situation would be eased by a reduction of the local population. The operative factors are almost certainly oceanographic, and independent of the birds. In the latter, by contrast, food is either super-abundant or (periodically) virtually non-existent. In either case, the size of the seabird population has little to do with availability of food. Although it would be illogical to doubt the possibility of density-dependent competition if populations rose high enough (and indeed, with inshore feeders this would certainly operate long before colonies reached the size of those characterising many offshore feeders) it seems highly likely on simple pragmatic grounds that in many seabirds the population is controlled independently of existing numbers, though intimately dependent on the availability of food.

References

Brown, L. H. and Urban, E. K. (1969). The breeding biology of the Great white pelican (*Onocrotalus pelecanus roseus*) at Lake Shala, Ethiopia. *Ibis,* **111**, 199–237.

Lack, D. (1954). *The Natural Regulation of Animal Numbers.* Oxford University Press, New York.

Lack, D. (1966). *Population Studies of Birds.* Oxford University Press, New York.

Lack, D. (1968). *Ecological Adaptations for Breeding in Birds.* Methuen, London.

Nelson, J. B. (1975). The breeding biology of frigatebirds: a comparative review. *The Living Bird,* **14**, 113 - 155.

Nelson, J. B. (1969). The breeding ecology of the Red-footed booby in the Galapagos. *J. Anim. Ecol.,* **38**, 181 - 98.

Pearson, T. H. (1968). The feeding biology of sea-bird species breeding on the Farne Islands, Northumberland. *J. Anim. Ecol.,* **37**, 521 - 552.

Poulin, J. M. (1968). Reproduction du Fou de Bassan (*Sula bassana*) Ile Bonaventure (Quebec) (Perspective Ecologique). M.Sc. Thesis, Laval University, 110 pp.

Simmons, K. E. L. (1967). Ecological adaptations in the life history of the brown booby at Ascension Island. *The Living Bird,* **6**, 187 - 212.

Snow, B. (1960). The breeding biology of the Shag (*Phalacrocorax aristotelis*) on the Island of Lundy, Bristol Channel. *Ibis,* **102**, 554 - 575.

Stonehouse, B. (1962). The tropicbirds (genus *Phaethon*) of Ascension Island. *Ibis,* **103b**, 124 - 161.

9 Coexistence and diversity in Heteromyid rodents

Dr M. L. Rosenzweig*

David Lack's interest in coexisting avian seed eaters was extraordinarily seminal. His work, *Darwin's Finches* (1947), suggested that several species of this family might be avoiding competitive exclusion by virtue of their dissimilar beak sizes. A likely hypothesis seemed to be that each beak size is specialised for a particular range of seed sizes (Bowman, 1961). Theories have since been developed that explain how this might come about (see, for example, MacArthur and Pianka, 1966; Emlen, 1966). In general they may be summarised by noting that they all require a trade-off in adaptiveness: the phenotype that is well adapted to utilising one portion of a resource spectrum is, perforce, poorly adapted for other portions of it. In the case of granivores, this might mean that, for a bird of given size, hulling over-large seeds is too costly in time or energy, and small seeds are too unrewarding, to be profitable. MacArthur and MacArthur (1972) have shown that this is probably true for certain American birds.

In the arid habitats of southwestern North America, there is a diverse assemblage of granivorous rodents. It is natural to wonder if their coexistence is based on mechanisms similar to those postulated for avian granivores.

My cursory inspection of typical reports of the mammals collected at particular localities quickly revealed an encouraging fact. At any one place, there tends to be an assortment of body sizes. When species of the same body size appeared to occur together, authors usually added that the two lived in different microhabitats. With the knowledge that available data were scarce and that a diverse and abundant assemblage of small mammals was awaiting investigation in a most pleasant set of habitats, my colleagues and I embarked on the past ten years of research. We hypothesised a coexistence based on body size and seed size — smaller species depending on smaller seeds, and so on, but it now seems likely that we were totally or substantially wrong. Evidence from several sources now points consistently to the conclusion that these rodents select different habitats, but investigations which

*Michael L. Rosenzweig received his Ph.D. in zoology from the University of Pennsylvania in 1966. Robert H. MacArthur and W. John Smith were his preceptors. His research has centred on population interactions, particularly on theoretical aspects of exploitation and competition in rodents. Having held positions at the University of New Mexico, SUNY-Albany, and Bucknell University, He is currently Professor of Ecology and Evolutionary Biology, at the University of Arizona.

searched for seed size preferences have, with one notable exception, failed to discover any. As we shall see, the exception, which is due to Brown and Lieberman (1973) and Brown (1975), may not have much to do with desert rodent coexistence.

Evidence for Habitat Selection

Our very first investigations surprised us. Desert rodents of various species had densities correlated with aspects of the structure of the foliage, however sparse (Rosenzweig and Winakur, 1969). Some species such as the desert pocket mouse *Perognathus penicillatus* weighing 18 – 20 g were associated with shrub vegetation. Others, such as the Merriam kangaroo rat *Dipodomys merriami,* appeared to live only in patches relatively free of vegetation of a certain sort; they avoid, for example, vegetation growing between heights of 75 and 500 mm. These vegetational specialisations, quite analogous to those of birds (see, for example, MacArthur, Recher and Cody, 1966), were confirmed by Brown and Lieberman (1973) using an entirely different technique of observation.

Not only do the rodents have densities which correlate well with structural features of the vegetation, but also the vegetation is at least one proximate cue which informs them of the acceptability of a habitat. This has been established by observing rapid alterations in rodent distribution following changes in foliage structure. On one occasion we induced change by cutting foliage (Rosenzweig, 1973), but effects of natural changes – for example growth after unusually heavy rain, were also observed (Schroder and Rosenzweig, 1975). In this second study we tested to see if abandonment of a habitat by one species was due to active displacement by a more aggressive species. It was not; removal of the aggressor did not encourage the re-establishment of the original species. A third test was performed independently on the same pair of species by Whitford *et al.* (unpublished): using a reversible herbicide treatment to modify vegetation, they came to the same conclusions.

In the past several years, C. Lemen and I have been investigating the habitat selection of two species which we originally thought were prime candidates for seed size selection. They are *D. ordii,* a kangaroo rat weighing 45 g which lives in semi-arid dunes and grasslands, and *P. flavus,* a tiny pocket mouse of only 8 g which is restricted to semi-arid grasslands. To examine their habitat selection, we soaked millet seeds with indium, cobalt or cadmium chlorides and scattered them in different habitat patches. In experimental plots, the different patches were next to each other. This prevented us from luring rodents from their proper patch with bait. In control plots, all three types of tagged millet were scattered in the same patches. The controls enabled us to adjust for differences in the palatability and physiological treatment of the salts. After several days of such treatment, rodents presumed to be consumers of the treated seeds were trapped alive in clean aluminum box traps and their faeces were analysed by atomic absorption spectrophotometry for concentrations of the three rare elements.

In these experiments, the habitat was classified into one of two types: grassy or open. This is not so arbitrary as it might seem, since there were many areas where fairly substantial clumps of grasses and forbs formed islands in virtually bare spots; more common were situations where the bare spots were like islands in the vegetation. Results were conclusive: in 22 pocket mice sampled, 68 per cent of feeding was in grassy and 32 per cent in open situations; comparable figures for the kangaroo rats (n = 12) were 2 and 98 per cent, showing clearly that the two species are separated by differences in microhabitat. That *D. ordii* gathered its millet seeds from open patches in a sea of grass agrees with our knowledge of its other habitats. It lives also in saltbush barrens (*Atriplex canescens*) and in mesquite-covered dunes (*Prosopis juliflora*) in central New Mexico.

Using the formulae of Levins (1968), I calculated estimates of α, the intensity of competition. These turn out to be α_{of} = 0.33 and α_{fo} = 0.56, values which according to May and MacArthur (1972) are sufficiently small to account for coexistence.

Seed Size Selection

Brown and Lieberman (1973) and Brown (1975) collected the seeds out of the pouches of snap-trapped heteromyids and sieved them into size groups. Their material came from a pair of transects hundreds of miles in length, and they concluded that heteromyids are selective in the size of seed which they take, partitioning supplies according to size. This conclusion is so appealing to a modern ecologist steeped in work on niche partitioning, that one must immediately fear it will be hard pressed to attract a 'loyal opposition'. Fortunately heteromyids themselves are also attractive, and other workers have amassed data relevant to the conclusion. Let us examine the significance of these data from the point of view of one interested in the Brown and Lieberman pattern.

First, if the pattern is accurate, no-one understands why. Present theories which might lead one to predict a pattern depend on larger animals having an energetic advantage when they seek out and take larger packages of food, and do not seem to apply in this case. For birds the crucial part of pursuit has been thought to be hulling and handling of seeds. But, unlike birds, rodents do not have to crack seeds open all at once. Instead, they can gnaw and nibble at their seeds. Consequently, even for a small rodent, a big desert seed is just as easily and efficiently (or even more efficiently) taken as is a small one. Thus larger rodents are stranded with no seed sizes left on which to specialise (Rosenzweig and Sterner, 1970).

Brown has suggested to me some ingenious modifications of existing theory which concentrate on mobility differences among rodents, for example, that rodents may find some sizes of seed more readily than others. Also shifting attention from pursuit activities to search activities, deserves careful investigation whether or not it applies in this particular case. However, other workers have been unable to discern the pattern of seed selection suggested by Brown's studies. Smigel and Rosenzweig (1974) examined seed size selection in the two species that were most often caught at the same trap stations in their study plots (*P. penicillatus* and *D. merriami*). Both species selected their seeds and in much the same way. Brown also found these species — despite their size difference — to be collecting similarly sized seeds and exploiting different microhabitats. On the other hand, in central New Mexico, Lemen (unpublished) studied three known habitat selectors which, according to Brown and Lieberman, might be expected also to have selected different seed sizes: *D. ordii* (weighing $\sim$ 45 g), *D. merriami*

(~ 40 g), and *P. flavus* (~ 8 g). Seeds collected from their cheek pouches over several seasons were weighed, not sifted, and mean seed size for each animal was obtained by averaging representative weights of each seed species. This technique tends to overestimate the average weight of seeds collected by an animal, because smaller seeds are more common, but it should detect differences between feeding habits if many individuals are sampled and the species really do prefer different sizes of seed. Lemen's findings were that 37 *P. flavus* had collected seeds averaging 154 mg, 38 *D. merriami*'s seeds averaged 294 mg, and 28 *D. ordii*'s seeds averaged 152 mg. Thus there was no significant difference in the size of seed taken by *D. ordii* and *P. flavus*, though *D. merriami* took significantly larger seeds than either of the other species. This result conflicts partly with that of Brown and Lieberman, but the difference is due solely to the fact that in the habitat of *D. merriami*, large seeds of *Larrea tridentata* (creosote bush) predominate and are the main food of the species, whereas in grassier places the smaller seeds of *Salsola kali* (Russian thistle) are prominent in the diet of both *D. ordii* and *P. flavus*.

In an unpublished work Lemen recently reviewed papers by Arnold (1942), Blair (1937), Dunham (1968), Franz *et al.* (1973), Monson (1943), Reynolds (1950), Reynolds and Haskell (1949) and Shaw (1934), relevant to seed selection by heteromyids. Samples of seeds mentioned in these publications were obtained and weighed. A synthesis of this information revealed no positive correlation of seed size and body size.

The paradox between these results and those of Brown and Lieberman (1973) and Brown (1975) has been resolved by Lemen who, through the kindness of Dr Brown, obtained the actual seeds used to obtain the results of these papers. Lemen reanalysed Brown's seeds using their weight as the criterion of their size. By this criterion the relationship between body-size and seed-size was random. Figure 9.1 shows these results for the Brown and Lieberman (1973) paper. Examination of the actual seeds provided a simple explanation. The smaller the heteromyid rodent, the more likely it was to have hulled or trimmed seed before pocketing it. This is a very interesting pattern in itself, possibly arising from a shortage of pouch capacity (Lemen, unpublished) and perhaps favoured because small heteromyids, trimming their seeds under dense cover, incur a smaller predation risk than large species which are less protected. Trimming, however, does not affect competitive exclusion; whether trimmed or not, the seed is in any case consumed by the animal which finds it.

It is too early to reject totally the concept that large and small heteromyids select matching sizes of seed as a basis for coexistence. Although trimming does not take place by mere weathering, the dispersal of seeds from fruits or heads and seed hulling does, and seeds may become naturally unavailable to the larger heteromyids as wind and dry conditions break them up and hull them. However, it is also true that heteromyids are clearly proficient, active hullers of seeds. The seed size pattern may yet prove the basis for the coexistence of the very largest of kangaroo rats (100–150 g) with smaller ones. No pattern of habitat selection has yet been discovered which would otherwise explain the coexistence of these two groups, and whole fruiting heads seem to appear more often in the pockets of the largest forms.

Habitat selection is not always to be expected as the basis of niche separation

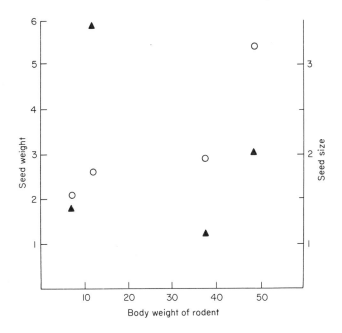

Figure 9.1 Body size (g body weight) against seed size. ○, Seed size averages obtained by Brown and Lieberman using their method of sieving seeds. The data plotted this way show an apparent strong correlation between body size and seed size. Only four out of six points are represented because the remaining two were derived from small samples. ▲, The same set of seeds analysed by Lemen according to seed weight (g × 10^{-1}); the correlation disappears. See text for explanation. (Figure from unpublished manuscript by Lemen.)

in granivorous rodents, as two cases of resource allocation in Heteromyidae show. Kenagy (1972) found that *D. microps*, by virtue of its behaviour and its chisel-like incisors, can depend on eating halophytic leaves. It strips the salty outer layers from the leaves and consumes the nutritious insides. Kenagy (personal communication) has considerable evidence that the species does in fact depend on this in nature. Another species, *P. baileyi* may specialise on jojoba (*Simmondsia chinensis*) seeds. This possibility was suggested by Rosenzweig and Winakur (1969), and has received support from Sherbrooke (in press). Sherbrooke discovered that *P. baileyi* virtually maintained its weight on a pure jojoba diet whereas other species died rather than even sample the food. It is not clear whether a toxin is involved or whether the unique whale-oil-like substance that forms half the seed's weight discourages (most) potential consumers.

Species Diversity

Considering the apparent dominance of vegetational structure in determining the niches of these mammals, it is not surprising that their species diversity can be predicted with a model of vegetational complexity (Rosenzweig *et al.*, 1975).

This model depends on subdividing the foliage into strata, but is quite different from that developed for birds by MacArthur *et al.* (1966). The strata are different, reflecting the different perspective of a nocturnal rodent, and the model incorporates the fact that some rodent species specialise on absence of foliage. Other factors also affect the diversity of these rodents. Brown (1975) and Brown and Lieberman (1973) carefully eliminated effects of vegetational structure. By selecting habitats of the same structure over a wide geographical area, they determined that species diversity also depends strongly on the dependable productivity of a place. Finally, as might be expected in a group of desert-adapted species, the diversity of the Heteromyidae depends positively on any reasonable index of aridity one might select (Rosenzweig, unpublished).

Why Habitat Selection?

The first criticism levelled at seed size selection — that is, that no one yet understands why it should generate selective advantages in desert rodents — can also be levelled at habitat selection. Perhaps habitat selection has been favoured because different predator defences are required in different foliage structures? Possibly foraging techniques must be different, or radically different search images are required.

Hoover *et al.* (unpublished) have suggested that the soil difference which separates *P. penicillatus* from *P. intermedius* (the latter is found only on rocky slopes), is based on different physiological requirements for success in the two habitats. The rocks permit a wider temperature fluctuation, which *P. intermedius* can tolerate much better than its congener. On the other hand, Ogsden (personal communication), observing behavioural differences between the species, suspects that the rocky substrate demands a strict quadrupedal gait, whereas the flat desert demands more bipedalism.

Physiological differences may indeed be responsible for this and other macrohabitat differences, but are unlikely to account completely for microhabitat preferences. Rosenzweig (1973) reported that *P. penicillatus* avoids a spot 8 m from the closest shrub, but does not avoid one only 4 m from it; it is unlikely that physiological conditions differ significantly between such points.

Intensity of Competition

The least pretentious formulation of the Principle of Competitive Exclusion is that two species with identical niches cannot coexist. More interesting has been the realisation that a degree of niche overlap which is less than complete may also result in competitive exclusion. The question then becomes, how much overlap is tolerable? This amount has been called limiting similarity (MacArthur and Levins, 1967; May and MacArthur, 1972). An error is introduced when one concludes that in a mature ecosystem, niches will overlap to the tolerable extent, and not less. The usual procedure for making predictions about mature ecosystems should be to use the theory of natural selection, and discover which phenotypes will enjoy maximum fitness. This should certainly be done when using the theory of limiting similarity to predict the competitive structure of mature ecosystems, because that which is tolerable is not necessarily optimal.

Rosenzweig (1974) explored the concept of optimal niche overlap by examining the question of optimal habitat selection in the case where there are two favourable habitats, and any number of unfavourable ones. A novelty of this treatment was the conclusion that habitat selection is an all-or-nothing phenomenon. If an individual should ignore one of the two habitats on some occasions, then it should always ignore this habitat. It may need to travel through, but selection should keep it from exploiting resources in the patch. Such habitat selection should occur if, and only if

$$\frac{\ln W_G}{\ln W_B} > 1 + \frac{t_M}{t_G} \tag{1}$$

where $\ln W_G$ and $\ln W_B$ are fitness (preferred patch and other favourable patch, respectively), t_G, the time actually spent exploiting the average preferred patch, from the time it is first encountered until the time it is abandoned owing to its depleted resources, and t_M is the average time spent travelling from one preferred patch to the next.

If there is only one species in the environment the inequality (1) is greatly affected by density. As the density of a habitat-selecting species increases, its fitness in the preferred patch falls, until finally the inequality is no longer satisfied. Fretwell and Lucas (1970) were the first to point this out, except that they tacitly assumed no travelling time. Thus, they predicted that the secondary habitat should receive individuals only when $\ln W_G$ had been lowered to equal $\ln W_B$. In (1), we see that habitat selection will be abandoned even before $\ln W_G$ has dropped to $\ln W_B$ (unless habitat selection is cost free; that is, $t_M = 0$).

Now we shall add a second species to our model ecosystem. Let us assume that it is a patch B specialist and needs to select patch B exclusively whenever its own population density is low. Finally let us define the proportional pressure of each population on the resources of its preferred patch as N_i/K_i where N_i is the density and K_i is the carrying capacity of the ith species in its preferred patch.

If the covariance of the two terms N_i/K_i is high, then high density of one species means high density of the other, and individuals cannot seek relief from high densities by abandoning habitat selection. For example, imagine both populations at equilibrium in both of their respective patches (figure 9.2). Individual $(K_2 + 1)$ which stays in patch B lowers the mean fitness of his species below zero. But it is not very far below zero. If on the other hand individual $(K_2 + 1)$ moves to patch G, it enters an environment in which its fitness is already substantially below zero. Thus it is best for that individual and others of its species not to abandon their habitat selection.

The importance of a high positive covariance is great. Without it, opportunities for profitable invasion of the poorer patch can be expected and will eliminate absolute habitat selection. With it, habitat selection remains powerful; neither species attempts to exploit the other's special world; and, even though they may be observed together, neither should evolve any sort of interspecific aggressiveness towards the other except perhaps as a basic species recognition display.

Under these circumstances, the optimal, mature amount of niche overlap and, consequently, of competition will often be zero. (An exception occurs if the resources are very mobile and move from patch to patch.) Yet competition

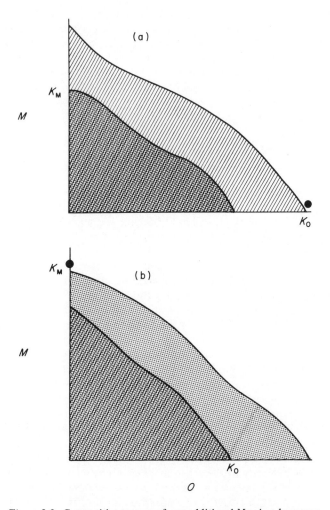

Figure 9.2 Competitive success of one additional Merriam kangaroo
rat in two habitats. (*a*), The competitive dynamics in patches of
grass; (*b*), the same in patches of open creosote bush. The represen-
tation is by zero isoclines: each axis is the density of one species;
the stippled area covers density compositions at which *D. merriami*
increases; the shaded, compositions at which *D. ordii* increases. In
(*a*) the additional Merriam rat has entered the grass, and the community
composition is represented by the large dot, which is very far from
the nearest rat isocline point. In (*b*) the extra rat stays in the open
creosote bush and the large dot is quite near the isocline.

remains important: natural selection to avoid competition is the mechanism that
has shaped the system. Moreover, the threat of competition hovers like a ghost
over the system and is necessary to its maintenence. Were this threat to disappear,
such as by uncoupling of N_i/K_i, then competition itself would return.

There is one case in which the α value of desert rodents has been measured experimentally. Schroder and Rosenzweig (1975) selected two species that are extremely similar morphologically, *D. ordii* and *D. merriami*. Although these two are found together over much of their geographical ranges, were actually caught at many of the same trap stations in our study areas, and had been reported to take somewhat different sizes of seeds by Brown and Lieberman (1973), they are absolute habitat selectors in our study area and do not compete there.

I am inclined to believe that the fact that our kangaroo rats fit in with the theory is not accidental. Consider where one might expect the covariance of N_i/K_i to be high. This should occur in a place where densities are quite stable and always near K_i; the tropics is commonly believed to be such a place. But it should also occur where one dominating abiotic variable determines environmental quality for the ecosystem as a whole, and is quite variable from year to year. Water supply in a desert is just such a variable. Went and Westergaard, (1949) showed that a good year for some desert ephemerals is generally a good year for all. Beatley (1969) showed the dominant effect of precipitation on the reproduction of many species of desert rodents.

It is highly speculative to suggest this, but one is tempted to account even for the loss of small mammal diversity in mesic habitats with this theory. After all, the covariance of N_i/K_i varies as the aridity. In environments with lowered covariance, mammal species should utilise a broader range of habitats. The resulting fine-grained competition should reduce the rodent diversity even in the presence of a vegetationally complex habitat such as a temperate deciduous forest. This application is speculative, but the two kangaroo rat species seem to agree with the theory, suggesting that at least it deserves further investigation.

Measuring α is often difficult and time consuming, and various formulae have been used to estimate it (for example, MacArthur and Levins, 1967; Levins, 1968), starting with an empirical measurement of the resource or spatial distributions of a pair of species. Dayton (1973) has criticised the formulae cogently, pointing out the dangers in using them when it is uncertain even that the objects of study are competing. The work on kangaroo rats enforces Dayton's criticism because the shortcut formulae fail to generate an accurate estimate of their competition. They fail because the kangaroo rats are habitat selectors and can be found in habitats which they do not exploit. Such behaviour need not be pathological, and we must be suspicious of any applications of these formulae to data involving spatial distributions *unless we also know what the animals are actually doing when observed in each place.* One must conclude that there are no shortcuts to painstaking field work.

Acknowledgements

Cliff Lemen supplied a considerable amount of unpublished data from his PhD thesis. This work benefited by criticism from my colleagues J. H. Brown, H. R. Pulliam and W. M. Schaffer. The US National Science Foundation supported the research.

References

Arnold, L. W. (1942). Notes on the life history of the Sand pocket mice. *J. Mammal.*, **23**, 339 - 341.

Beatley, J. C. (1969). Dependence of desert rodents on winter annuals and precipitation. *Ecology*, **50**, 721 - 724.

Blair, W. (1937). Burrow and food habits of the Prairie pocket mouse. *J. Mammal.*, **18**, 188 - 191.

Bowman, R. I. (1961). Morphological differentiation and adaptation in the Galapagos finches. *Univ. Cal. Publ. Zool.*, **58**, 1 - 302.

Brown, J. H. (1975). Geographical ecology of desert rodents, in *"Ecology and Evolution of Communities"*. (eds. M. L. Cody and J. M. Diamond.) Harvard University Press, Cambridge, Massachusetts, 315 - 341.

Brown, J. H. and Lieberman, G. A. (1973). Resource utilisation and coexistence of seed eating desert rodent in sand dune habitats. *Ecology*, **54**, 788 - 797.

Dayton, P. K. (1973). Two cases of resource partitioning in an intertidal community: making the right prediction for the wrong reasons. *Am. Nat.*, **107**, 662 - 670.

Dunham, M. (1968). A comparative food habit study of two species of kangaroo rats, *Dipodomys ordii* and *Dipodomys merriami.* M.S. thesis, University of New Mexico, Albuquerque. 25 pp.

Emlen, J. M. (1966). The role of time and energy in food preference. *Am. Nat.*, **100**, 611 - 617.

Fretwell, S. D. and Lucas, H. L., Jr (1970). On territorial behaviour and other factors influencing habitat distribution in birds. I. Theoretical development. *Acta Biotheor.*, **19**, 16 - 36.

Franz, C. E., Reichman, O. J. and Van de Graeff, K. M. (1973). Diet, food preferences and reproductive cycles of some desert rodents. Research Memorandum, RM 73 - 74.

Lack, D. L. (1947). *Darwin's Finches.* Cambridge University Press, London.

Levins, R. (1962). Theory of fitness in a heterogeneous environment. I. The fitness set and adaptive function. *Am. Nat.*, **96**, 361 - 373.

Levins, R. (1968). *Evolution in Changing Environments*, Princeton University Press, Princeton, New Jersey.

Kenagy, G. J. (1972). Saltbush leaves: excision of hyper-saline tissue by a kangaroo rat. *Science*, **178**, 1094 - 1096.

MacArthur, R. H. and MacArthur, D. (1972). Efficiency and preference at a bird feeder. *J. Ariz. Acad. Sci.*, **7**, 3 - 5.

MacArthur, R. H., Recher, H. and Cody, M. (1966). On the relation between habitat selection and species diversity. *Am. Nat.*, **100**, 319 - 332.

MacArthur, R. H. and Levins, R. (1967). The limiting similarity, convergence and divergence of coexisting species. *Am. Nat.*, **101**, 377 - 385.

MacArthur, R. H. and Pianka, E. R. (1966). On optimal use of a patchy environment. *Am. Nat.*, **100**, 603 - 609.

May, R. M. and MacArthur, R. H. (1972). Niche overlap as a function of environmental variability. *Proc. natn. Acad. Sci. U.S.A.*, **69**, 1109 - 1113.

Monson, G. (1943). Food habits of the Bannertailed Kangaroo Rat in Arizona. *J. Wildl. Mngmt.*, **7**, 98 - 102.

Reynolds, H. G. (1950). Relation of Merriam Kangaroo Rats to range vegetation in southern Arizona. *Ecology*, **31**, 456 – 463.

Reynolds, H. G. and Haskell, H. S. (1949). Life history notes on Price and Bailey Pocket Mice of southern Arizona. *J. Mammal.*, **301**, 150 – 156.

Rosenzweig, M. L. (1973). Habitat selection experiments with a pair of coexisting heteromyid rodent species. *Ecology*, **54**, 111 – 117.

Rosenzweig, M. L. (1974). On the evolution of habitat selection. In *Proc. 1st Int. Congr. Ecology*, 401 – 404.

Rosenzweig, M. L., Smigel, B. and Kraft, A. (1975). Patterns of food, space and diversity, in *Rodents in Desert Environments.* (eds. I. Prakash and P. Ghosh), Monographiae Biologicae, 28, Dr W. Junk, bv, The Hague. 241 – 268.

Rosenzweig, M. L., and Sterner, P. (1970). Population ecology of desert rodent communities: body size and seed-husking as bases for heteromyid coexistence. *Ecology*, **51**, 217 – 224.

Rosenzweig, M. L. and Winakur, J. (1969). Population ecology of desert rodent communities: habitat and environmental complexity. *Ecology*, **50**, 558 – 572.

Schroder, G., and Rosenzweig, M. L. (1975). Perturbation analysis of competition and overlap in habitat utilisation between *Dipodomys ordii* and *Dipodomys merriami. Oecologia*, **19**, 9 – 28.

Shaw, W. T. (1934). The ability of the giant kangaroo rat as a harvester and storer of seeds. *J. Mammal.*, **15**, 275 – 286.

Sherbrooke, W. C. (1976). Differential acceptance of toxic jojoba seed (*Simmondsia chinensis*) by four Sonoran desert heteromyid rodents. *Ecology*, **57**, 596 – 602.

Smigel, B. W. and Rosenzweig, M. L. (1974). Seed selection in *Dipodomys merriami* and *Perognathus penicillatus. Ecology*, **55**, 329 – 339.

Went, F. W. and Westergaard, M. (1949). Ecology of desert plants. III. Development of plants in the Death Valley National Monument, California. *Ecology*, **30**, 26 – 38.

10 Courtship-feeding and clutch size in *Common terns* Sterna hirundo

Dr I. C. T. Nisbet*

Two of David Lack's most original and important contributions to the study of breeding biology in birds were his discussions of the significance of courtship-feeding (Lack, 1940) and the determination of clutch size (Lack, 1947). He originally argued in general terms that clutch size is adjusted, through the action of natural selection, to correspond to the maximum number of young that the parents can raise successfully (Lack, 1954). However, he agreed in his later writings that in some species clutch size may be limited below this optimum number by the nutritional reserves available to the female at the time of laying (Lack, 1963; 1966; 1968). If the nutritional state of the female is indeed limiting in this way, the phenomenon of courtship-feeding takes on special significance, because the food provided by the male may contribute directly to the reserves required for egg laying (Lack, 1966 p. 23; 1968 p. 287). There is now evidence in several species that courtship-feeding contributes significantly to the nutritional intake of the laying female (Royama, 1966; Brown, 1967; Krebs, 1970; Nisbet, 1973). This paper reports recent observations of the relationships between courtship-feeding, clutch size, egg size, and the weight of laying females in Common terns *Sterna hirundo*.

The Context of Courtship-feeding in Common Terns

Courtship in Common terns takes place in three main phases (Tinbergen, 1938; Palmer, 1941; Cullen, 1956; Nisbet, 1973). In the first, or 'pair-formation' phase, males carry fish around the colony and display with them to prospective mates. At first, most of the display is aerial, and both males and females sometimes perform display flights with several potential mates in quick succession. The displaying males attempt to lead the females down to territories established within the nesting area, where they display to them with fish. At first the males are very reluctant to give up their fish, but as pair bonds become established they start to feed the females. The first copulations take place late in this phase of courtship, usually immediately following feedings.

*Ian C. T. Nisbet was born in Britain, and obtained his Ph.D. in physics at Cambridge University in 1958. After working in departments of engineering and physics in the United States, Britain, and Malaysia, he returned to the United States in 1968 and is now Director of Scientific Staff, Massachusetts Audubon Society. He is now specialising in Environmental biology and in studies of the fate and effect of toxic chemicals. He has been studying the behaviour and ecology of terns in Massachusetts since 1970.

The second, or 'honeymoon' phase starts after the pairs are formed and lasts for roughly 5 - 10 days. The pairs then spend much of the daytime on the feeding grounds, where the male feeds the female frequently, usually with little accompanying display. They visit the nesting territory intermittently during this period, but appear to spend less time there than in the later part of the pair-formation phase.

In the third, or 'egg-laying' phase the female spends most of her time in the nesting territory, leaving it primarily to drink and bathe: the male brings fish to the territory and feeds her there. After 1 - 6 days the first egg is laid; the frequency of feeding drops off rapidly after the second and third eggs are laid (Nisbet, 1973).

In 1971 I made a detailed study of twelve pairs during the egg-laying phase of courtship: I found that the amount of food fed to the female by the male was correlated with the total weight of her clutch (Nisbet, 1973). This supports the hypothesis that the food furnished by the male contributes significantly to the nutritional reserves of the female and hence to the weight of her clutch. However, at least three questions were left unresolved by this study:

(1) The results did not distinguish between the possible effects of courtship-feeding on the number of eggs and on the size of the eggs. The performance of the male was associated with both (Nisbet, 1973: table 1), but not significantly with either independently of the other.

(2) The study was limited to the third phase of courtship, when the male has to spend much of his time flying to and from the feeding grounds. In the 'honeymoon' period, when the female is usually near him, he has the opportunity to provide food much more rapidly.

(3) No measurements were made of the condition or performance of the females. I have since found that 'high quality' males (as measured by their performance in bringing food to their growing young) tend to be mated to 'high quality' females. Thus it is possible that the correlation observed in 1971 between the performance of the males and the weight of their mates' clutches may have been due in part to their selection of mates with comparable performance.

Since 1971 I have continued observations on courting and laying Common terns in attempts to throw further light on these questions.

Feeding Ecology of Common Terns in Two Areas

In 1970, 1971 and 1972 I studied Common terns in and around a colony at Bird Island, Marion, Massachusetts (41°40'N, 70°43'W). In 1974 and 1975 I worked primarily near a colony at Monomoy Island, Chatham, Massachusetts (41°38'N, 69°58'W), with occasional visits to Marion.

The feeding behaviour of the terns in the two areas differed markedly, apparently in relation to differences in topography and marine conditions. The shoreline around Marion is highly convoluted, including a mixture of sandy and rocky shores with extensive salt marshes. Many of the Common terns from Bird Island defended feeding territories along the shoreline and appeared to spend a substantial fraction of their time foraging there. Alternatively, during the rising and falling tides many terns fed in shallow water within 1 km of the colony, primarily by picking shrimps and other invertebrates from near the surface of the water. During the egg-laying period the Bird Island terns spent little time feeding in flocks over schools of fish.

In contrast, at Chatham, most of the shoreline is straight, open and sandy. The Monomoy terns did most of their feeding in flocks, especially on the rising and falling tides when schools of small fish were swimming over shallow sand bars. Comparatively few appeared to spend much time in feeding territories along the shore. Although the Monomoy terns often caught and ate invertebrates on the feeding grounds, they rarely fed them to their mates.

Courtship-feeding, Egg Size and Clutch Size in Two Colonies

In May 1975 with three assistants I studied courtship-feeding during the egg-laying period in the Monomoy colony, using the methods employed in 1971 at Bird Island (Nisbet, 1973). A hide was set up in the colony overlooking an area where 12 – 15 female Common terns were being fed by their mates. The birds were watched for a total of 24.5 hours between May 22 and 26. Each item of food fed to the females was observed from a range of 4 – 12 m. Small sea herrings (*Clupea harengus*) were the commonest food: their lengths were estimated relative to the birds' bills and their weights were then estimated using length – weight curves derived from samples of herrings taken near the colony.

Table 10.1 summarises the results of this study and compares them with the results of the 1971 study at Bird Island. The figures given for feeding rates in each study are limited to observations made on females within 3 days before and 2 days

Table 10.1 Comparison of courtship-feeding and egg-laying performance at two colonies of Common terns (*Sterna hirundo*)

	Bird Island (Nisbet, 1973)	Monomoy (this work)
Year of study	1971	1975
Courtship-feeding in the egg-laying phase*:		
Dates of study	May 21 – 28	May 22 – 26
No. of pairs watched	4 – 11	11 – 13
Total length of watches (h)	76	24.5
No. of feeds per female per hour	1.57	1.34
Percentage of shrimps among items fed	81 per cent	0 per cent
Percentage of small fish (< 1 g)	1 per cent	98 per cent
Percentage of large fish (1 – 12 g)	18 per cent	2 per cent
Estimated mean feeding rate (g per female per hour)	1.9	0.7
Egg and clutch characteristics†:		
Mean clutch size	2.95	2.78
Mean weight of first egg (g)	21.1	20.2
Median laying date	May 25	May 28

*Data limited to feeding of females within 3 days before and 2 days after laying of their first egg (see text).

†Clutch sizes based on 100 clutches at Bird Island and 250 at Monomoy; egg weights and laying dates based on 35 clutches at Bird Island and 45 at Monomoy. The figures refer only to the peak of nesting; late laying birds are not included.

following the laying of their first eggs, since the 1971 study had shown that their attendance at the territory and the males' feeding, are most consistent in this period (figure 1 of Nisbet, 1973).

The figures in table 10.1 indicate that the laying females at Bird Island received 2 - 3 times as much food as those at Monomoy. Although the number of feeds per hour was slightly higher at Bird Island, the principal reason for the difference was the much larger number of large fish fed at Bird Island. About half of these fish consisted of mature silversides (*Menidia menidia*), together with a variety of other fish including sandlaunce (*Ammodytes americanus*), common mummichogs (*Fundulus heteroclitus*), and cunners (*Tautogolabrus adspersus*). Most of these fish were estimated to be in the weight range 3 - 12 g, larger than any fish fed at Monomoy. At Bird Island most of these fish were probably brought from distant feeding territories, since terns were not seen catching them near the island and individual males brought them only at infrequent intervals (40 - 60 min or more between feedings). At Monomoy, however, terns rarely caught silversides along the shore, and despite regular seining for fish in 1973 and 1974 I did not catch any silversides until June 15. Thus a principal reason for the better performance of the male terns at Bird Island appears to have been the earlier availability of silversides inshore there in 1971.

The last three lines in table 10.1 show that the female terns at Bird Island laid earlier than those at Monomoy, and laid larger clutches with larger eggs; each of the differences is statistically significant ($P < 0.01$). These differences between the two colonies have been greater in other years of the study (1972 - 75); mean clutch sizes at Bird Island have been consistently in the range 2.81 - 2.96, whereas those at Monomoy varied from 2.47 to 2.57 between 1972 and 1974; mean egg weights have been in the range 21.1 - 21.8 g at Bird Island, as against 20.3 - 20.5 g at Monomoy; the median laying date at Bird Island was about 14 days earlier in 1972 than that at Monomoy, and 9 - 10 days earlier in 1973, 1974, and 1975.

It is unlikely that these differences in laying performance between the colonies can be explained by differences in age composition. Trapping of banded birds of known age in 1975 showed that 4 - 5-year-old birds tend to lay smaller clutches and to lay later than older birds in each colony. However, these young birds are well represented in the Bird Island colony, and in any case they did not lay smaller eggs than older birds in 1975. Hence it seems likely that at least some of the differences in clutch size, egg size, and laying date may be attributable to the greater food intake of the females at Bird Island. As shown in the next section, the females at Bird Island probably received more food in the 'honeymoon' period, as well as in the egg-laying period for which data are given in table 10.1.

Courtship-feeding During the 'Honeymoon' Period

In each of the six years of this study I have spent some time watching Common terns during the 'honeymoon' phase of courtship. Although these observations were made during gaps in other work and were not planned systematically, they cover a number of different feeding areas and most hours of the day and states of the tide.

I was unable to catch terns before egg laying and to make extensive observations of marked individuals or pairs. However, by watching regularly in the same feeding territories it was possible to observe pairs on several successive days: in many cases the same perches were used regularly and it could reasonably be presumed that the same pairs were involved. It soon became clear, however, that the pairs spent only a fraction of their time in the feeding territories. At Marion, during 85 hours' observations in actively used feeding-areas, the presumed territory-owners were present for only 37 hours. At Chatham, during 42 hours' observation, the presumed territory-owners were present for only 14 hours. I was not able to determine where the birds spent the rest of the day, although from observations in the nesting area I believe that they did not spend much time there.

When a 'honeymooning' pair is in the feeding territory the female usually stands on the beach, on a rock, on a boat, or some other prominent place, while the male catches and brings fish to her. Occasionally she may fly to another perch, but she rarely takes part in defending the territory (except to posture and call when intruders fly past), and rarely seeks food for herself. In the course of 51 hours' observation, birds known to be females were seen to dive only eight times (although for part of this period I did not know the sex of the bird or birds present).

During May 1975 I also spent some 13 hours among flocks of Common terns feeding over sand bars at Chatham. Although the majority of birds were eating the fish that they caught, a number (presumably males) carried fish away towards the colony. A few carried fish to nearby beaches and fed females there. On seven occasions I saw females being fed by males in a feeding flock: on each occasion the female begged to the male in the air and alighted briefly on the water to receive the fish. Although it is difficult to follow individual birds for long in a flock, I was able to watch five of these females for up to 4 min after they were fed. None of them made any attempt to catch a fish for herself; instead each stayed close to the male, begging when he dived. Although these observations are very limited, they suggest that even in a flock-feeding situation some females rely primarily or entirely on their mates for food during this period.

In the feeding territories the males hunt intermittently, sometimes perching for up to 15 – 20 min after a successful bout of feeding. Their activity may be responsive to the behaviour of the females, whose begging varies markedly in intensity. However, I have rarely seen a female refuse food brought to her; the few refusals that I have observed followed several feedings in quick succession. It appears that the females are occasionally satiated, but this may occur only at favourable times in the day, or in good territories. For much of the time the females eagerly solicit food, so it seems likely that they would accept more food if the male could bring more. Thus, although the males do not work all the time, it is likely that the more successful males would supply more food than those with less competence or with poorer territories.

During 42 hours' observation at Marion, the males under observation brought 151 food items to their mates. The average frequency of feeding (3.6 items per female per hour) was thus 2 – 3 times larger than that observed in 1971 during the egg-laying phase (table 10.1). In addition the average size of the items fed on the territories was significantly greater. Of the 107 food items identified there, 84 per cent were fish with an estimated median weight of about 2 g, whereas in the nesting-area over 80 per cent of the food items brought to the females were shrimps, with an estimated median weight of only 0.5 - 0.6 g (Nisbet, 1973). Although many of the feedings took place too far away to observe the size of the fish with any accuracy, I have estimated the rate of food intake by the females on the feeding territories as roughly 6 - 7 g/h, compared to less than 2 g/h in the egg-laying phase (Nisbet, 1973; figure 1; table 10.1). Thus, when the females were in the feeding territories, it appeared to be economical for the males to hunt for fish there, since they had to spend no more than 30 s flying to the females. However, when the females were in the colony a reliable supply of fish was more distant and the males concentrated for much of the day on catching shrimps near the island.

At Chatham also the average rate of feeding observed in the feeding territories (33 fish in 14 hours, or 2.4 fish per female per hour) was substantially higher than that observed in the nesting area (table 10.1). In addition, the median size of the fish caught on the feeding territories (estimated as roughly 1 - 1.5 g) appeared to be much larger than that of the fish brought to the colony (estimated as about 0.5 g). Again the commuting distance appeared to be a critical factor influencing the foraging strategy of the birds: along the shore larger fish were available, but when the females were in the colony the most economical food source appeared to be schools of very small herring fry running over sand bars much closer to the colony.

These observations were not very systematic, *inter alia* because individual pairs could not be followed throughout the day. However, they seem sufficient to show: (*a*) that during the 'honeymoon' phase the females depend on the males for most of their food; (*b*) that the males are able to provide food much more rapidly at this time than in the egg-laying phase, when they have to commute to and from the colony.

The estimate derived for the average feeding rate at Chatham (roughly 3 g per female per hour) appears substantially smaller than that estimated for birds during the 'honeymoon' period at Marion (6 - 7 g per female per hour). The significance of this difference should not be overstated, because each figure is based on only a limited number of pairs. However, it is at least consistent with the hypothesis advanced in the previous section, that the smaller clutches and smaller eggs laid by the females at Monomoy are a consequence of their smaller food intake.

Weights of Laying Females

Since 1971 I have caught 29 female Common terns during the egg-laying period, in most cases on the day on which their first or second egg was laid. The weights of these females and of their clutches are listed in table 10.2; the weights of the females are further summarised in table 10.3 and are compared with weights of males and females trapped later in the incubation period.

There was almost no overlap in weight between the females which subsequently laid two more eggs and those which were to lay only one more egg (table 10.2). The former group weighed 18 - 20 g more than the latter group, and 36 - 38 g more than birds trapped late in incubation (table 10.3). Since the eggs usually weigh 19 - 24 g (table 10.2), these differences indicate that immediately after laying their first egg, the females already have most of the reserves (as measured by total weight) that they need to lay the remainder of the clutch. So far I have been unable to catch any females before they laid their first eggs. However, since many of the birds listed in table 10.2 were caught within a few hours of laying their first eggs, it is reasonable to presume that they had most of the weight reserves required for the entire clutch immediately before laying. This in turn implies that the most important feeding period for them was before laying. If their food intake immediately before and after laying the first egg in fact contributes to the total weight of the clutch (as suggested in my 1971 study; Nisbet 1973), it probably constitutes no more than the icing on the cake.

Table 10.2 Weights of female Common terns (*Sterna hirundo*) caught during the egg-laying period, related to the size and weight of their clutches

Colony	Date	Weight of female (g)	Weight of clutch (g)*
After laying 1 egg of a clutch of 3			
Bird Island	22 May 1971	169	61.8
Bird Island	25 May 1971	166.5	67.6
Bird Island	27 May 1971	153.5	59.4
Bird Island	27 May 1971	158	66.9
Bird Island	21 May 1972	172	71.5
Bird Island	21 May 1972	157.5	Not weighed
Bird Island	2 June 1973	160.5	63.2
Monomoy	26 May 1974	162	60.8
Monomoy	31 May 1974	164	64.1
Monomoy	4 June 1974	153.5	58.6
Monomoy	5 June 1974	166	66.4
Monomoy	6 June 1974	148.5	60.9
After laying 2 eggs of a clutch of 3			
Bird Island	22 May 1972	129	63.0
Bird Island	10 June 1973	146	72.7
Monomoy	31 May 1974	142.5	60.8
Monomoy	4 June 1974	150.5	64.2
Monomoy	21 June 1975	137	Broken
After laying 1 egg of a clutch of 2			
Bird Island	1 July 1972	151	39.2
Bird Island	1 July 1972	145.5	41.8
Monomoy	4 June 1974	154.5	41.6
Monomoy	4 June 1974	149	44.1
Monomoy	4 June 1974	138.5	42.0
Monomoy	5 June 1974	134	38.9
Bird Island	17 June 1975	126	38.7
After laying 2 eggs of a clutch of 2			
Bird Island	2 June 1973	118.5	42.7
Monomoy	31 May 1974	123	39.8
Monomoy	31 May 1974	129	43.1
Monomoy	31 May 1974	106	40.3
Monomoy	21 June 1975	118	38.6

*Estimated fresh weight. In cases where the eggs were not weighed on the day of laying, a correction has been incorporated based on measurements of the average rate of weight loss during incubation.

Table 10.3 Weights of female Common terns (*Sterna hirundo*) during and after
the egg-laying period

Time when weighed	Number	Mean weight ± s.e. (g)	Range (g)
After laying 1 egg of c/3	12	161 ± 2.0	148.5 – 172
After laying 2 eggs of c/3	5	141 ± 3.7	129 – 150.5
After laying 1 egg of c/2	7	143 ± 3.9	126 – 154.5
After laying 2 eggs of c/2	5	119 ± 3.8	106 – 129
Late in incubation (both sexes)*	116	125 ± 0.7	106 – 147
Late in incubation (known females)	21	123 ± 1.7	109 – 134

*Note: There are significant differences between years and/or between colonies. The mean
weight of 82 Common terns caught late in incubation in three Massachusetts colonies in 1975
was 126.7 ± 0.9 g (range 108 – 147 g) compared to 122.7 ± 1.2 g (range 106 – 143.5 g) for 34
birds from the same colonies in 1973 – 74. The mean weight of 265 Common terns caught at
Great Gull Island, New York, in 1970 was 120.4 g (range 103 – 145 g) (LeCroy and LeCroy,
1974).

Among the females trapped after laying one egg of a clutch of three (first
section of table 10.2), the weights of the birds were correlated with the weights
of their clutches (Spearman rank correlation coefficient = + 0.670, $P < 0.05$).
This supports the hypothesis that much of their excess weight consisted of
material subsequently incorporated into their eggs.

Conclusions and Discussion

The principal findings of this study are:

(1) That the females already have most of the weight reserves required for
their clutches when they start to lay.

(2) That the most important period of food intake by the females is during
the 'honeymoon' period.

(3) That the females depend primarily on their mates for food even during
this period, a number of days before egg laying.

(4) That the amount of food provided by the males at the two colonies
differed substantially, apparently as a result of local differences in food availability.

(5) That this difference was associated with differences in average egg-size,
clutch-size, and laying date.

Findings (1) – (3) suggest that feeding by the males makes a substantial contribution to the nutritional requirements of the laying female. Findings (4) – (5)
suggest further that the amount of food which the males can provide is one of
the factors limiting the number and size of the eggs. This conclusion, drawn
from a comparison between colonies, adds to the evidence that the performance
of individual males is associated with the number and/or size of the eggs laid by
their mates (Nisbet, 1973).

Nevertheless, these observations do not exclude the possibility that clutch
size and egg size are limited also by characteristics of the female. The fact that
males do not forage all the time, even when the females' requirements are
greatest, suggests that the males' performance is not the sole limiting factor. It
remains possible that the clutch size is determined to a lesser or greater extent

by the female, and that her behaviour then determines how assiduously the male forages for her.

If the availability of food indeed limits clutch size and egg size, it might be asked why the female stops feeding for herself at the time when her requirements appear greatest. I suggest that this may be because she is then too heavy. Before laying a clutch of three eggs, the average female probably weighs about 180 g, 50 per cent more than her normal weight. When foraging, terns have to cruise for long periods at low speeds; catching fish requires precise control of their flight during dives. Even in good conditions, Common terns typically have to dive three or more times to catch each fish. It may well be that the heavy load carried by gravid females would be a major handicap during foraging, so that it is more economical for them to rest and rely on the male for food.

Acknowledgements

This chapter reports studies that were inspired directly and indirectly by questions posed originally by David Lack, and is dedicated with gratitude and respect to his memory. Among a number of field assistants who have helped with this study, I thank especially William A. Broad, A. Gordon Brown, Julia W. Davies, Elizabeth P. Mallory, and Karen J. Wilson. Fieldwork since 1973 has been supported by the Frederick W. Beinecke Fund. This is Contribution no. 143 from the Scientific Staff, Massachusetts Audubon Society.

References

Brown, R. G. B. (1967). Courtship behaviour in the Lesser Black-backed Gull, *Larus fuscus. Behaviour*, **29**, 122 - 153.

Cullen, J. M. (1956). A study of the behaviour of the Arctic Tern (*Sterna macrura*). D. Phil. thesis, University of Oxford.

Krebs, J. R. (1970). The efficiency of courtship feeding in the Blue Tit *Parus caeruleus. Ibis,* **112**, 108 - 110.

Lack, D. (1940). Courtship feeding in birds. *Auk*, **57**, 169 - 178.

Lack, D. (1947). The significance of clutch-size. *Ibis*, **89**, 302 - 352.

Lack, D. (1954). The Natural Regulation of Animal Numbers. Oxford University Press, Oxford.

Lack, D. (1963). Cuckoo hosts in England. *Bird Study*, **10**, 185 - 201.

Lack, D. (1966). Population Studies of Birds. Oxford University Press, Oxford.

Lack, D. (1968). Ecological Adaptations for Breeding in Birds. Methuen London.

Lecroy, M. and Lecroy, S. (1974). Growth and fledging in the Common Tern (*Sterna hirundo*). *Bird-banding*, **45**, 326 - 340.

Nisbet, I. C. T. (1973). Courtship-feeding, egg-size and breeding success in Common Terns. *Nature*, **241**, 141 - 142.

Palmer, R. S. (1941). A behavior study of the Common Tern (*Sterna hirundo hirundo* L.). *Proc. Boston Soc. nat. Hist.,* **42**, 1 - 119.

Royama, T. (1966). A re-interpretation of courtship feeding. *Bird Study*, **13**, 116 - 129.

Tinbergen, N. (1938). Erganzende Beobachtungen über die Paarbildung der Flussseeschwalbe (*Sterna h. hirundo* L.). *Ardea*, **27**, 247 – 249.

Section 3 Breeding adaptations and reproductive rates

David Lack's book on *Ecological Adaptations for Breeding in Birds* was, in his own words, primarily '. . . an interpretation of those adaptations of birds to the external environment which affect the number of young raised' and secondarily '. . . a review of nesting dispersion, the pair bond, clutch size, egg size, the incubation and fledging periods and the age of first breeding in birds'. He considered that '. . . the adaptations concerned are closely interrelated and have been evolved through natural selection in the natural habitat of the species.' The close interrelation of breeding adaptations in the achievement of successful reproductive strategies is the common theme of the chapters in this section.

In chapter 11, Dr Ian Newton, formerly a research student at the Edward Grey Institute, examines the ways in which Arctic geese have evolved towards a short breeding cycle – one which enables them to raise young in an environment where the period suitable for breeding and rearing is itself extremely short. Two chapters on aspects of co-operative breeding follow. Dr Hilary Fry of Aberdeen reviews a lengthening literature on birds which breed co-operatively, including those in which several adults combine to tend young ('helpers at the nest'), those which feed the young of several nests communally, and those in which several females contribute eggs and care to a common nest. The evolutionary implications of thse strategies for breeding are only just becoming apparent, and link with some of the questions on altruism raised in the opening chapters. Professor Gordon Orians and his ornithologically-minded family describe co-operative breeding in Argentinian blackbirds (chapter 13), examining its significance in other temperate species and possible mechanisms for its evolution. In chapter 14, Professor H. N. Kluyver and two of his colleagues from the Institute of Ecological Research, Arnhem, consider the opportunities taken by individual pairs of Great tits, arising from small differences in habitat and food supply, to reduce the time interval between first and second broods; these enable them to raise their second broods quickly, and so increase the likelihood of their survival.

The four following chapters are centred on a topic which concerned David Lack throughout his working life – the factors controlling clutch size. In chapter 15 Dr Dennis Owen, former research assistant at the Edward Grey Institute, develops Lack's theory of latitudinal gradients in clutch size, suggesting that the small clutches typical of tropical regions are due to the wide diversity of food species and predators to be found there. Dr Christopher Perrins, formerly a student of Lack's and now his successor as Director of the Institute, discusses the effects of predation on evolution of clutch size and associated nesting strategies, and Dr Robert E. Ricklefs develops, in chapter 17, a model which describes how predation, competition for food among nestlings and environmental variability interact to influence the optimum size of avian clutches. Finally in this section Drs Levin and Turner of the University of Texas consider some of the implications of clutch size (interpreted as number of ovules per flower or per head of flowers) in one family of dicotyledons. As in birds, clutch size in the Compositae shows variation with latitude, and convergence between taxa within the family as a whole.

11   Timing and success of breeding in tundra-nesting geese

Dr I. Newton*

This chapter is concerned with the factors that influence breeding output by tundra-nesting geese, and is a review of published information. In recent years, many detailed studies on individual species have been made but, to my knowledge, no synthesis of ideas. The main ideas favoured here are (1) that breeding in such geese depends largely on reserves of body fat and protein accumulated at wintering and migration areas, and (2) that its timing and success is influenced by the date in spring when snow melts and nest sites become available. The first idea is an extension of David Lack's (1967; 1968) last contribution on the evolution of clutch size in birds, while the second interested him many years earlier (Bertram *et al.*, 1934). Regarding geese, major contributors to the field discussed here have included Cooch (1958), Barry (1962), Hanson (1962), MacInnes (1962, 1966), Ryder (1970) and F. Cooke (papers in preparation).

Background

Some 12 species of goose in the genera *Anser* (including *Chen*) and *Branta* are involved; for Latin names of species mentioned see table 11.1. These are large, robust waterfowl, which inhabit relatively open landscapes and feed entirely, or almost entirely, on plant material. They nest at high latitudes, in tundra and forest tundra, and migrate south for the winter. Only the Canada goose in North America and the Greylag in Eurasia also breed south of the Boreal Zone.

North of the treeline, geese are restricted to damp fertile areas where sedges (*Carex* spp.) predominate, such as river deltas or near sea coasts. They nest either solitarily, or in loose colonies, which in some species usually contain only a few pairs and in others hundreds or thousands of pairs. The same species may breed solitarily in one area and colonially in another (Nyholm, 1965; Uspenski, 1965a; King, 1970). The largest concentration of Snow geese (on Wrangel Island, Siberia) numbered more than 400 000 birds (Uspenski, 1965), and several other colonies in northern Canada contained more than 300 000 (R. H. Kerbes, *in*

*After reading zoology at Bristol University, Ian Newton joined the Edward Grey Institute, Oxford, in 1961 and worked for six years on the ecology of British finches. This included a study of the problem of bud-eating by Bullfinches in fruit orchards. He then moved to the Nature Conservancy in Scotland to study wildfowl, and spent a year in North America again working on wildfowl. In recent years he has worked on sparrowhawks and other raptors, including the problems created for these birds by agricultural pesticides.

litt.). Within such colonies, the ganders defend territories of a few metres radius around a nest, and remain there while the females incubate (Cooch, 1958; Lemieux, 1959; Ryder 1972).

The actual nest sites of geese are often on islets surrounded by water (MacInnes, 1962; Ryder, 1967; King, 1970), on cliffs or bluffs (Kessel and Cade, 1958; Blurton Jones and Gillmor, 1959; Kumari, 1971), near nests of birds of prey (Uspenski, 1965; Cade, 1964), or in other places which supposedly offer protection against mammalian predators. Birds nesting solitarily or in small groups are usually in less accessible sites than those in large colonies, where the sheer numbers and synchrony of breeding confer protection (Kerbes, Ogilvie and Boyd, 1971). Nesting usually occurs in habitats which differ from, but are adjacent to, good feeding areas. The commonest clutch sizes are 3 – 5 eggs.

After hatching, the family may move many kilometres to good feeding areas, where they may live on their own or in groups of families. While the young grow, the adults moult their flight feathers, starting about 2 weeks after the eggs have hatched, so for a time the whole family is flightless. Some Greylag broods walked and swam more than 25 km in the early brood-rearing period (Paakspuu, 1972), and some broods of Snow geese more than 100 km (MacInnes and Lieff, 1968), but these were exceptionally far. Such movement was necessary because local food supplies were insufficient for all the birds that nested.

Adaptations to Arctic Nesting

For geese nesting in the tundra, the main constraint is the shortness of the season. To produce young that are strong on the wing in the period between spring melt and autumn freeze up, such geese show three main adaptations:

(1) They conform to the widespread trend among birds that the further north they breed, the shorter their individual breeding cycles, and the more they are synchronised with one another (Newton, 1972). This is shown by comparing tundra nesting populations with temperate zone ones, and among tundra ones by comparing the northern Brent, Barnacle and Snow geese with the rest (table 11.1). The main shortening of breeding seasons is due to a reduction in the spread of the starting dates of individuals, rather than in the shortening of individual cycles. In most tundra species, individuals need three months to complete a breeding cycle, but the three most northerly ones only require 2.5 months.

(2) Courtship and copulation occur either on winter quarters or at stopping places on northward migration (Barry, 1962, MacInnes, 1962; Ryder, 1967; Kistchinski, 1971). The birds arrive in pairs, the females already fertilised; if conditions are right they can lay within a week (Barry, 1962; Ryder, 1967; 1971). The males' testes have already begun to regress (Barry, 1962; Ryder, 1967). If the clutch is destroyed after completion, a second is not laid, but if some eggs are destroyed before the clutch is completed, the remaining eggs may be laid in a new nest, with no longer break than usual between eggs (Barry, 1962; MacInnes, 1962). The completion of the pre-nesting phase of breeding before arrival may be usual in other northern waterfowl, but is apparently not the norm in other birds.

Table 11.1 The duration of the breeding cycle (days) in different species of geese. As far as possible, species (or races) are listed according to their breeding distributions north to south.

Species	Locality	Spread in start of laying	Usual clutch size	Incubation period	Fledging period	Total egg* period of population	Breeding period of individual	Total breeding period of population	Source
Brent *B. bernicla*	Southampton Island	6	4	24	45-50	34	73-78	83	Barry, 1962
Greater Snow Goose *A. hyperboreus*	Bylot Island	7	4	23-25	44	36	71-73	79	Lemieux, 1959
Snow Goose *A. caerulescens*	Eastern Arctic Canada	7-14	4	22-3	42	34-43	68-69	75-84	Cooch, 1964; Ryder, 1971; Ryder, 1967; 1972
Ross's Goose *A. rossii*	Central Arctic Canada	6-11	4	22	—	32-37	—	·	
Barnacle *B. leucopsis*	Spitzbergen	—	4	24	49	—	78		Norderhaug et al., 1964
Canada Goose *B.c. hutchinsii*	Eastern Arctic Canada	8-10	4	24-5	—	37-39	—	—	MacInnes, 1962
Pinkfooted Goose *A. brachyrhynchus*	Iceland	—	4-5	26	58	—	88-90	· —	Witherby et al., 1938; Scott et al., 1955; Norderhaug et al., 1965
Emperor Goose *A. canagicus*	Northeast Siberia	13	5-6	25	—	45	—	—	Barry, 1964; Kistchinski, 1971
White-fronted Goose *A. albifrons*	Alaska	21	5	27	42-49	54	75-82	102	Lack, 1968; Dzubin et al., 1964
Bean Goose *A. fabalis*	Northern Europe	—	5	28	—	—	—	—	Lack, 1968
Canada Goose *B.c. maxima*	Missouri	38	5-6	28	65-70†	73	99-104	141	Brakhage, 1965 Yocum and Harris, 1965
Greylag *A. anser*	Temperate Europe	49	6-7	27-28	56	86	90-93	141	Hudec and Rooth, 1970 Newton and Kerbes, 1974

*From first egg laid to last hatched by population, assuming each bird lays one egg every 1.5 days (Barry, 1962; Brakhage, 1965; Kossoch, 1950; Klopman, 1958; Lemieux, 1959; MacInnes, 1962; Yocum and Harris, 1965; Ryder, 1967; 1971).
†Calculated for *B.c. moffitti*.

(3) All breeding activities before the eggs hatch are accomplished almost entirely on reserves accumulated at wintering and migration areas (see below). Successful females evidently lay down enough reserves for migration (or at least for the last stage of it), for producing the eggs and for their own maintenance during incubation (Ryder, 1967; 1970; Harvey, 1971). The males expend less energy than the females (see below), but are more active, establishing and defending the nest area until hatch. These demanding activities together take a minimum of 4 – 6 weeks in each successful bird. The dependence of migration on body reserves is well known among other bird families at all latitudes (Moreau, 1972), and at least a partial dependence of breeding on reserves is known in some other waterfowl (Milne, 1974) and other birds (Fogden, 1974) further south. But northern geese (and swans) seem extreme in the number and duration of activities which they accomplish on reserves, with little opportunity for replenishment.

Evidence for Dependence on Reserves

The evidence for such heavy and prolonged dependence on reserves is of two types. First, successful birds feed little or not at all until after the eggs hatch (Harvey, 1971; MacInnes *et al.*, 1975). When the geese arrive, the ground may be snow covered and they do not start nesting until bare patches have appeared (Lemieux, 1959; Barry, 1962; Uspenski, 1965); solitary nesters wait until ice around nesting islets has melted, thus making them safer from mammalian predators (MacInnes, 1962). The amount of feeding by geese depends partly on the interval between arrival and nesting; this is shortest at the highest latitudes, and in any one area is shortest in the warmest springs. It depends also on whether food is available, that is, on snow cover and habitat, and on the condition of the female, which is partly dependent on local weather (Harvey, 1971). Feeding can occur only on places clear of snow, and at first the food is of poor quality because new growth does not begin for another 2 – 3 weeks, well after incubation has started. Also, the birds often nest in areas of low scrub or heath, which are deficient in food plants. During incubation the female spends nearly all her time on the nest (Salomonsen, 1950; Harvey, 1971), but she may leave occasionally for short periods to obtain what little food is available nearby, and very rarely she may fly to feed elsewhere, accompanied by her mate. At this stage the male is more free to feed than the female, but is still restricted to a small area near his mate.

Spring migration of geese is normally accomplished in a series of long flights, each of several hundred kilometres, between traditional stopover points (Cooch, 1964; Dzubin, 1965; MacInnes, 1966; Prevett and MacInnes, 1972). The extent to which birds can feed on migration varies between populations and whether they migrate over favourable terrain, but in general the nearer to their destination, the less their chance of getting fresh protein-rich vegetation. The Pinkfooted geese, which fly the 1100 km over water between Scotland and central Iceland in spring, have no access to good feeding for several weeks after setting off. Other populations that move over land feed at stopping places (MacInnes, 1966; Drobovtsev, 1972; Kumari, 1971), but the bulk of the reserve needed for breeding is almost certainly laid down earlier.

Secondly, the weights of breeding birds show a massive loss between the dates of arrival and hatching, irrespective of weather and snow cover. Female Ross's geese lost about 25 per cent of their initial weight over the few days of egg laying, and another 25 per cent during the 22 days of incubation, then gained rapidly when they began to feed again after hatching (Ryder, 1967). The males lost about 25 per cent of their weight between arrival and hatching and then, like the females, gained weight again. Female Snow geese lost about 25 per cent of their weight between arrival and hatching, males about 17 per cent (Cooch, 1964). Also, while nesting female Brent geese lost 25 per cent of their weight, non-nesting females lost only about half as much as this. The enormous fat reserves present in geese on arrival on their breeding areas have been noted in Brent geese (Barry, 1962), Snow geese (Cooch, 1958), White-fronted and Canada geese (Hanson *et al.*, 1964). Carcass analyses of nesting Canada and Snow geese showed that weight loss resulted from metabolism of both fat and protein stores, the latter mainly in the pectoral muscles (Hanson, 1962; Harvey, 1971). Female Snow geese became extremely emaciated. Their breast muscles were resorbed until "only vestiges of stringy gelatinous tissues remained", and the keel protruded sharply. Their necks also became thin and their bill linings blackened and dried (Harvey, 1971).

Timing and Success of Breeding

The arrival of geese on their breeding areas does not always coincide with the dates of snow melt, which can vary by up to three weeks from year to year. Generally, arrival dates are much more constant from year to year than melt dates (Goodhart and Wright, 1958; Barry, 1962; Kerbes, 1969). In years when the ground has begun to clear at the time of arrival, geese begin nesting within a week, but in other years they remain on the breeding areas and wait (Barry, 1962; Uspenski, 1965), all the time depleting their reserves. The more of the reserves used in prior maintenance, the less that can be allocated to eggs (Ryder, 1970). In these circumstances, breeding output might be influenced both by the initial size of the reserves (hence, on conditions further south) and also on how long the geese have to wait between accumulating their reserves and laying their eggs. Peak laying dates of several populations have been up to two weeks later in late than in early years (Hardy, 1967; Uspenski, 1965); after even more delay, the birds did not attempt to nest (Barry, 1962; Cooch, 1964).

To my knowledge, the only attempt to relate breeding output in tundra nesting geese with previous conditions on winter quarters is that of Cabot and West (1973), with the Greenland Barnacle that winter on the Inishkea Islands, off western Ireland. Using data for 13 years, they found a significant positive correlation ($P < 0.05$) between mean brood sizes in any one winter and mean temperatures over the preceding winter. Temperature had a major influence on the production of winter food (grass) and thus, it was presumed, on the condition of the geese. MacInnes *et al.* (1974) also provided indirect evidence that conditions south of breeding areas were partly responsible for annual variations in production by two populations of Canada geese. Further work of this type is much needed.

The relationship between the date of snow melt and the timing and success of breeding in geese has been well documented in White-fronted, Canada and Brent geese (Barry, 1962), Snow geese (Cooch, 1964; Kerbes, 1969; Uspenski, 1965*b*), Ross's geese (Ryder, 1970), Pinkfooted and Barnacle geese (Goodhart and Wright, 1958), and by casual observations on other species. In general, the later that nest-sites become available, the smaller the proportion of birds that lay, the smaller the clutches and (at least in Snow geese) the poorer the hatching success (see later). Availability of nest sites in spring is influenced not only by temperatures, but also by snow depth, flooding and other factors.

Resorption of ova is evidently the mechanism by which clutch sizes are reduced, or, in very late years, breeding is prevented (see photographs in Barry, 1962). The importance of this is not so much the nutrients the female reclaims at the time, but rather those she saves through not committing herself to a larger clutch and a later incubation. In some female Brents killed off nests in a late year, 60 per cent of ova had been resorbed, compared with 95 per cent in birds killed from non-breeding flocks (Barry, 1962). The few eggs laid by the latter were probably "dumped", that is, laid in another bird's nest or on the ground.

In all species in which it was studied, mean clutch size was smaller in late than in early years (Barry, 1964; Cooch, 1964), and it declined as each season progressed, sometimes by more than 50 per cent in a few days. Thus in 1967 on Baffin Island, the mean clutch size of Snow geese was 5.0 for birds starting on June 17, 3.9 for birds starting on June 18, 3.5 on June 19, 2.6 on June 20 and 2.1 on June 21 (Kerbes, 1969). Seasonal declines of similar magnitude were noted in other populations of Snow geese (Ryder, 1971), and in Ross's geese (Ryder, 1967); less marked ones were noted in *A.c. atlantica* (Lemieux, 1959), and Brent geese (Barry, 1962). In Snow geese they also occurred within particular age groups, and could not be wholly attributed to younger birds (which laid smaller clutches) nesting later than older birds (Finney and Cooke, in press). The depletion of reserves may not be the only factor involved, moreover, for a late bird laying a small clutch will also hatch its eggs earlier than if it laid a large clutch, and in a short season the few days saved could be crucial (see later).

Once they have bred, individual marked female geese have usually returned to breed in the same area year after year (Barry, 1962; Cooch, 1958; MacInnes and Lieff, 1966; Cooke, MacInnes and Prevett, 1975). In late years, competition for what snow-free ground was available led to much fighting, as seen in Brent geese, Snow geese (Barry, 1962) and White-fronted geese (Dzubin *et al.*, 1964). The youngest potential breeders (22 and 34 months old) among Snow geese nested chiefly in early springs, when nest sites were superabundant from the start of the season (Cooch, 1964). Young birds nested a few days later, on average, than did older, experienced breeders, even when not constrained by shortage of sites; they also laid smaller clutches than older birds nesting at the same date (Finney and Cooke, in press).

Hatching Success

Whether a bird's reserves will last through incubation depends not only on their extent, but also on their rate of loss, which in turn depends on the weather and its effect on heat loss (Harvey, 1971). In one cold year at the McConnell River, some female Snow geese were found dead on their eggs, which were near to hatching. On average, they weighed 43 per cent less than did breeders on arrival. Other geese interrupted incubation to feed, thus exposing their nest to predation. Among individuals that were closely watched from a hide, predation was highest among birds which interrupted incubation to feed. Most losses occurred late in incubation, when there were fewer predators present, but when females left their nests most often. Usually the whole clutch was taken, even though covered with down. Hence, a bird's condition influenced not only the number of eggs it laid, but also its chance of hatching them.

In some studies, the mean hatching success was found to be better in early than in late springs when availability of nest sites was delayed. Cooch (1964) estimated the usual success of Snow goose nests to be around 81 per cent in early seasons, 64 per cent, while Kerbes (1969) found 76 per cent in an average year and ting the losses to predation; L. Maltby (*in litt.*) found equivalent figures of 97, 79 and 16 per cent; while Kerbes (1969) found 76 per cent in an average year and 26 per cent in a late year. On the other hand, Ryder (1971) found no reduction in nesting success of Ross's geese in a late season. MacInnes (1962) gave figures of 75 and 90 per cent for *B.c. hutchinsii* in two early years.

Problems arise in assessing predation rates, however, because the main predators on the eggs of arctic-nesting geese are skuas (mainly Arctic skuas) and gulls (mainly Herring gulls). These are mainly opportunists, unable to drive a goose off a nest, but able to get in quickly and take unguarded eggs. They are thus greatly assisted in their efforts by the presence of a biologist working in the area. Hence, predation rates recorded in the literature are probably higher than they would be in natural conditions, and they do not usually distinguish clutches that were deserted before being eaten. MacInnes and Misra (1972) thought that the partial depletion of clutches in *B.c. hutchinsii*, which accounted for 55 per cent of all eggs lost, occurred only under human disturbance, and that an average of 0.65 eggs was lost from each nest at every visit by the observer. Otherwise predation rates would have been around 10 per cent, would not have varied much in the five years studied, and would have been restricted to whole clutches.

Arctic foxes *Alopex lagopus* have locally accounted for a large number of clutches. In one study, a single fox was seen to destroy all the 200 Snow goose clutches in 3 km^2 in 5 days. The fox systematically removed the eggs from one nest after another and buried them (MacInnes and Misra, 1972). The heavy destruction of Snow goose and Brent goose eggs in another area was associated with the presence of an active fox den in the colony (Barry, 1964). These effects were local and, unless foxes were numerous and short of normal food over a wide area, they would be unlikely to have any major impact on goose populations or breeding success, especially where geese nest in a more dispersed manner. Grizzly Bears *Ursus horribilis* have been known to have a heavy local impact: two ate the contents of 295 goose nests in 6 days (Barry, 1964).

Fledging Success

This has usually been assessed by comparing the mean brood sizes in a population at different dates between hatching and fledging, ignoring pairs without young. The survival of young *B.c. hutchinsii* in two early seasons near McConnell River was thus estimated at 85 and 90 per cent (MacInnes, 1962), that of young Ross's geese in two early seasons near Perry River at about 67 per cent and that of young Pinkfooted geese between hatching in Iceland and arrival in Britain in October at 67 per cent or more (Scott *et al.*, 1955). All these figures imply good survival, but are almost certainly overestimates because they take no account of broods which failed entirely. In general, losses were greatest in the week after hatching, when the young were still small. The main predators were large gulls (especially Herring gulls), which in *B.c. hutchinsii* took any young that strayed more than about 3 m from their parents (MacInnes, 1962). On Baffin Island, survival among Snow geese was poorer in a late year than in an average year (Kerbes, 1969), a finding that needs more investigation.

Breeding Success and Population Trends

The result of varying seasons is that production of young has varied enormously from year to year. This was especially so in high Arctic nesters, like Brent and Snow geese, in which the proportion of young in the autumn populations varied from < 1 to > 50 per cent (table 11.2), and in which poor years were most frequent. In general, there was a good correlation between mean brood size and percentage of young in the population, but the latter was also influenced by the number of immature birds, and hence on breeding success in previous years.

Table 11.2   Annual variations in production of young, judged from brood counts in winter quarters. As far as possible, species are listed according to their breeding distributions, north to south

	Breeding/ wintering area	No. of years of observation	Range of values from different years in		Source
			Percentage young in population	Mean brood size	
Brent goose B. bernicla	Northern Siberia/ Britain, France	15	< 1 – 53	–	Ogilvie and Matthews, 1969
Snow goose A. caerulescens	Northeast Canada/ Southern United States	11	2 – 55	1.6 – 2.7	Lynch and Singleton, 1964
Barnacle goose B. leucopsis	Northeast Greenland/ Western Scotland	9	8 – 31	1.4 – 2.8	Boyd, 1968
	Northeast Greenland/ Western Ireland	12	1 – 14	1.5 – 2.4	Cabot and West, 1973
Pinkfooted goose A. brachyrhynchus	Iceland/ Britain	19	11 – 49	1.3 – 3.3	Boyd and Ogilvie, 1969
Greylag A. anser	Iceland/ Britain	12	6 – 44	1.3 – 3.8	Boyd and Ogilvie, 1972
White-fronted goose A. albifrons	Northern Russia/ England	17	11 – 47	2.5 – 3.9	Boyd, 1965

Good production was often associated with big increases in population and vice versa; several good years in succession led to a massive increase in numbers, and several poor years to a decline. Differences also occurred within a breeding population wintering in different areas. The population of Brent geese that winter on the Inishkea Islands remained stable for 12 years, with a mean adult loss of 8.3 per cent and a mean production of 7.8 per cent young, while the Islay ones increased continuously for at least 18 years, with a 10.1 per cent adult loss and 16.5 per cent production of young (Cabot and West, 1973; H. Boyd, *in litt.*). Both populations breed in eastern Greenland. Such increases are presumably possible only until populations become limited by shortage of breeding or wintering habitat.

Icelandic populations of Pinkfooted and Greylag geese more than doubled in size in 20 years and, while production of young varied greatly from year to year, on average the proportion of mature birds producing young declined (Boyd and Ogilvie, 1969; 1971). This implied some steadily increasing restriction in production of young as numbers rose. One way in which this might have occurred is through competition for limited areas of breeding habitat, an idea supported by the recent finding that nesting and feeding habitat of Pinkfooted geese was occupied at very high density (Kerbes, Ogilvie and Boyd, 1971).

Non-Breeders

Probably most geese nest for the first time at 3 years (34 months), but some at 4 years (46) or 2 years (22). This is the case with *Anser* and *Branta* species in captivity (Scott *et al.*, 1955), and has been confirmed in the wild in Canada geese (Craighead and Stockstad, 1964; Martin, 1964), Brent geese (Barry, 1962) and Snow geese (Finney and Cooke, in press); it has not been confirmed in European White-fronted geese, most of which bred at 4 yr and only a few at 3 years (Boyd, 1965). These generalisations may of course be upset by seasonal phenology, by the intensity of competition for breeding places and other factors (Cooch, 1958; Finney and Cooke, in press); even during its breeding life, a bird may not nest every year (MacInnes *et al.*, 1974).

Geese that do not breed moult their feathers 2 – 3 weeks earlier than those that do. They may remain in the breeding area, in flocks separated from breeders (Hardy, 1967; Newton and Kerbes, 1974), or migrate up to several hundred kilometres northwards to traditional areas where they gather in flocks containing up to several thousand birds. They are ready for autumn migration, and in some species leave well before the breeders. Moult migrations are regular in all species of geese (see Salomonsen, 1968 for Greylag, White-fronted, Lesser white-fronted, Bean, Pinkfooted, Canada, Brent and Snow geese, and Kistchinski (1971) and Blurton Jones (1972) for Emperor geese).

Discussion

Breeding season

The short and more synchronised breeding cycles of arctic geese, compared with those of the temperate zone have parallels in other bird families that breed in both regions (Newton, 1972), though geese in general have longer breeding cycles than most other land birds at similar latitudes, largely because their young take longer to reach the flying stage. This may in turn be linked with their almost wholly vegetarian diet in which (with swans) geese seem unique. The completion of the pre-nesting phase of breeding before arrival may be usual in other northern water-fowl, but it is apparently not usual in other tundra nesting birds. Also geese seem extreme in the number and duration of activities that they accomplish almost entirely on reserves (though again swans *Cygnus* spp. may provide parallels). Both features could be interpreted as adaptations to save time.

The period needed to complete a breeding cycle is probably a major feature preventing any one goose species from extending its range northwards, for the further north a species breeds, the smaller the proportion of summers in which it can raise young (Lensink, 1973). It is not clear to me what prevents a species from breeding further south, unless perhaps some competition is experienced from related species. Reduced competition was probably also involved in the evolution of moult migrations. By leaving the nesting areas, non-breeding geese avoid com-peting for food with families of their own species and, by flying north, they also gain the advantage of longer days. Some populations also get two spring flushes of plant-growth, the first on the breeding area and the second on the moulting area. The flightless period of non-breeding geese lasts less than half as long as a success-ful breeding cycle, so individuals can moult well to the north of where they can breed.

Clutch size and non-breeding

Lack (1967) proposed that clutch size in waterfowl had evolved in relation to the average amount of food available to the female at the time (and place) of laying, modified by the relative size of the eggs. Starting with this idea, Ryder (1970) proposed that clutch size in Ross' geese had evolved to correspond with the maxi-mum that the female could normally produce on her reserves, that resorption of ova allowed adjustment of clutch size to depleted reserves, and that the female usually retained a certain reserve to enable her to incubate effectively. The female was thus largely independent of food supplies at the time and place of laying. Not only should this view now be extended to other species of geese, but annual variations in spring body reserves may also be considered partly responsible for annual variations in production.

When reserves are falling, resorption allows the transfer of food materials from the ova back to the female. Further, it prevents the female from committing her-self to extra breeding effort, and from nesting so late that she would have little chance of raising young. Selection operates against late and repeat nesting; Barry (1962) found 21 Brent geese that had hatched in a late spring, frozen in the ice the following spring. They were well preserved, and had nothing wrong with them except that their feather development was 4 - 5 days short of allowing them to fly. When birds fail to nest, they do not suffer a big energy drain, are able to feed all summer and moult earlier, and thus presumably enhance their own chances of surviving to attempt to breed again. Since individual geese nest in several seasons, they might produce most young by skipping a year in return for greater success the following year.

Breeding success

Greater egg loss in late years results from less sustained incubation (facilitating greater predation from skuas and gulls), and perhaps also from shortage of sites at laying time forcing more birds than usual to nest in unsuitable places (Cooch, 1964). The greater numbers of avian predators often seen around large goose colonies in late years (Barry, 1964; Kerbes, 1969) has usually been attributed to shortage of the predator's normal food. The alternative view, that they are attracted there by more vulnerable (exposed and dumped) eggs, has seldom been considered. The survival of young after hatching may also be poorer in late years, irrespective of any being frozen in, but needs more study.

The relationship between the timing and success of breeding also holds in waterfowl nesting in temperate regions. Thus the Greylag geese resident in the snowfree environment of the Outer Hebrides Islands, off northwest Scotland, nested earlier and more successfully in warm than in cold springs and, within each season, the first birds to nest were also the most successful (Newton and Kerbes, 1974). Here nest sites were unrestricted and the differences between years were due mainly to differential predation, which occurred only when the nest was left unattended. The same association within a season also held in at least four species of duck nesting at Loch Leven, in east-central Scotland (Newton and Campbell, 1975). Disastrously poor seasons for production of young have also occurred in temperate zone waterfowl. They were frequent in Scottish Eiders *Somateria molissima,* (Milne, 1974) and Shelducks *Tadorna tadorna* (Jenkins *et al.*, 1975), but in these species the failures were due to heavy mortality of small young, rather than to widespread failure to lay, as in arctic geese.

Further research

The model presented here, based on links between the level of body reserves accumulated on winter quarters, on the availability of nest sites in spring, and the timing and success of breeding, is probably oversimplified and likely to be modified by further work. Whether events before or after arrival on breeding areas have most effect on production of young may also vary between populations and between years in the same population. A major research need is for more study of the relationship between winter food supplies and the spring condition and breeding success of the birds. Future studies on breeding areas could also include some good measure of progressive nest site availability each year. Different years from the same and different studies could then be more precisely compared with one another, and the relationship between timing and success of breeding examined more rigorously. In studies of gosling survival, some account must also be taken of broods depleted to nil, and some assessment made of the quality of summer grazings.

Acknowledgements

I thank H. Boyd, F. Cooke and C. D. MacInnes for giving me the opportunity to study breeding geese in arctic Canada, H. Boyd, F. Cooke, J. P. Dempster, G. Finney, M. Harris, D. Jenkins, R. H. Kerbes and L. Maltby for many constructive comments on the manuscript, and F. Cooke, G. Finney, C. D. MacInnes and L. Maltby for allowing me to quote from their unpublished work.

References

Barry, T. W. (1962). Effect of late seasons on Atlantic Brant reproduction. *J. Wildl. Mgmt.*, **26**, 19 - 26.

Barry, T. W. (1964). Brant, Ross' Goose and Emperor Goose, *Waterfowl Tomorrow* (ed. J. P. Linduska), US Government Printing Office, Washington, pp. 145 - 154.

Bertram, G. C. L., Lack, D. and Roberts, B. B. (1934). Notes on East Greenland birds, with a discussion of the periodic non-breeding among arctic birds. *Ibis*, **4**, (13) 816 - 31.

Blurton Jones, N. G. (1972). Moult migration of Emperor geese. *Wildfowl*, **23** 92 - 93.

Blurton Jones, N. G. and Gillmor, R. (1959). Some observations on wild geese in Spitsbergen. *Wildfowl Trust Ann. Rep.*, **10**, 118 - 132.

Boyd, H. (1965). Breeding success of White-fronted geese from the Nenets National Area. *Wildfowl Trust Ann. Rep.*, **16**, 34 - 40.

Boyd, H. (1968). Barnacle geese in the West of Scotland, 1957 - 67. *Wildfowl*, **19**, 96 - 107.

Boyd, H. and Ogilvie, M. A. (1969). Changes in the British-wintering population of the Pinkfooted goose from 1950 to 1975. *Wildfowl*, **20**, 33 - 46.

Boyd, H. and Ogilvie, M. A. (1972). Icelandic Greylag geese wintering in Britain in 1960 - 1971. *Wildfowl*, **23**, 64 - 82.

Brakhage, G. K. (1965). Biology and behaviour of tub-nesting Canada geese. *J. Wildl. Mgmt.*, **29**, 751 - 771.

Cabot, D. and West, B. (1973). Population dynamics of Barnacle geese, *Branta leucopsis*, in Ireland. *Proc. R.I.A.*, **73B**, 415 - 443.

Cade, T. (1960). Ecology of the Peregrine and Gyrfalcon populations in Alaska. *Univ. Calif. Publ. Zool.*, **63**, 151 - 290.

Cooch, F. G. (1958). The breeding biology and management of the Blue goose (*Chen caerulescens*). PhD thesis, Cornell University, Ithaca, New York.

Cooch, G. (1964). Snows and Blues, in *Waterfowl Tomorrow* (ed. J. P. Linduska), US Government Printing Office, Washington, pp. 125 - 133.

Cooke, F., MacInnes, C. D. and Prevett, J. P. (1975). Gene flow between breeding populations of Lesser snow geese. *Auk*, **92**, 493 - 510.

Craighead, J. J. and Stockstad, D. S. (1964). Breeding age of Canada geese. *J. Wildl. Mgmt.*, **28**, 57 - 64.

Drobovtsev, V. (1972). Character of the spring and autumn passage of geese in the North-Kazakhstan Region, in *Geese in the USSR* (ed. E. Kumari), Tartu. pp. 132 - 138. Russian, with English summary.

Dzubin, A. (1965). A study of migrating Ross geese in western Saskatchewan. *Condor*, **67**, 511 - 534.

Dzubin, A., Miller, H. W. and Schildman, G. V. (1964). Whitefronts, in *Waterfowl Tomorrow* (ed. J. P. Linduska). US Government Printing Office, Washington, pp. 135 - 143.

Finney, G. and Cooke, F. (in press). Reproductive strategies in the Snow goose: the influence of female age. *Ibis*.

Fogden, M. P. L. (1974). The seasonality and population dynamics of Equatorial forest birds in Sarawak. *Ibis*, **114**, 307 - 342.

Goodhart, J. and Wright, T. (1958). North-east Greenland expedition 1956. *Wildfowl Trust Ann. Rep.*, 9, 180 - 192.

Hanson, H. C. (1962). The dynamics of condition factors in Canada geese and their relation to seasonal stresses. *Arct. Inst. Tech. Pap.*, 12, 68 pp.

Hardy, D. E. (1967). Observations on the Pink-footed Goose in Central Iceland, 1966. *Wildfowl Trust Ann. Rep.*, 18, 134 - 141.

Harvey, J. M. (1971). Factors affecting Blue goose nesting success. *Can. J. Zool.*, 49, 223 - 234.

Hudec, K. and Rooth, J. (1970). *Die Graugans* (Anser anser *L.*). Die Neue Brehm-Bucherei, Wittenberg Lutherstadt, 148 pp.

Jenkins, D., Murrary, M. G. and Hall, P. (1975). Structure and regulation of a Shelduck population. *J. Anim. Ecol.*, 44, 201 - 231.

Kerbes, R. H. (1969). Biology and distribution of nesting Blue geese on Koukdjuak Plain, N.W.T. MSc thesis, University of Western Ontario, London, Ontario, Canada.

Kerbes, R. H., Ogilvie, M. A. and Boyd, H. (1971). Pink-footed geese of Iceland and Greenland: a population review based on an aerial nesting survey of Thjorsarvar in June 1970. *Wildfowl.*, 22, 5 - 17.

Kessel, B. and Cade, T. J. (1958). Birds of the Colville River, Northern Alaska. *Biol. Papers Univ. Alaska*. No. 2.

King, J. G. (1970). The swans and geese of Alaska's arctic slope. *Wildfowl*, 21, 11 - 17.

Kistchinski, A. A. (1971). Biological notes on the Emperor goose in north-east Siberia. *Wildfowl*, 22, 29 - 34.

Klopman, R. B. (1958). The nesting of Canada geese at Dog Lake, Manitoba. *Wilson. Bull.*, 70, 168 - 183.

Kossack, C. W. (1960). Breeding habits of Canada geese under refuge conditions. *Am. Midl. Nat.*, 43, 627 - 649.

Kumari, E. (1971). Passage of the Barnacle goose through the Baltic area. *Wildfowl*, 22, 35 - 43.

Lack, D. (1967). The significance of clutch-size in waterfowl. *Wildfowl Trust Ann. Rep.*. 18, 125 - 128.

Lack, D. (1968). *Ecological Adaptations for Breeding in Birds*. Methuen, London. 409 pp.

Lemieux, L. (1959). The breeding biology of the Greater Snow Goose on Bylot Island, Northwest Territories. *Can. Field Nat.*, 73, 117 - 128.

Lensink, C. J. (1973). Population structure and productivity of Whistling swans on the Yukon Delta, Alaska. *Wildfowl*, 24, 21 - 25.

Lynch, J. J. and Singleton, J. R. (1964). Winter appraisals of annual productivity in geese and other water birds. *Wildfowl Trust Ann. Rep.*, 15, 115 - 126.

MacInnes, C. D. (1962). Nesting of small Canada geese near Eskimo Point, Northwest Territories. *J. Wildl. Mgmt.*, 26, 247 - 256.

MacInnes, C. D. (1966). Population behaviour of eastern Arctic Canada geese. *J. Wildl. Mgmt.*, 30, 536 - 535.

MacInnes, C. D., Davis, R. A., Jones, R. N., Lieff, B. C. and Pakulak, A. (1974). Reproductive efficiency of McConnell River Small Canada geese. *J. Wildl. Mgmt.*, 38, 686 - 707.

MacInnes, C. D. and Leiff, B. C. (1968). Individual behaviour and composition of a local population of Canada geese, *Canada Goose Management* (ed. R. L. Hine and C. Schoenfeld), Dembar, Madison, 93 - 101.

MacInnes, C. D. and Misra, R. K. (1972). Predation on Canada goose nests at McConnell River, Northwest Territories. *J. Wildl. Mgmt.*, **36**, 414 - 422.

Martin, F. W. (1964). Behaviour and survival of Canada geese in Utah. *Utah State Dept. Fish Game, Dept. Inform. Bull.*, 89 pp.

Milne, H. (1974). Breeding numbers and reproductive rate of Eiders at the Sands of Forvie National Nature Reserve, Scotland. *Ibis*, **116**, 135 - 154.

Moreau, R. E. (1972). *The Palaearctic-African Bird Migration Systems*. Academic Press, London and New York, 384 pp.

Newton, I. (1972). *Finches*. Collins, London, 288 pp.

Newton, I. and Campbell, C. R. G. (1975). Breeding of ducks at Loch Leven, Kinross. *Wildfowl*, **26**, 28 - 102.

Newton, I. and Kerbes, R. H. (1974). Breeding of Greylag Geese (*Anser anser*) on the Outer Hebrides, Scotland. *J. Anim. Ecol.*, **43**, 771 - 783.

Norderhaug, M., Ogilvie, M. A. and Taylor, R. J. F. (1964). Breeding success of geese in West Spitzbergen. *Wildfowl Trust Ann. Rep.*, **16**, 106 - 110.

Nyholm, E. S. (1968). Ecological observations on the geese of Spitzbergen. *Ann. Zool. Fenn.*, **2**, 197 - 207.

Ogilvie, M. A. and Matthews, G. V. T. (1969). Brent geese, mudflats and man. *Wildfowl*, **20**, 119 - 125.

Paakspuu, V. (1972). Present-day status of the Greylag goose populations in the Matsalu Bay, in *Geese in the USSR*, (ed. E. Kumari), Tartu. pp. 13 - 19 Russian, with English summary.

Prevett, J. P. and MacInnes, C. D. (1972). The number of Ross' geese in central North America. *Condor*, **74**, 431 - 438.

Ryder, J. P. (1967). The breeding biology of Ross's goose in the Perry River region, Northwest Territories. *Can. Wildl. Service Rep. Ser.*, **3**, 56 pp.

Ryder, J. P. (1970). A possible factor in the evolution of clutch size in Ross's goose. *Wilson Bull.*, **82**, 5 - 13.

Ryder, J. P. (1971). Distribution and breeding biology of the Lesser snow goose in Central Arctic Canada. *Wildfowl*, **22**, 18 - 28.

Ryder, J. P. (1972). Biology of nesting Ross's geese. *Ardea*, **60**, 185 - 215.

Salomonsen, F. (1950). *The Birds of Greenland*. Munksgaavd, Copenhagen.

Salomonsen, F. (1968). The moult migration. *Wildfowl*, **19**, 5 - 25.

Scott, P., Boyd, H. and Sladen, W. J. L. (1955). The Wildfowl Trust's second expedition to Central Iceland, 1953. *Wildfowl Trust Ann. Rep.*, **7**, 63 - 98.

Uspenski, S. M. (1965a). The geese of Wrangel Island. *Wildfowl Trust Ann. Rep.*, **16**, 126 - 129.

Uspenski, S. M. (1965b). *Die Wildganse Nordeurasians*. Wittenburg Lutherstadt, Ziemren.

Witherby, H. F., Jourdain, F. C. R., Ticehurst, N. F. and Tucker, B. W. (1938). *The Handbook of British Birds*. Vol. 3. Witherby, London.

Yocum, C. F. and Harris, S. W. (1965). Plumage descriptions and age data for Canada goose goslings. *J. Wildl. Mgmt.*, **29**, 874 - 877.

12 The evolutionary significance of co-operative breeding in birds

Dr C. H. Fry*

Although the phenomenon had barely been hinted at before Skutch's review of 1961, co-operation by three or more adult birds in some aspect of reproductive effort is remarkably widespread as a usual feature of breeding biology, and is now known to involve at least 1.5 per cent of the world's avifauna. I hold co-operative breeders to include the many birds with 'helpers at the nest', and the several species which breed in pairs but are group-territorial and provision the young communally, as well as the few in which several females lay in a common nest. For reasons given below we should exclude the half-dozen species that co-operate in building a supernest in which each pair rears its young unaided — those are best designated communal nesters. I include, however, birds in which later broods in the same season are regularly fed by their siblings of earlier broods, since this seems similar to their retaining family bonds and, when adult, helping parents at the nest in subsequent years. Details of social organisation and the nature of the co-operative effort vary considerably with species; they typically involve pre-breeding adult individuals helping a breeding pair with nest building, incubation, territorial defence, feeding nestlings and attending fledglings.

There is a twofold special interest in co-operative breeding. First, Lack (1968) and others have demonstrated the environmental correlates of various mating systems. We need to explain in terms of its adaptive advantages *why* co-operative breeding has evolved. Secondly, we have yet to understand the precise nature of the selection pressures and of the units on which they have operated, to explain *how* co-operative breeding has evolved. For ecologists, the challenge of co-operative breeding in birds is to demonstrate its environmental correlates, and quantify the disadvantages and advantages accruing to the helpers. For evolutionists, the challenge is to refine theory accounting for its natural selection.

Helpers seem to provide a classic example of altruistic behaviour in animals, and hence evidence for the operation of group or kin selection. On the face of it we may think that a bird is altruistic if by "helping" it materially improves the

*Dr Hilary Fry became interested in tropical ornithology on leading an expedition from Cambridge to West African islands in 1959. On graduating he returned to Nigeria, where he took his Ph.D. degree under the supervision of David Lack, studying the biology of bee-eaters. At Aberdeen University since 1967, he has retained that interest with frequent visits to Africa and other parts of the tropics.

survival of the brood, while prejudicing its own fitness by abstaining from reproduction and exposing itself to increased risk. But against such personal disadvantages to the helper should be set experience likely to benefit it in a subsequent
season's role of breeding bird, improving its individual fitness in the Darwinian
sense. Whether helping and related "altruistic" behaviours in birds are the product
of classical individual selection or of kin selection is thus still in some doubt.

However, evidence for group selection as at least a plausible force in evolution
is accumulating, especially where the group is a small deme having a high coefficient of kinship. Indeed, among social insects with sterile castes which sacrifice
themselves to the common good by defending the colony (often with their lives)
and feeding its reproductive oligarchy, there is no doubt that such castes have
evolved by means of selection operating above the level of the individual, and in
a sense counter to individual selection, as Darwin himself perceived.

Among other animals, birds closely resemble social insects in the evolution of
social organisations. Entomologists distinguish communal and co-operative
insects, restricting the terms as I have above for birds; either organisation can lead
to advanced sociality. Insects and birds also run parallel in the high frequency of
social (brood) parasitism, and reasons for these evolutionary similarities are discussed by Wilson (1975). We may speculate that, given time, birds might well
evolve sterile helper castes, like the insects.

The Species

Skutch showed that about 50 species are regular co-operative breeders, and in a
review in 1972 I extended the list to about 80 birds. Great interest in the phenomenon was generated at the 16th International Ornithological Congress in
Canberra in 1974, where it was demonstrated that about 30 species in Australia,
40 in Africa, 10 in Eurasia and 40 in the Western Hemisphere are certainly involved, and there may also be some 60 additional species. It was formerly widely
overlooked because research in the tropics, where the majority of co-operative
nesters occur, has lagged far behind ornithological investigation elsewhere, and
also because the social roles of individuals can be discerned only by protracted
field observation of marked birds, an approach which has become widespread
only within the past 15 years. The total of over 120 species is likely eventually to
be doubled or tripled.

The species are systematically diverse. In Australia co-operative breeding
occurs in a species of goose, a flightless grazing gallinule, the Kookaburra *Dacelo
gigas*, a parrot and many songbirds of Australasian and palaeotropical families; in
the Americas, cuckoos, a hawk, swifts, woodpeckers, a toucan, swallows, jays, nut-
hatches, wrens and tanagers are co-operative nesters. In Africa cooperative
breeding has arisen in bee-eaters, the Ground Hornbill *Bucorvus cafer*, a kingfisher,
shrikes, babblers and others. The species involved altogether belong to about 28
families of landbirds in seven orders, so we may be sure that cooperative breeding
has arisen independently many times. But if, as that suggests, the habit originates
easily, it also has some evolutionary latency as evidenced from its wide occurrence
in some families that have undergone considerable adaptive divergence (for example,
American jays (Corvidae) *Aphelocoma, Gymnorhinus, Calocitta, Cissilopha,*

Psilorhinus and *Cyanocorax,* and Australian honeyeaters (Meliphagidae) *Manorina, Myzantha, Melithreptus, Conopophila* and *Anthocaera.*

Ecological diversity is as great. The co-operative breeding species inhabit tropical rainforest, deciduous and coniferous woods, swamps, mangroves, lakes and desert steppe. Some have generalised diets, others are specialised insectivores (bee-eaters), predators of lizards or fish, fruit-, grain-, plant-eaters, and nectivores (honeyeaters), and their number includes a sapsucker and a mast-storer (the wood-peckers *Melanerpes formicivorus* and *M. cruentatus*). Mostly these birds are altricial, but a few have precocious, nidifugous young, for example the Magpie goose *Anseranas semipalmatus,* and four species of gallinule. In view of all these differences it is not surprising that details of the social organisation and breeding biology of the co-operative breeders vary between species as much as they do among pair-nesting birds, making it difficult to determine the adaptive bases and survival value of such social organisations. Analysis has, however, revealed several correlates, as will be discussed later.

Social Systems

A few examples will suffice to indicate the range of social organisation. Red-throated bee-eaters *Merops bulocki,* are common and gregarious birds of the wooded bush and farmland of Africa's northern savannas, feeding on venomous wasps caught in flight. From January to March they nest in compact colonies in stream banks, excavating metre-long tunnels in cliffs. Flocks of 30 to 100 birds occupy traditional sites of wooded sandy watercourse 1 km long, with shade trees nearby, and are highly sedentary. They are not territorial, except for some weak defence of the nest entrance. Most nests are attended by the parents only, but about a quarter have up to three adult helpers which are usually, if not always, last year's progeny of the parents they now assist. There are one and a half times as many males as females in the population. Pair bonds are for life, and bonds forged between a breeding pair and their helpers may endure two years. Helpers participate in excavation, incubation and provisioning the young. They are generally male, and sometimes copulate with the female so that the paternity of the clutch can only be surmised at. There is a single clutch, of a mean 2.6 eggs, and average production is about 8 per cent greater from nests with helpers than from nests without them. Studies are now being made to discover any difference in fitness of the classes of young. After fledging the young depend on the adult group for food for about six weeks while they perfect their aptitude for devenoming the bees and wasps which form the main part of their diet. Even when independent for food, young birds remain in close association with the adult group as a family party within the larger social unit of the colony, and in the following year surviving one-year-olds will either find a mate and reproduce, or more likely attach themselves as helpers to their parents.

Long-tailed shrikes *Corvinella corvina,* have a similar distribution and habitat, are insectivorous, but feed mainly on the ground. In Ghana, Grimes (1975) found that they breed *only* in groups; the young are never reared by an unaccompanied pair. Each group has only one nest; the clutch may be laid by a single matriarch for several successive years, and there is a maximum of three successive broods per year. A group consists of from 4 to 23 birds, usually 10 to 15, comprising the breeding pair, their offspring of several previous nestings and unrelated non-breeding adults. The species is sedentary and territorial, and co-operative effort includes territorial defence, mobbing predators, nest building and feeding the incubating female as well as the young before and after fledging.

American jays have attracted much interest (Brown, 1974). Western populations of the Scrub jay *Aphelocoma coerulescens*, are pair-breeding with classical, all-purpose territories; they are sedentary, single-brooded, permanently monogamous and very long lived. In the Florida population, however, half the breeding pairs have helpers, and young birds remain in their parents' territory and participate in many breeding activities the following year (Woolfenden, 1975). Some, mainly males, are still helping when 2 years old, some even when they are three or four. In the Mexican jay *A. ultramarina,* in contrast, several pairs nest as a loose colony with a pool of helpers; they flock all year round within a well-defined (but not vigorously defended) group territory.

Much the same may be said for the Kookaburra, a mouse- and insect-eating giant kingfisher inhabiting woodland in Queensland and New South Wales (Parry, 1973). In this species territory size varies in proportion with the size of the group; since helpers form about a third of the adult population and occupy the same amount of territory as a breeding bird, and since territories are contiguous in all but the most marginal habitats, it means that the breeding potential of the population is one-third less than would be the case if young birds were recruited directly into breeding instead of becoming helpers. *Prima facie* this is a good instance of the voluntary breeding restraint that Wynne-Edwards (1962) has argued to be a group selected adaptation to density-dependent population homeostasis.

The Groove-billed ani *Crotophaga sulcirostris*, is a non-parasitic relative of the cuckoo found in tropical areas of the New World. From one to four pairs co-operate in constructing a common nest of sticks in which the females lay one clutch each. All birds in a group incubate and care for the young. Comparing pairs with larger groups in respect of foraging efficiency, provisioning, nest defence and fitness, Vehrencamp-Bradbury (1975) has found an apparent balance of costs and benefits for each adult individual.

Lastly, the granivorous Sociable weaver *Philetairus socius,* of southern Africa is a communal and also a co-operative breeder. Its huge supernest consists of a communally-constructed water-shedding thatch beneath which a matrix of grass bents contains numerous discrete chambers in one, two or three storeys. Sociable weavers are monogamous and during the rainy season a pair produces up to four broods successively from one nest chamber, averaging three young per brood. The young help to feed the next brood, so a fourth brood can be attended by the parents with up to nine helpers, the youngest of which may be only 25 days of age (Maclean, 1973).

Correlates

The diversity of social systems is even greater than is suggested by these few examples. Some characteristics of co-operative breeding species may, however, be identified:

Geography and habitat

Most of the species are found in the warmer parts of the world, between latitudes 30°N and 30°S. Since the biology of tropical and subtropical avifaunas is relatively poorly known, this majority is likely to increase. Southern Australia and western USA provide the principal exceptions. In the temperate Palaearctic some breeding collaboration has been found only in the Moorhen, *Gallinula chloropus* (Wood, 1974), Long-tailed tit *Aegithalos caudatus*, (Nakamura, 1975) and House-martin *Delichon urbica* (Bryant, 1975), whereas in the hotter regions of the Palaearctic it is known in the Arabian babbler *Turdoides squamiceps* in Israel (Zahavi, 1974) and in one or two other species.

Nearly all of the co-operative breeders are land birds no larger than corvids, and although they occupy a diversity of niches and habitats, many live in some-what arid areas where rainfall is sparse. That applies to much of Australia and Mesoamerica and to most of the non-forest areas of lowland Africa – the wooded savannas, thorn bush and desert steppe. The seasonality of such environments surely decrees seasonal variation in the quality and quantity of food resources and it generates a great deal of migration. Yet the only co-operative breeders which migrate are some of the bee-eaters, the swifts, hirundines and wood swallows. The others are all markedly sedentary; individual Red-throated bee-eaters seldom move further than 2 km from the place where they were born. White-throated bee-eaters *M. albicollis,* in contrast, are strongly migratory in the northern tropics of Africa, moving up to 1000 km between desert steppe breeding grounds and forested wintering regions; this species has up to six helpers at the nest.

Approximately half of the species have well-protected nests, usually holes in trees or the ground. The proportion is greater than among non-helping species, but in view of our present poor knowledge of the incidence of co-operation in tropical birds (which have a greater frequency of inaccessible nests than have non-tropical birds) it is unwise to speculate on the significance of this point.

Food distribution

Clearly there is no common diet among the co-operative breeding species. One ecological guild is well represented, the aerial insectivores (bee-eaters, swifts, swallows, wood swallows). Other guilds are faintly discernible, and they suggest that co-operative breeding may be one device to cope with widely dispersed food. This hypothesis has been explored extensively – and environmental determinants of mating systems and breeding dispersion have been demonstrated – by Lack (1968). With so many of the species being sedentary birds of the seasonal tropics and subtropics, seasonal changes of food dispersion or foraging strategy are likely also to promote co-operation. No intensive studies of the foraging biology of a co-operative species have yet been made.

Social and demographic attributes

The sex ratios of many of the species are biased towards the males. Disparate sex ratios also occur in other bird populations, usually in association with polygamy. The preponderance of males seems to arise after hatching and fledging rather than at fertilisation, and in the few co-operative-breeding species for which there are data it is thought to result from the propensity of females to disperse earlier and further than males, thereby incurring greater mortality. Male helping is sometimes disproportionate with the sex ratio; Dow (1975) found that one brood of Noisy miners *Manorina melanocephala,* in Queensland were fed by their mother and no fewer than 20 males.

Commonly, birds of all post-fledging ages and both sexes have plumages almost exactly alike, which is another reason for co-operation being long over-looked. Sexual maturity is frequently deferred, with the age of initial breeding varying widely within a single group. A further trait of co-operative breeders is that each group keeps together all year round, in small parties of its kind. Indeed it is a striking feature of bird life in Africa, for example, that numerous species are perennially gregarious, keeping in cohesive flocks of from three to a dozen indivi-duals. Mean flock size is a specific characteristic, hence in Australia the name Apostle-bird for the babbler *Struthidea cinerea* (a timaliid organised somewhat like the ani).

Many co-operative breeders – and land birds of tropical and stable climates in general – are proving to be extremely long lived. For most small passerines in the temperate zone, the annual survival rate is about 50 per.cent (Perrins and Moss, 1975 and references therein), but for the House-martin it is 64 per cent (Rhein-wald and Gutscher, 1969), the highest temperate-zone value known for asmall passerine. In subtropical Queensland the co-operative babbler *Pomatostomus temporalis,* has an adult survival rate of 78 per cent (Boehm, 1974). Florida scrub jays are remarkably long lived and so is the 'magpie' *Gymnorhina tibicen* (Cracti-cidae), a group-territorial species which in New South Wales reaches 16 years of age or even more (Carrick, 1972).

In the north temperate zone, survival curves for small land birds are linear and decline steeply so that only a tiny fraction of populations survive to 7 years (then largely under suburban patronage). For passerines excluding corvids, re-coveries show individuals of only 34 species to have achieved that age in Europe, and 31 in North America; seven species have been identified with at least one individual living to 13 – 20 years in Europe, and seven others have reached 11 – 14 years in North America (Rydzewski, 1973). But in Australia, where passerines have been ringed in substantial numbers only since the early 1960s, 320 individual birds of 50 species have already been recovered 7 years after being ringed, with several at 10 and one at 13 years (D. Purchase, personal communication). Rather more than one million birds have been ringed in Australia, as against a hundred million in Europe and North America. Comparing these values is beset with pitfalls, but the evident higher longevity of Australian birds compared with Holarctic species can probably be related to the harsher climates of high latitudes.

Tropical longevity data are fragmentary, but the following records are relevant. Several Red-throated bee-eaters that I ringed in northern Nigeria in 1964 were still alive in 1975, though at 24 g the species is far from robust. Neotropical manakins (a promiscuous lek species belonging to the family Pipridae) live to 14 years and have an adult annual survival rate between 80 and 90 per cent (Snow and Lill, 1974). Several southern African species survived 10-12 years and robins, *Pogo-nocichla* and *Cossypha,* in Natal lived at least 14 and 17 years, and in a small patch of outlying forest in Ivory Coast savanna, 40 per cent of the small birds survived at least into their fifth year, while in Malaysian rainforest small songbirds have known lifespans in excess of 12 years (references in Fry, in press). These data need to be evaluated in relation to the paucity of long-term field studies and to the still poor knowledge of bird life outside Europe and North America.

Discussion

Clearly co-operative breeding is a heterogeneous phenomenon, and debate will continue for some time as to the socioecological frameworks from which it has originated. Brown (1974) proposed three main routes to this type of sociality among jays — from colonies, from a surplus of males and by retention of young in the family. Other authors have proposed additional routes (see, for example, Koenig, 1974).

The last of Brown's bases is of crucial importance, as follows. Selection operates on populations to produce life history characteristics at or between extremes known as r- and K-selection. r-selected species are adapted to life in unpredictable environments. They have high rates of population turnover and are opportunists that quickly colonise a patch of suitable habitat as it becomes available. Local extinction is commonplace; individuals are short lived, and there is not much overlap between generations. K-selection occurs in stable habitats. Individual and group survival is good; mortality and fecundity are low, and individuals are long lived with consequent wide generation overlap (Pianka, 1974). By comparison with most insects, birds are strongly K-selected; but most land birds living in middle and high latitudes are far from the K end of the selection spectrum and have relatively r-adapted social organisations. Their high mortality, resulting from overwintering or from the alternative of long-distance migration, is balanced by high productivity and rapid turnover of the population. Low-latitude (tropical) birds mostly adopt K-strategies, both within the highly stable rainforest habitat with its uniform photoperiod, temperature and rainfall, and in those non-forest biomes lacking great seasonal adversity (of which protracted freezing temperatures are evidently far more hostile to bird life than seasonal drought).

Co-operative breeding is found mainly in birds which are (or which one can on other grounds strongly infer to be) markedly K-selected. Most of the species are tropical, sedentary and long lived. Adult survival rates of at least some of the extra-tropical co-operative nesters are exceptionally high. In all these birds, whatever the level of annual production of young, the problem arises of assimilating more young than initially there are vacancies for in an adult population which is, in K-selected species, intensely competitive. One adaptive solution is to integrate the first-year birds as non-breeders, when under some circumstances kin selection will favour the evolution of co-operative breeding (Ricklefs, 1975).

I conclude with a small irony. It is that on the one hand kin selection is being increasingly accepted as the operative mode in the evolution of co-operative breeding and related altruistic behaviours (Eberhard, 1975), while, on the other hand, more and more people are challenging the notion that helping at the nest is altruistic at all. Zahavi (1974) has argued the case for 'helping' being the product of individual selection (but see Brown, 1975), which is in part endorsed by the

observations of Carrick (1972), Parry (1973), Woolfenden (1975) and others; as
Wilson (1975) put it, 'the theory of group selection has taken the good will out
of altruism'.

References

Boehm, E. F. (1974). Results from banding Chestnut-crowned Babblers. *Aust.
Bird Bander*, **12**, 76 - 78.

Brown, J. L. (1974). Alternate routes to sociality in jays — with a theory for the
evolution of altruism and communal breeding. *Am. Zool.*, **14**, 63 - 80.

Brown, J. L. (1975). Helpers among Arabian Babblers *Turdoides squamiceps.
Ibis*, **117**, 243 - 244.

Bryant, D. M. (1975). Breeding biology of House Martins *Delichon urbica* in relation to aerial insect abundance. *Ibis*, **117**, 180 - 216.

Carrick, R. (1972). Population ecology of the Australian Black-backed Magpie,
Royal Penguin, and Silver Gull in *Population Ecology of Migratory Birds*. US
Department of Interior Bureau of Sport Fisheries and Wildlife, Wildlife
Research Report 2, 41 - 99.

Dow, D. D. (1975). Breeding biology and behaviour of *Manorina* (*Myzantha*)
melanocephala, a communally nesting honeyeater. *Emu,* **74** (1974), Supplement,
308.

Eberhard, M. J. W. (1975). The evolution of social behaviour by kin selection.
Q. Rev. Biol., **50**, 1 - 39.

Fry, C. H. (1972). The social organisation of bee-eaters (Meropidae) and co-
operative breeding in hot-climate birds. *Ibis*, **114**, 1 - 14.

Fry, C. H. (in press). Survival and longevity among tropical land birds, in *Proc.
4th Pan-Afr. Ornith. Congr.* (1976), South African Ornithological Society.

Grimes, L. (1975). Co-operative breeding behaviour and family relationships
within groups of the long-tailed Shrike *Corvinella corvina. Emu,* **74** (1974),
Supplement, 309 - 310.

Koenig, W. D. (1974). Communal breeding in birds. Privately circulated manuscript 41 pp.

Lack, D. (1968). *Ecological Adaptations for Breeding in Birds*. Methuen, London,
409 pp.

Maclean, G. L. (1973). The Sociable Weaver, part 3: breeding biology and moult.
Ostrich, **44**, 219 - 240.

Nakamura, T. (1975). Co-operative breeding in the Japanese Long-tailed Tit.
Emu, **74** (1974), Supplement, 311.

Parry, V. (1973). The auxiliary social system and its effect on territory and
breeding in Kookaburras. *Emu*, **73**, 81 - 100.

Perrins, C. M. and Moss, D. (1975). Reproductive rates in the Great Tit. *J. Anim.
Ecol.*, **44**, 695 - 706.

Pianka, E. R. (1974). *Evolutionary Ecology*. Harper and Row, New York, 356 pp.

Rheinwald, G. and Gutscher, H. (1969). Das Alter der Mehlschwalbe (*Delichon
urbica*) in Riet. *Vogelwarte*, **25**, 141 - 147.

Ricklefs, R. E. (1975). The evolution of co-operative breeding in birds. *Ibis*,
117, 531 - 534.

Rydzewski, W. (1973). Longevity records. *Ring*, 74, 7 - 10; 75, 40 - 41; 76, 63 - 70; 77, 91 - 96.

Skutch, A. F. (1961). Helpers among birds. *Condor*, 63, 198 - 226.

Snow, D. W. and Lill, A. (1974). Longevity records for some neotropical land birds. *Condor*, 76, 262 - 267.

Vehrencamp-Bradbury, S. (1975). Adaptive advantage of communal nesting in Groove-billed Anis. *Emu*, 74 (1974), Supplement, 312.

Wilson, E. O. (1975). *Sociobiology*. Belknap Press, Cambridge, Massachusetts, 697 pp.

Wood, N. A. (1974). The breeding behaviour and biology of the Moorhen. *Br. Birds*, 67, 105 - 115 and 137 - 158.

Woolfenden, G. E. (1975). Florida Jay helpers at the nest. *Auk*, 92, 1 - 15.

Wynne-Edwards, V. C. (1962). *Animal Dispersion in Relation to Social Behaviour*. Oliver and Boyd, Edinburgh, 653 pp.

Zahavi, A. (1974). Communal nesting by the Arabian Babbler. A case of individual selection. *Ibis*, 116, 84 - 87.

13 Helpers at the nest in some Argentine blackbirds

Professor G. H. Orians, C. E. Orians and K. J. Orians*

Systems of communal or cooperative breeding in which several adults help to feed one brood (Lack, 1968, p. 72) have been considered to be relatively rare among birds, but recent evidence suggests strongly that the habit may be very widespread especially in tropical and subtropical regions (Fry, 1972; Harrison, 1969). An excellent review is provided by Brown (1974) who also discusses several theories for the evolution of this and related forms of breeding. He points out that known communal breeders are nearly all permanent residents in relatively stable habitats where their populations are likely to be at or close to the carrying capacity of the environment most of the time. He postulates that a period of K-selection, followed by a period of kin selection are the normal prerequisites for the evolution of communal breeding.

Despite extensive studies on a number of species and casual observations on many others, helpers at the nest were not reported among the American blackbirds (Icteridae) until Fraga's (1972) brief report on helpers in the Bay-winged cowbird *Molothrus badius*. We were therefore surprised to discover helpers at the nest in four of the six species of blackbirds we studied in Argentina and Bolivia during the 1973–74 breeding season. Because of the unusual features of the social organisation of the Bolivian blackbird *Oreopsar bolivianus*, that species will be dealt with in a separate publication (Orians, Erckmann and Schultz, in preparation). Here we present our field data on three Argentine species, the Brown-and-yellow marshbird *Pseudoleistes virescens*, the Bay-winged cowbird *M. badius* and the Austral blackbird *Curaeus curaeus*, together with a conceptual framework for considering problems of the evolution of offering and accepting help at the nest.

*Gordon H. Orians is Professor of Zoology at the University of Washington. Seattle. His research on the ecology of vertebrate social systems began while he was a graduate student at the University of California, Berkeley and has involved many species of blackbirds in North and South America. He was a Fulbright Fellow with David Lack at the Edward Grey Institute of Field Ornithology from September 1954 until August 1955.

Carlyn E. Orians, a student at Bowdoin College, Brunswick, Maine and Kristin J. Orians, a student at the Overlake School, Redmond, Washington, have participated in studies of blackbirds for several years. They discovered helpers at the nest in the Brown-and-yellow marshbird while participating in an alternative education programme of the Shoreline School District, Seattle, Washington.

Brown-and-yellow Marshbird P. virescens

The non-migratory Brown-and-yellow marshbird, the more southerly of two species in its genus, breeds from extreme southern Brazil south through Uruguay and eastern Argentina to southern Buenos Aires Province. The most dense populations are found in the temperate wet pampas of eastern Argentina, a flat region of wet grasslands interspersed with extensive marshes and lagoons (Vervoorst, 1967). The region has a mild, temperate climate with a July mean temperature of about 8°C and a January mean of about 20°C. Extremes are somewhat greater away from the Atlantic Ocean but nowhere do the marshes freeze in the winter.

Our field studies were carried out in the general vicinity of Pinamar, on the coast south of Cabo San Antonio (Weller, 1967). Most of our observations were made along a paved, all-weather road leading inland from Pinamar to the town of Madariaga. The tall clumps of Pampas grass, *Cordateria argentea*, which grew on the shoulders of the road were favoured nesting sites of the marshbirds, especially early in the spring when other nesting cover was scarce. Studies began on October 8, 1973 and continued daily until November 28. Unfortunately males and females are difficult to distinguish in the field, though the males are slightly larger. Some sexing was possible by observing the behaviour of colour banded birds.

Because we knew that emerging aquatic insects are the most important food source supporting breeding of blackbirds in western North America (Orians, 1966; Orians and Horn, 1969; Willson, 1966), we carried out a sampling programme for emergence in four different marshes using 51×89 cm wire mesh cages covered with plastic screening. A total of 1243 trap days during October and November 1973, yielded a total of 512 emerging insects, only 0.41 per trap per day. Emergence on productive lakes in western North America is 50 – 100 times as great (Orians and Horn, 1969; Orians, in press). Therefore, the aquatic food base supporting breeding blackbirds in eastern Argentina is much less than in North America. Not surprisingly, the populations of blackbirds are much less dense, and there are other differences in social organisation and types of food exploited (Orians, in preparation).

During the winter, marshbirds tend to be gregarious (Hudson, 1923; Weller, 1967) though they are not migratory. Large flocks are occasionally found even during the breeding season. On November 1 we found flocks of dozens of birds in drier grasslands southwest of Mar de Ajó, about 40 km northwest of our main study area, but we never saw flocks of this size around Pinamar. Nest building commenced at Pinamar in mid-September and continued, though at reduced intensity, throughout November. During the first week of the study we discovered helpers at one nest, and thereafter we made special efforts to determine the number of birds attending as many nests as we could observe.

We discovered no evidence of territoriality among the Brown-and-yellow marshbirds. Nests were often situated as close to one another as 4 m, and several were regularly found in one small group of pampas grass clumps. Other nests were isolated more than several hundred metres from their nearest neighbours. Even at clumped nests there were rarely signs of agonistic interactions among the birds. The only clear case we observed involved chasing of a group of five birds from a *Scirpus – Typha* clump by a pair, one of which was carrying nesting material. Feeding grounds are also undefended. Birds regularly feed within a few centimetres of one another, and birds from many different

Table 13.1 Number of adults visiting nests of the Brown-and-yellow marshbird, Pinamar Argentina. For the nesting stage the number refers to the birds actually bringing food.

	Nest characteristics			Building			Incubation			Nestling		
Nest no.	Clutch size	No. of young fledged	Distance of nearest nest (m)	No. of birds	Observation time (h)	No. of birds marked	No. of birds	Observation time (h)	No. of birds marked	No. of birds	Observation time (h)	No. of birds marked
2881	5	3	24							2	3:54	0
2854	5	0	24							1	2:38	1
F	3	2	103							8	5:06	0
L	4	0	500							2	1:13	0
D	3	0	7				4	4:24	0			
2811	?	0	30	2	2:40	0						
2826	4	0	30				2	3:45	0			
2815	4	0	38				2	2:40	0	2	1:03	0
2819	5	0	9	2	1:20	0	3	2:05	0			
2845	?	0	20	2	3:16	0						
2865	5	3	11				3	1:33	1			
2874	4	2	4				2	1:33	1			
2842	4	3?	21							3	0:40	0
2841	3	2	66							4	1:32	1
2850	4	3	9				3	2:05	0	5	2:00	0
2839	4	0?	32							4	1:30	0
2838	4	1	12							5	2:05	2
2897	5	4	12				2	2:00	1		0:20	0
2885	5	4	83				3	1:10	0			
2882	4	?	7				4	0:39	0	6	0:40	0
Acreage	4.2	1.4	52.1	2.0			2.8			3.8		

nests often gather at one general feeding area.

Because of high nest predation rates we were unable to follow any one nest all the way from nest building to fledging of the young, but many nests were observed in two of the three major stages, for example, building–incubation or incubation–nestlings. Data on all nests at which we made estimates of the number of adults in attendance are summarised in table 13.1. The total time we watched the nest at each stage, and the number of colour-banded birds available to aid in estimating the number present, are indicated. All estimates represent the minimum number of birds that could account for the visitation pattern we observed. Since the country was open we were able to follow many individuals on their foraging trips away from the nest, and could make reasonable estimates of the number involved even when none was colour-banded.

We never saw more than two birds attending a nest under construction and never observed males bringing material to the nest. Typically the male accompanies the female on gathering trips, returning with her to the nest vicinity without bringing material. The three copulations we observed took place on gathering trips when the nest was nearly completed. During casual observations at many other nests under construction we never saw more than one pair of birds.

As far as we could determine only one bird incubated, though we have good data for only three nests at which the incubating bird was colour-banded. One of those three was seen copulating and known to be a female, and we suspect that incubation is done by the female alone. At most of the nests, the incubating bird was fed on the nest, often by more than one other bird. At one nest where the incubating female was not fed, the male usually sat near the nest and regularly flew to it, hovering above it. This usually induced the female to leave and the two flew together to a foraging area. We found no evidence that more than one female laid in a single nest, and there is no correlation between clutch size and number of birds attending. None the less, clutch size was rather variable, and a surprising number of eggs failed to hatch (22.5 per cent). Whether this indicates involvement of other birds during laying cannot be determined without more intensive field work with a colour-banded population.

Another increase in the number of birds attending a nest occurred after the eggs hatched. The maximum estimate of eight birds was at the first nest to hatch (and to be observed), and the count was made without the aid of colour-banded birds. After these initial observations, we colour-banded four of the group, but our estimate of its size did not change.

The sexes and ages of the extra birds were unknown, and we know nothing of their origins or relationships to the birds they helped. They do not seem to have been derived from other nests in the vicinity because we never observed a bird known to have been in attendance at one nest bringing food to another, but our sample is not large (38 individuals).

Indirect evidence suggests that the birds within these nest-flock units may be closely related. Rohwer (unpublished) has suggested that birds associated in groups with high coefficients of relationship are more likely to scream when removed from mist nets than individuals from migrant species, particularly nocturnal migrants, that travel in flocks of unrelated individuals in the winter. This relationship was predicted because screaming is probably either an altruistic act or one which evokes altruistic acts from those that hear the signal. The Brown-and-yellow marshbirds were loud and vigorous screamers, and their calls induced mobbing behaviour from other members of the nest flock and birds from nearby nests.

Nests were built in dense cover near wet pastures which were the major foraging sites for the birds at all stages in the breeding cycle. Of the 73 nests we found, 53 were situated in Pampas grass, the others being distributed among cattails *Typha latifolia,* sedges *Cyperus,* and thistles *Circium,* in order of decreasing frequency. In October the pampas grass clumps provided the best cover and contained nearly all the nests. In November, the growth of herbaceous plants and the leafing out of shrubs increased the total amount of cover available and nest sites were more variable. Apparently, suitable nesting sites did not seem to be in short supply because only a small fraction of the available clumps of pampas grass were occupied in the 7 km stretch of road where most of our work was done. Nevertheless, our evaluation of suitability could have been in error.

Young blackbirds were very difficult to follow after they fledged, and we have data from only one brood which was fed for at least 2 weeks after fledging by the same birds that brought food to the nest (F see table 13.1). There is no information available on the permanence of family units during the autumn and winter, the coefficients of relationship in winter flocks, or the extent of movement of these flocks.

Bay-winged Cowbird *M. badius*

The Bay-winged cowbird is common in grasslands and savannas from coastal southern Brazil, Paraguay and Bolivia south to central Argentina. Except in Bolivia, where it is the common icterid in valleys on the east side of the Altiplano up to an elevation of 3350 m, it is a bird of the lowlands. The species is resident throughout its breeding range, but there are local movements, and during the autumn and winter birds travel in medium to large flocks. It is unique among the cowbirds in incubating its own eggs and feeding its own young in nests built by other species, particularly members of the family Furnariidae (Friedmann, 1929). Nests which have been abandoned are regularly used, but eviction of the builders is not uncommon.

The Bay-wing tends to be a late breeder in much of its range. Friedmann (1929) reports a peak in breeding in late January and early February in the province of Tucumán, Argentina, where the climate is subtropical with a long, intense rainy season extending from November to April. Our main study area was southwest of Tucumán in and around the town of Andalgalá, Province of Catamarca; this lies in the creosote bush (*Larrea*) desert zone that extends north to about latitude 25°S in valleys west of the first front ranges of the Andes. These mountains, which rise to over 3000 m at the latitude of Tucumán, intercept the moisture-bearing winds from the Atlantic Ocean. Consequently, the valley systems lying between them and the main range of the Andes are much drier, and receive only light rainfall primarily during the months of January, February and March.

At Andalgalá, Bay-wings were found breeding within the town itself and also along the small washes with shrubs and small trees of *Prosopis alba, P. torquata, Acacia aroma, A. furcatispina, Bulnesia retama* and *Cassia aphylla* which are interspersed among flats dominated by creosote bush (*Larrea cuneifolia*). During the non-breeding season the cowbirds are found principally around the towns and agricultural areas in this vegetation zone, and they gradually spread into the surrounding countryside during the rainy season. At Andalgalá, the principal nest selected is that of the Brown cachalote *Pseudoseisura lophotes*, an ovenbird about the size of a jay that builds a large, sturdy domed nest with a tubular entrance on one side which normally lasts for many years. In addition, nests were found in abandoned holes of the Golden-breasted woodpecker *Chrysoptilus melanolaimus* in the Giant cardon cacti *Trichocereus terscheckii,* or in crevices at the bases of palm fronds.

During his intensive studies at Concepción, Friedman always observed Bay-wings in monogamous breeding pairs. He reported that the nest site was selected by the pair acting together and that they both fought for its possession if the owners were still in residence. After the nest had been occupied, a singing tree was selected by the adults, and defence of the nest site and its immediate environs began. It is uncertain whether the Bay-wing actually defends territories, but agonistic interactions are not restricted to the immediate vicinity of the nest, and the loud, musical song carries long distances.

The first evidence of helpers at the nests of the Bay-winged cowbird was obtained by Fraga (1972) during a study of these birds near Lobos, Buenos Aires Province. He found that the nests were selected primarily by the females, who slept in them during egg laying. If a nest was actually built by the birds, as sometimes occurs, the female did most of the building with some help from the male. During incubation he stood guard near the nests but then joined the female in bringing food to the nestlings.

At Lobos, most pairs were assisted by one or sometimes two helpers which brought food to the nestlings and participated in mobbing predators. In the 1970-71 breeding season there were helpers at 5 of 8 nests that reached the nestling stage. Fraga found no evidence that laying, incubating or brooding was performed by more than one female. Of four colour-banded helpers, two were clearly males, and others were suspected of being males though no clear evidence is provided. In every case, when the young fledged, extra individuals began to attend them, feeding them and mobbing predators. On one occasion 18 scolding birds were found with one fledgling. Fledglings of the parasitic Screaming cowbird *Molothrus rufoaxillaris* are similarly attended and protected. One of the colour-banded females in Fraga's study area changed mates for her second nest and began to lay a new clutch when her first mate was still feeding the fledglings from the first nest.

We found five nests of Bay-winged cowbirds at Andalgalá, two in old Cachalote nests, two in woodpecker holes in cacti and one in a palm tree. One of the cactus nests was either quickly destroyed or abandoned, and no useful data were obtained from it. Our most intensive observations were made at a nest 6 km west of Andalgalá in an old Cachalote nest in a small *Acacia aroma*. This nest was watched a total of 913 min, 505 of which occurred after two of the birds had been mist-netted and colour-banded. Four adults regularly brought food to the young. One of the marked birds was judged to be a male on the basis of its larger size and its regular habit of perching directly above the nest and singing before and after delivery of food. The other colour-banded bird was smaller and usually entered and left the nest without stopping to sing. Each of the colour-banded birds associated with an unbanded bird which exhibited behaviour typical of the opposite sex. Therefore, though the birds were not collected for anatomical verification of sex, we believe that two pairs were in attendance. It was not uncommon to have all four birds at the nest area simultaneously, and two or three were sometimes in the nest together. We observed no agonistic interactions among them.

On February 26, in contrast to the situation of the previous week, the two colour-banded birds regularly arrived and left together, often in association with a third, unbanded individual. On February 28 we captured a recently fledged young bird whose screaming evoked mobbing by three adults, two of them unbanded, the missing bird presumably being the banded male. On March 2 the fledglings were found about 450 m up the wash from the nest. Again we were mobbed by only three birds, and the banded male was not among them. Since this nest was not watched during the incubation period we do not know the roles of any of the birds at that time.

Four kilometres west of Andalgalá we watched a nest in a hole in a cactus for 160 min during the incubation period. At this nest we never saw more than one pair in attendance. One bird, presumably the female, did all the incubating while the male sat in the top of the cactus or in a nearby bush most of the time. Usually the pair left to forage together. A second pair of birds approached the nest on two different occasions during the observation periods and both times they were driven away by the male who sang vigorously and flew around the intruders. Unfortunately, this nest had not hatched by the time we left in late March.

The third nest, high in a palm tree in town was watched for 135 min when it had nestlings. Four birds were in attendance, but we could not obtain positive evidence that more than two of them were actually bringing food. When four birds were present, only two had food and one pair regularly came in to the tree with nothing in their bills even though they did show considerable interest in the nest site.

The final nest, situated in an old Cachalote nest in a small *Celtis* tree at the edge of an abandoned field, was watched for a total of 233 min when it had nestlings. None of the adults at this nest was colour-banded and none could be sexed, but a minimum of three birds and possibly four regularly brought food.

Thus, there were more than two adults at all three nests we watched during the nestling stages, but only a pair at the one nest watched during the incubation period. These results are similar to those of Fraga (1972) suggesting that breeding of the Bay-wing may be similar to that of the Brown-and-yellow marshbird.

The Austral Blackbird *Curaeus curaeus*

The Austral blackbird is the sole representative of its family in the beech (*Nothofagus*) forests of southern Chile and Argentina. Almost nothing has been written of its ecology or behaviour, and our data, are, unfortunately, more fragmentary on this species than the others, primarily because young had already fledged at the three locations where we were able to observe the birds.

Observations were made on a flock of birds at Llao Llao, Rio Negro Province, on December 11 and 12, 1973. A careful search revealed at least four fledglings in the bushes at the site and the screams of a captured individual evoked mobbing by six adults. As one fledgling attempted to fly across the road from the area where it was first found, six adults immediately approached and flew with it to the trees on the opposite side. More than two adults were seen with food in their bills, but we were unable to observe fledglings being fed.

Additional observations were made on several groups of birds at Lago Futalaufquen, Chubut Province, December 15 - 18, 1973. At one site a single fledgling was fed by at least three adults, and four birds were definitely in attendance. At another site the fledglings could not be discovered, but at least three adults flew into the general area with food.

Birds were relatively easy to observe feeding in open pastures at Lago Futalaufquen and we obtained data on two different pairs that were not gathering food for offspring. In both cases the pair maintained close contact and flew from site to site together. A third bird, loosely associated with each pair, gave a much higher frequency of contact calls than did either member of the pair. One of the hangers-on was judged to be a male on the basis of the glossiness of his plumage.

A final flock was observed on December 27, 1973 at Lapataia, Parque Nacional de Tierra del Fuego. This group, foraging in a logged forest of *Nothofagus*, was attending at least one fledgling. We were mobbed by four adults but could not determine how many of them actually were feeding or how many fledglings were present.

None of our data on Austral Blackbirds is complete enough to exclude the possibility that, after fledging, young from two or more nests come together with their parents to form a small flock in which attendance and care of the young is shared but helping does occur at least at the fledgling stage. If the foraging pairs we observed were birds with nests under construction or with eggs, the presence of a third bird would suggest that potential helpers may associate with nests early in the breeding cycle.

Discussion

Our data do not provide any evidence about the relationships among helping and helped individuals in any of these three species of blackbirds. Therefore, they cannot be used to evaluate the importance of individual and kin selection on the evolution of these breeding systems. Nevertheless, the apparent widespread incidence of helpers at the nest in South Temperate icterids, compared with the total absence of any reports for North American species, (some of which have been intensively studied) suggests that a general consideration of factors favouring helping and accepting help might throw some light, however tentative, on the different course of evolution of blackbird social systems on the two continents.*

Recent evidence that helpers at the nest are more common among birds than previously supposed suggests that the conditions favouring this type of behaviour

*Since writing the above we have observed a nest of Brewers' blackbirds *Euphagus cyanocephalus*, to which food was being delivered by one female and two males. The frequency of such helpers in this species is unknown.

may not be highly restrictive. There are several theories to explain helping that find support among some students of the subject. The first postulates that helping evolved for the good of the species. This explanation requires some form of selection at the level of the species, the problems of which have been adequately reviewed several times (Brown, 1966; Maynard Smith, 1964; 1965; Lewontin, 1970; Williams, 1966) and need not be discussed here. Since all suggested mechanisms at this level require stringent conditions unlikely to be met by the birds we have studied, we prefer to seek explanations based on less restrictive evolutionary processes.

A second theory postulates that helping evolved by kin selection. The balances between risks, benefits and relationships among individuals that results in the spread of genes which cause their bearers to help close relatives, are generally known (Hamilton, 1964; 1972). The importance of kin selection in the evolution of helping among birds is strongly suggested by the fact that, in those species for which adequate data are available, helpers are often close relatives, especially young of the previous breeding season (Brown, 1970; 1972; Ligon, 1970; Parry, 1973; Skutch, 1935; 1960; Rowley, 1965; Woolfenden, 1975; Zahavi, 1974). In fact, most recent authors have assumed that helping is altruistic and that kin selection must be involved.

The third theory postulates that the helping individuals gain experience which is of later benefit, or obtain some kind of protection of more immediate benefit. Future benefits may involve improved access to territories, knowledge of foraging techniques, improved foraging, tending skills, and so on. These benefits are most likely to result in a net gain in total fitness when opportunities to obtain the same benefits from independent breeding are poor (Selander, 1964).

Fourthly, helping could evolve through parental manipulation of the offspring even if the offspring do not benefit from the behaviour (Alexander, 1974; Trivers, 1974). This appears unlikely, because if the probability of success of independent breeding by potential helpers is high the inclusive fitness of the parents, as well as that of the helpers, should be increased if the yearlings attempt to rear their own young.

Finally, helping may result from an imbalance in the sex ratio among potential breeders in an area. This may, in fact, be the cause of most of the occasional helping observed in normally monogamous species, but it cannot explain the most highly evolved systems that are our concern here.

None of the theories as currently developed gives adequate consideration to the potential risks and benefits to *both* partners in a helping situation, the helpers *and* those being helped. Helping is a two way interaction. Not only must there be individuals that will help, but there must also be individuals that accept help. As is usually the case with two-party interactions, the interests of the participants are not necessarily the same.

Accepting help

The immediate benefits potentially accruing to individuals accepting help are improved care of the offspring (Parry, 1973), increased future reproductive success of the parents resulting from their opportunity to reduce current investment in reproduction (Gaston, 1973), enhanced detection of predators, and more individuals to mob and drive away predators and competitors. A longer-term benefit might be the opportunity to hand down the territory to a close relative, as occurs, for example, in the Mexican jay (Brown, 1974), the Florida race of the Scrub jay (Woolfenden, 1975), and possibly in the Tufted jay (Crossin, 1967).

Risks potentially associated with accepting help include cuckoldry, the attraction of predators by the higher density and increased activity of birds around the nest, and greater reduction of food supplies in the vicinity of the nest by the larger group of individuals (Zahavi, 1974). Also, there is a potential risk that the extra individuals might destroy the eggs or young of the breeding pair. Though this possibility has not, to our knowledge been considered in previous discussions of communal breeding, it is not difficult to imagine conditions in which extra individuals could improve their fitness by eliminating the off-spring of others in the group. Such behaviour has been described on the part of dominant individuals in a number of species of birds and mammals (Ghiselin, 1974; Trivers, 1972), but advantages could also accrue to a subordinate if it could kill the offspring of the dominant without being detected.

Helpers should be permitted near the nest only if the inclusive fitness of the breeders is increased. Most birds may not allow extra individuals on their territories because accepting help is too risky. On the other hand, the existence of occasional helpers at nests of many species normally lacking them could mean that accep-tance would be general if there were more willing helpers. On the basis of the risks and benefits identified above, we can suggest conditions that would tend to reduce risks and increase benefits. For example, in larger birds, predation risks are usually lower and additional individuals are more effective in driving away predators. The risk of nest destruction by the helpers should be reduced or eliminated if the helpers are close relatives since an individual is less likely to im-prove its inclusive fitness by killing close relatives. Also, cuckoldry reduces inclusive fitness less if it is performed by a close relative.

Offering help

The potential benefits to a helper from helping include gaining experience in foraging and nest building, especially in circumstances where opportunities are available to learn by observing tactics and successes of older and more experienced individuals. Being a member of an experienced flock may also afford protection from predation, though it is not necessary to help to gain this benefit. Member-ship of a group may enhance the probability of gaining the territory on the death of one of the dominant birds.

A disadvantage of helping is failure to produce one's own offspring. Also, as suggested previously, individuals in groups may suffer higher predation and grouped individuals must forage in areas with higher densities of conspecifics. Of these risks, probably the most significant is failure to reproduce. Significantly, most birds do breed in their first year even though they are less successful than older birds (Coulson and White, 1956; Lack, 1954). However, if helpers are in-dividuals unable to obtain a mate because of, for example biased sex ratios, then *any* benefit to helper suffices to favour helping. Helping close relatives is pre-ferable to helping unrelated individuals, but producing one's own offspring, if there is a reasonable chance of success, should be better than helping. Benefits may accrue to helpers regardless of their relationship to those they are helping.

Experience and access to territories in the future are not dependent on close relationships and, if help is given randomly with respect to adult genotypes, it does not affect gene frequencies among the adult population. Therefore, neither altruism on the part of helpers nor kin selection is necessary for the evolution of helping. Nevertheless, helping *is* more likely among close relatives because, for a helping individual, the point at which it is better to help rather than attempt to breed occurs at a higher probable breeding success from independent reproduction than does helping non-relatives. Furthermore, helping is more likely to be accepted if it is offered to close relatives.

Since individuals can be expected to offer help only when the average benefits from such behaviour are greater than the average benefits from independent reproduction, helping should evolve only when expected success from independent reproduction is very low, as argued by Selander (1964). Failure to find a mate is one obvious cause of this but many species are known in which helpers of both sexes are found in the same area, as occurs in the Bay-winged cowbird. Alternatively, when resources are in short supply the poorer performances of inexperienced individuals reduce breeding success more than when resources are more plentiful (Brown, 1974). For example, if there are insufficient nest sites and older birds are better competitors for them, younger birds may have to settle for a poor site or nothing at all. If there is only a small pulse in the food supply at the time of breeding, younger individuals may be at a competitive foraging disadvantage with older ones. Similarly, if there is a shortage of suitable territories, younger individuals may be forced to occupy inferior sites, which, when combined with their inexperience, could result in very low average breeding success. Therefore, relative resource shortages are probably the most common condition favouring the evolution of helping and probably account for the prevalence of helping in larger passerines of low latitudes (see also Selander, 1964).

The prevalence of non-helping

Any theory to explain the evolution of helping should also be able to explain why most species do not have helpers at the nest. A number of these species would seem to fulfil the conditions we have specified as favouring the evolution of helping. For example, many seabirds do not begin to breed until they are several years old, and in several cases the young individuals are known to be less efficient foragers than older individuals (Ashmole and Tovar, 1968; Orians, 1969). In addition many nest in relatively predation-free environments where the help of additional individuals should pose little or no predation risk. Nevertheless, helping at the nest has not been reported in any of these species, though many have been studied intensively.

Similarly, helping at the nest is not known in other blackbirds. A number of species are polygynous and have substantial non-breeding populations of first year and older males, but helping has never been observed even though the potential benefits we have postulated should be operative. The major reason for lack of helping in these birds may be that the extra males pose too severe a genetic threat to the territorial male to be permitted on the territory. This is especially critical among the polygynous blackbirds where additional females may be attracted over a long time interval and females are available to be fertilised over a correspondingly long period. In these circumstances males accepting helpers may actually pass fewer genes than those excluding them. In the case of many of the seabirds, especially gulls, extra individuals may eat eggs or young of the breeding pair. These birds are regularly active nest predators on their neighbours, apparently finding it to their advantage to inhibit the reproduction of other birds in the colony while gaining resources for themselves.

Because neither seabirds nor temperate zone blackbirds occupy nesting areas outside the breeding season, their helpers do not benefit from holding territories continuously; helpers in resident species may gain in this way (Brown, 1974). Also, in migratory species, the probability that breeding birds can recognise the identity of potential helpers may be lower unless close relatives remain together in flocks during the non-breeding season.

Helping among blackbirds

Because of our limited knowledge of the South American icterids with helpers at the nest, we do not know which of the above conditions they fulfil. We lack any information about the relationships among helpers and those they help and do not know the sexes of most helpers. None of the three Argentina species is sedentary within its breeding area but in all three, parents and their offspring may remain together during the autumn and winter so that individual identification cannot be ruled out.

Suitable nest sites might be limited for the Bay-winged cowbirds that must take over nests of other species, but our general impression is that old nests of a number of species of furnariids are common in the breeding range of the Bay-wing in Argentina. At Andalgalá many such nests were unoccupied by cowbirds. Early in the spring there are few good nesting sites in our study area for the Brown-and-yellow marshbird but even then many clumps of Pampas grass were unoccupied. However, we may be poor judges of nest-site quality and, given the high rate of nest predation at Pinamar, subtle differences may be very important to breeding success. Helpers at early nests of the Brown-and-yellow marshbird are consistent with an assumption of a shortage of nest sites, but inconsistent with an interpretation that helpers are individuals whose nests have been destroyed.

Close relatives may remain together in the Austral blackbird, but there are no data. There is some altitudinal migration in this species, flocks descending to the Patagonian plains as snow accumulates in the shrub–forest interface where they breed, but the distances are short and parents and offspring could remain together. Nest sites seem abundant in the breeding areas.

Data on the food supply in the breeding season are meagre for Argentine blackbirds, but in general, south temperate South America is characterised by maritime climates with mild winters and cool summers. Our emergence sampling procedures at Pinamar did not adequately sample the food utilised by breeding Brown-and-yellow marshbirds because they foraged in wet pastures and the edges of marshes and primarily dug lepidopteran and coleopteran larvae from the moist turf. Nevertheless, the maritime climates of temperate South America probably result in less widely fluctuating food supplies than in temperate North America. Suggestive evidence comes from the fact that the passerines at Pinamar have

Table 13.2 Clutch size of passerine birds at Pinamar, Argentina,
spring (October – November), 1973

Species	Mean clutch size	No. of clutches
Brown-and-yellow marshbird	4.1	37
Scarlet-headed blackbird	3.1	9
Yellow-winged blackbird	3.0	19
Wren-like rush bird, *Phleocryptes melanops*	2.8	9
Many-coloured rush tyrant, *Tachuris rubrigastra*	3.0	1
Spectacled tyrant, *Hymenops perspicillata*	2.0	12
Rufous-collared sparrow, *Zonotrichia capensis*	3.0	2

smaller clutches than their North American counterparts (table 13.2).

Two associated marsh breeding icterids at Pinamar, the Scarlet-headed black-bird, *Amblyramphus holosericeus,* and the Yellow-winged blackbird *Agelaius thilius* do not have helpers at the nest. Of the two, the Scarlet-headed Black-bird is strongly territorial, and no extra birds are allowed on the large territories. Attempts to take over territories by pairs of adults were seen throughout the breeding season and there is also a distinctive first-year plumage further indicating the existence of non-breeders potentially available to help. The Yellow-winged is non-territorial and extra birds regularly move about close to occupied nests and may even perch on the rims. Both species occur in flocks during the winter, often large ones in the Yellow-winged. Helping, if it does occur must be rare at least in the Pinamar area or we would have observed it during the many hours we watched nests of both species.

Available evidence on helpers at the nest suggests that it is most prevalent among passerines at low latitudes and more likely to occur among larger species than smaller ones. This pattern is consistent with our suggestion that helping is most likely where resources are relatively constant so that the birds face a limited food and/or nest site supply during most breeding seasons. Most, though not all of the species known to regularly have helpers are resident and, as pointed out by Lack (1968), this would facilitate identification of close relatives. Even so, helping apparently has not evolved in most tropical species so that pre-sumably a combination of the risks in helping or accepting help are strong enough to select against it. Possibly many species have high enough nest predation rates that the number of birds unable to obtain reasonable territories is not sufficient to provide helpers for most nests. Also, if Skutch (1949) is correct in his view of selection for clutch size in tropical birds, there would be no advantage to accepting help, and help should in fact be refused when offered except among larger species where predators can be driven off by the combined efforts of breeders and helpers.

Lack of data prevents a more thorough interpretation of our observations, but we feel that a careful dissection of the helping process into its two basic components will focus attention on some neglected problems and suggest additional interpretations to the evolution of offering and accepting help.

Acknowledgements

Support for the field work in Argentina was provided by funds from the John Simon Guggenheim Memorial Foundation and a sabbatical from the University of Washington. The hospitality of our many friends and landowners in the Pinamar area made the field work possible and enjoyable. John C. Schultz helped with mist netting and nest watching of Brown-and-yellow marshbirds. We appreciate comments on the manuscript by J. L. Brown, D. W. Foltz, L. Erckmann, C. M. Perrins, S. Rohwer, R. L. Trivers and S. Vehrencamp.

References

Alexander, R. D. (1974). The evolution of social behavior. *A. Rev. Ecol. Systematics,* **5**, 325 – 383.

Ashmole, N. P. and Tovar, H. (1968). Prolonged parental care in royal terns and other birds. *Auk,* **85**, 90 – 100.

Brown, J. L. (1966). Types of group selection. *Nature,* **211**, 870.

Brown, J. L. (1970). Cooperative breeding and altruistic behavior in the Mexican Jay, *Aphelocoma ultramarina. Anim. Behav.,* **18**, 366 – 378.

Brown, J. L. (1972). Communal feeding of nestlings in the Mexican Jay (*Aphelocoma ultramarina*): interflock comparisons. *Anim. Behav.,* **20**, 395 – 402.

Brown, J. L. (1974). Alternate routes to sociality in jays – with a theory for the evolution of altruism and communal breeding. *Am. Zool.,* **14**, 63 – 80.

Coulson, J. C. and White, E. (1956). A study of colonies of the kittiwake, *Rissa tridactyla* (L.) *Ibis,* **98**, 63 – 79.

Crossin, R. S. (1967). The breeding biology of the tufted jay. *Proc West. Found. Vert. Zool.,* **1**, 265 – 297.

Fraga, R. M. (1972). Cooperative breeding and a case of successive polyandry in the bay-winged cowbird. *Auk,* **89**, 447 – 449.

Friedmann, H. (1929). *The Cowbirds.* Thomas, Springfield, Illinois.

Fry, C. H. (1972). The social organization of bee-eaters (Meropidae) and co-operative breeding in hot climate birds. *Ibis,* **114**, 1 – 14.

Gaston, A. J. (1973). The ecology and behaviour of the Long-tailed Tit. *Ibis.* **115**, 330 – 351.

Ghiselin, M. T. (1974). *The Economy of Nature and the Evolution of Sex.* University of California Press, Berkeley.

Hamilton, W. D. (1963). The evolution of altruistic behavior. Am. Nat., **97**, 354 – 357.

Hamilton, W. D. (1964). The genetical evolution of social behavior. Parts I and II. *J. Theor. Biol.,* **7**, 1 – 52.

Hamilton, W. D. (1972). Altruism and related phenomena, mainly in the social insects. *A. Rev. Ecol. Systematics,* **3**, 193 – 232.

Harrison, G. J. O. (1969). Helpers at the nest in Australian birds. *Emu,* **69**, 30 – 40.

Hudson, W. H. (1923). *Birds of La Plata.* J. M. Dent & Sons, London.

Lack, D. (1954). *The Natural Regulation of Animal Numbers.* Oxford University Press, Oxford.

Lack, D. (1968). *Ecological Adaptations for Breeding in Birds.* Methuen, London.

Lewontin, R. C. (1970). The units of selection. *A. Rev. Ecol. Systematics,* **1**, 1 - 18.

Ligon, J. O. (1970). Breeding and breeding biology of the Red-cockaded Woodpecker. *Auk,* **87**, 255 - 278.

Maynard Smith, J. (1964). Group selection and kin selection. *Nature,* **201**, 1145 - 1147.

Maynard Smith, J. (1965). The evolution of alarm calls. *Am. Nat.,* **99**, 59 - 64.

Orians, G. H. (1969). Age and hunting success in the brown pelican. *Anim. Behav.,* **17**, 316 - 319.

Orians, G. H. and Horn, H. S. (1969). Overlap in foods of four species of blackbirds in the Potholes of central Washington. *Ecology,* **50**, 930 - 938.

Parry, V. (1973). The auxiliary social system and its effect on territory and breeding in Kookaburras. *Emu,* **73**, 81 - 100.

Rowley, I. (1965). The life history of the superb blue wren *Malurus cyaneus. Emu,* **64**, 251 - 297.

Selander, R. K. (1964). Speciation in wrens of the genus *Campylorhynchus. Univ. Calif. Pub. Zool.,* **74** 1 - 305.

Skutch, A. F. (1935). Helpers at the nest. *Auk,* **52**, 257 - 273.

Skutch, A. F. (1949). Do tropical birds rear as many young as they can nourish? *Ibis,* **91**, 430 - 455.

Skutch, A. F. (1960). Life histories of Central American birds. II. *Pacific Coast Avifauna,* **34**, 1 - 593.

Trivers, R. L. (1972). Parental investment and sexual selection, in *Sexual Selection and the Descent of Man,* (ed. B. Campbell) Aldine, Chicago, 136 - 179.

Trivers, R. L. (1974). Parent - offspring conflict. *Am. Zool.,* **14**, 249 - 264.

Vervoorst, F. B. (1967). *La Vegetación de la República Argentina. VII. Las comunidades vegetales de la Depresión del Salado.* Serie Fitogeográfica No. 7. Instituto Nacional Technología Agropecuaria, Secretaria de Estado de Agricultura y Ganadería de la nación. Buenos Aires.

Weller, M. W. (1967). Notes on some marsh birds of Cape San Antonio, Argentina. *Ibis,* **109**, 391 - 411.

Williams, G. C. (1966). *Adaptation and Natural Selection.* Princeton University Press, Princeton, New Jersey.

Willson, M. F. (1966). The breeding ecology of the Yellow-headed Blackbird. *Ecol. Monogr.,* **36**, 51 - 77.

Woolfenden, G. E. (1975). Florida Scrub Jay helpers at the nest. *Auk,* **92**, 1 - 15.

Zahavi, A. (1974). Communal nesting by the Arabian Babbler. A case of individual selection. *Ibis,* **116**, 84 - 87.

14 The occurrence of time-saving mechanisms in the breeding biology of the Great tit, Parus major

Dr H. N. Kluyver, Dr J. H. van Balen and Dr A. J. Cavé*

Introduction

Studies by Perrins (1965; 1970) have suggested that in several bird species there is a strong selection pressure favouring early breeding, that is as early as the feeding conditions for the female in the period of egg formation permit. This conclusion is confirmed by the recent finding that the provision of additional food in the pre-laying period can advance egg laying in the Great tit *Parus major* (Källander, 1974).

In the past, attention has been paid mainly to the early part of the breeding season – the period in which the first brood is reared. In many populations of Great tits many second clutches are produced, the proportion varying with, for instance, habitat and population density (Kluyver, 1951; 1963; 1971; Lack, 1955).

Although the young of second broods are generally at a disadvantage compared with those of first broods, as judged by the smaller contribution to the local breeding population in the next year, the tendency towards an increased survival associated with early fledging holds not only for first, but also for second broods (*cf.* Kluyver, 1966, figure 2). Hence, there must be a selection pressure favouring the early fledging of the young in both first and second broods.

These considerations imply a genetic component, but no data on this point are available and further study is badly needed.

There are several ways in which early fledging of second-brood young can be achieved. Time can be saved in several stages of the breeding cycle. Apart from such an obvious factor as the earliness of the first brood (early first broods are followed by early second broods and late first by late second broods), the first stage to be examined is the transition period between first and second brood. Usually, there is an interval of several days between the fledging of a first brood and the start of laying of the second clutch, but in a number of cases this does

*The authors are on the staff of the Institute for Ecological Research, of which Kluyver was the first director (1954 – 67). Kluyver specialised in behavioural and population ecology of passerines, and initiated a long-term study on the population ecology of the Great tit. Since 1958 this study has been continued by van Balen, working with A. J. Cavé, J. A. L. Mertens (physiological aspects) and P. J. Drent (behavioural aspects). van Balen specialised in the population dynamics of the Great tit, while Cavé has made a study of the population ecology of the kestrel.

not occur, that is the first egg of the second clutch is laid when the young of the first brood are still in the nest. These cases are termed overlapping broods (in German: *Schachtelbruten*).

The phenomenon of overlapping broods has been observed in several species. Heinroth (1909) discovered its occurrence in the European nightjar *Caprimulgus europaeus*. Later, Lack (1930*a*; 1930*b*; 1957) showed it to be a common phenomenon in this species, where it is the male's task to rear the first brood while the female lays and incubates a second clutch. Other species in which overlapping broods have been found include several Columbids, also *Gallinula chloropus, Acrocephalus arundinaceus* (Stresemann 1927 – 1934, p. 377), *Certhia familiaris* (Vollbrecht, 1938), *Passer montanus* (Deckert 1968, p. 52) and *Regulus regulus* (Palmgren, 1932).

Among the Paridae, several authors give observations of overlapping broods in the Great tit (Kingsley Siddall, 1909; Musselwhite, 1930; Berndt, 1938; Delmée, 1940; Rheinwald, 1971). Löhrl (1974, p. 63) and Winkel (1975) mention the occurrence of overlapping broods in the Coal tit *Parit ater*.

Other phenomena contributing to the early fledging of second broods occur at a slightly later stage. Second clutches are smaller than first clutches, which reduces the laying period by several days. Whereas in first clutches there is often an interval between the cessation of laying and the start of incubation (Zink, 1959), the incubation of second clutches usually starts several days before the last egg is laid. This gives the impression of very short incubation periods (9 – 12 days, calculated from the laying date of the last egg), compared with the 14 days usual for first clutches. A further time reduction might be achieved by shortening the fledging period, but we know of no observations pointing in this direction.

In this chapter the duration of the interval between first and second brood, defined below, is considered in relation to several factors in an attempt to study the causation of the phenomenon. Moreover, the consequences of a short interval for the survival of the young of both first and second broods will be analysed and the ultimate significance discussed.

Methods

Populations of Great tits were studied in two habitats — mature oak forests on fertile soil (figure 14.1), and pine plantations mixed with some deciduous stands on poor sandy soil (figure 14.2). As examples of the first type, which is an optimal breeding habitat for the Great tit (van Balen, 1973), Liesbosch near Breda and Oosterhout near Arnhem were taken. The second type, which is a marginal habitat for breeding Great tits, was studied in the Hoge Veluwe area near Arnhem. For a detailed description of these areas and for studies on the breeding ecology of the tits, the reader is referred to van Balen (1973). Some additional data were collected on the island of Vlieland, where the breeding habitat consists mainly of pine plantations.

Data on the breeding chronology of the tits were collected from 1955 to 1967 by weekly inspections of nestboxes. In addition daily inspections during two years (1961 and 1962) provided detailed data on the sequence of nesting events, thus permitting the exact calculation of the incubation and nestling periods.

Second broods were identified by the ring numbers of both parents. Since the

Figure 14.1 Liesbosch, fertile soil. Mature oak wood. Dense population of Great tits, many pairs with short intervals.

dates of hatching and fledging were usually not known exactly, calculation of the interval between first and second brood could not be based on the fledging or hatching dates, and another exactly known date had to be taken. Thus, for each pair the interval was taken as running from the laying date of the last egg of the first clutch to the laying date of the first egg of the second clutch (assuming that one egg is laid a day). This interval includes the incubation and fledging period of the first clutch and brood. No allowance could be made for any difference in the duration of the interval between laying and the start of incubation, but there are no indications for a difference between the two habitats.

In 1964, some of the first clutches in the Hoge Veluwe area were reduced to 4 eggs by the removal of eggs in the laying period. In some years the first clutches on Vlieland were reduced to 2 – 4 eggs (*cf.* Kluyver, 1971). The results for these years were kept separate from those of normal years. For general purposes, only the results from normal years were used.

The food availability was investigated by sampling caterpillars feeding in the canopy (van Balen, 1973). For Oosterhout, data covering several years were available. Liesbosch data were collected only in 1961/62, but these findings correspond closely with the Oosterhout data. In the Hoge Veluwe area the feeding conditions were studied in pine plantations, whereas the data on the breeding chronology were collected over the whole area, including pine plantations and mixed deciduous woods.

The rank correlation methods applied in this study were according to Kendall (1955). It was necessary to consider correlations within years, eliminating the variation between years. This was done by adding the values of the sums of ranks (S) for each year separately and testing the significance of the whole set against the sum of variances of the individual rankings (Kendall 1955, p. 65). A correlation coefficient can be constructed by dividing the sum of the S values of each

Figure 14.2 Hoge Veluwe, poor soil. *a*, Mature scots pine with young fir at right. Unfavourable for Great tit, no pairs with short intervals. *b*, Border between young birches and scots pines. Rather unfavourable for Great tits, hardly any pairs with short intervals. *c*, Crooked grown oaks, some scots pines in the background. Moderately favourable for Great tits, more pairs with short intervals.

year by the sum of their maximum values, giving a combined correlation coefficient

$$r_c = \frac{S_1 + S_2 + S_3 + \ldots}{S_{max_1} + S_{max_2} + S_{max_3} + \ldots}$$

in which $1, 2, 3, \ldots$ = year 1, year 2, year 3, ...

Results

General aspects

The frequency distribution of the calculated intervals between first and second broods is shown in figure 14.3. It is evident that the intervals for the Hoge Veluwe are larger than those for Liesbosch (36.7 and 33.4 days, respectively). The means differ significantly (*t*-test; $P<0.01$). The mean interval for Oosterhout corresponds closely with that for Liesbosch (32.9 days, $n = 17$), and the mean interval for Vlieland is the largest of all (mean = 39.4 days, $n = 123$).

These mean intervals can be used in combination with the daily observations from 1961 and 1962 to obtain an approximate picture of the time between fledging and the start of a second brood. For Liesbosch a mean incubation period (defined as the number of days from the laying of the last egg until the hatching of the first young) of 14.2 days was found, and a mean nestling period of 19.1

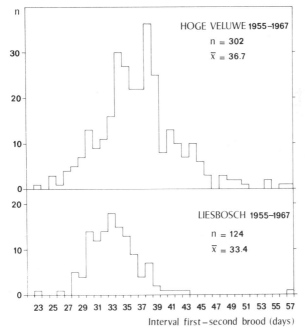

Figure 14.3 Frequency distribution of the calculated intervals between first and second broods in two areas. (For method of calculation, see section headed *Methods*.)

Evolutionary ecology

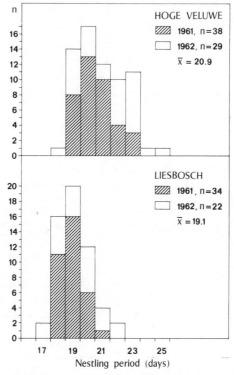

Figure 14.4   Frequency distribution of nestling periods in two areas.

days (see figure 14.4). If these values are representative for all the years for which data are available, the mean interval of 33.4 days (figure 14.3) implies that in Liesbosch the start of the second clutch coincides on average with the fledging of the first brood. The values for the Hoge Veluwe were 14.7 days for the incubation and 20.9 days for the nestling period (see figure 14.4). A mean interval of 36.7 days (figure 14.3) indicates that on average one day elapses between fledging and laying.

Variation in interval between years

At first sight, the time difference between the mean intervals in the two areas seems rather unimportant, but some interesting trends emerge from a more detailed analysis.

Figure 14.5 shows that annual differences in mean interval occur in both forests. A year-by-year comparison shows that in six of the years the difference between the two areas is significant (asterisks in figure 14.5). For five years (1958, 1961, 1962, 1965 and 1967) the number of data is too small for a valid comparison, and in 1964 and 1966 the mean intervals in the two areas do not differ significantly. Generally, it is clear that over the period 1955–63 the mean intervals for the Hoge Veluwe are distinctly larger than those for Liesbosch, whereas the means for 1964 are virtually equal. The situation after 1964 is less clear.

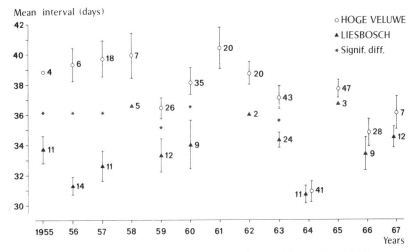

Figure 14.5 Mean intervals over a series of years. Mean and standard error are given where *n*> 5. Significant differences between the areas are indicated by asterisks.

To take a different approach, the mean interval for the Hoge Veluwe in 1964 is significantly shorter than the means for all other years, suggesting that a special situation occurred in this year. The fact that in 1964 the mean interval for Liesbosch is also rather short, suggests that (at least in this year) a factor common to both areas influenced the length of the intervals. Generally, however, the mean intervals in both areas are not correlated (Kendall's rank correlation: τ = 0.121, P>0.6). This could mean that the annual mean intervals are not greatly affected by a common factor, such as climate.

Among the factors possibly affecting the length of the interval, the feeding conditions during the nestling period could have an important role. The effect of the feeding conditions can be direct, that is influencing the energy available for egg production, but also indirect, influencing the general condition and behaviour of the parents during the energy-consuming period of nestling care. First, the direct effect will be considered.

Feeding conditions in the breeding season can be assessed by sampling the frass of caterpillars, the primary source of food for nestling Great tits (van Balen, 1973). It has been found that both the abundance of caterpillars and the timing of the caterpillar peak vary from year to year. As a consequence the feeding conditions in the period, critical for the start of a second clutch, can vary widely.

Measurements of caterpillar frass fall in the Hoge Veluwe area (pine plantations) are available for the years 1957-65, in Liesbosch for 1961-62, and in Oosterhout for 1957-66 (figure 1 and table 2 of van Balen, 1973). From these measurements the average weight of caterpillar frass falling daily during the critical period was calculated. The critical period was taken from days 25 to 35 after the average laying date of the last egg of the first clutches in a given year, that is approximately the period in which the nestlings of these broods were 11 to 21 days old.

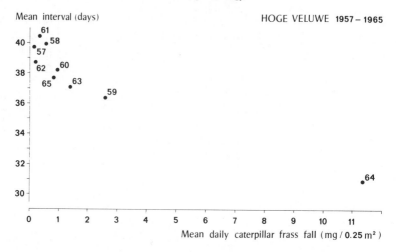

Figure 14.6   Relationship between the annual mean interval between
first and second brood in the Hoge Veluwe area, and the feeding conditions in pine wood during the critical period. Numbers denote years.

Comparison of the annual mean intervals for Liesbosch with the food availability in Oosterhout did not produce a significant correlation.

The results for the Hoge Veluwe are given in figure 14.6. Here, a highly significant negative correlation was found between the annual mean interval and the
amount of caterpillar frass fall during the critical period ($\tau = -0.722, P<0.01$).
The correlation still holds when the aberrant year 1964 (clutch reduction; see
Methods) was excluded. Apparently, when feeding conditions are good the tits
can shorten the interval between the first and the second brood considerably. The
results shown in figure 14.6, which refer to intervals measured in pine and mixed
deciduous forests and to measurements of caterpillar abundance in pure pine
plantations, are corroborated by the observation that in 1964, more than in
the other years, *Tortrix viridana* larvae caused heavy defoliation of oaks in large
parts of the country, including the study area (Doom, 1965). Thus, in 1964 both
oak and pine provided an abundant food supply.

The conditions in which the young grow up can be assessed further from the
survival of the nestlings and from the breeding success. Annual means were calculated for the survival, measured as the percentage of young fledged from the
eggs laid, and for the breeding success, the number of young fledged per first
brood. Those for Liesbosch appeared to be totally unrelated to the length of the
interval, but the annual means for the Hoge Veluwe were negatively correlated
with the interval. Figure 14.7 shows the correlation between interval and nestling
survival, which is highly significant ($\tau = -0.692, P<0.01$). These correlations still
hold when the 1964 results are excluded. These findings strengthen the conclusion
that in the Hoge Veluwe area short intervals are related to good feeding conditions
for the first brood.

The annual mean intervals were also compared with the breeding density and
with the mean laying dates of the first clutches, but neither in the Hoge Veluwe
nor in Liesbosch was a significant correlation obtained.

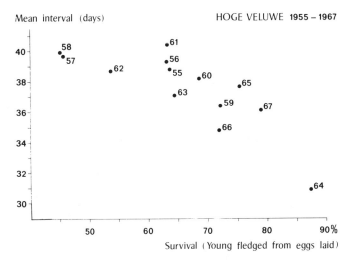

Figure 14.7   Relationship between the annual mean interval between first and second brood in the Hoge Veluwe area, and the nestling survival in the first brood. Numbers denote years. A similar picture was obtained when the intervals were compared with the mean number of young fledged per brood.

Variation in interval within years

So far, the phenomenon of varying intervals between first and second brood has been analysed by comparing results from different years, which led to the conclusion that the feeding conditions during the nestling period have an important role. A more detailed comparison can be made by analysing the variability within each year.

For this purpose, the intervals of the individual pairs were compared with the laying dates and the number of young fledged in each brood. This was done by applying a rank correlation method (see *Methods*), first for each year and next for all years combined, according to Kendall (1955).

For Liesbosch, no significant correlation between the length of the interval and the number of fledged young was found for either the separate years or the combined data. The data for the Hoge Veluwe showed, however, a positive correlation for most of the years, but for only two years (1960, 1963) was this significant. Combining the data for all years gave a small, but significant, positive correlation ($\tau_c = +0.142, P<0.01$). This implies that, in this habitat, the interval between first and second brood, considered within a single year, increases with increasing numbers of young reared. This result seems to contradict the conclusion reached earlier that years with high nestling survival and high breeding success are years with *short* intervals. Although the between-year relationship is probably related to annual differences in the general level of feeding conditions, the within-year relationship is probably based on differences in the condition of the females. It seems logical to assume that, when feeding conditions in the Hoge

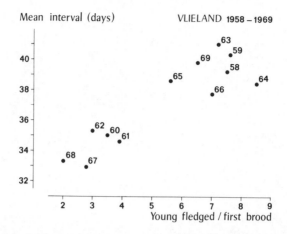

Figure 14.8 Relationship between the annual mean interval between first and second brood on Vlieland, and the mean number of young fledged in the first brood. Numbers denote years. Further explanation in text (see under *Variation in interval within years*).

Veluwe area are similar within a given year, females rearing few young maintain relatively good condition and are able to produce a second clutch without delay.

Further support for this conclusion is provided by a comparison of the annual mean intervals for Vlieland with the mean number of young fledged (figure 14.8). In this area, where the number of fledged young varied widely due to experimental clutch reduction (*cf*. Kluyver, 1971), this parameter does not give an indication of the external conditions. It is more likely that the distinct differences in mean interval between normal and experimental years, which are apparent from figure 14.8, are due to differences in the condition of the females, those in experimental years having reared far fewer young than they set out to rear.

The relationship between the duration of the interval and the date of laying of the first clutch of each pair was analysed in a similar way. Once more, no correlation was found in the Liesbosch data. In data from the Hoge Veluwe material, however, the length of the interval tended to be negatively correlated with the laying date in 2 out of 12 years and in the combined data ($\tau_c = -0.156$, $P<0.01$).

This tendency for females breeding later in the season to have somewhat shorter intervals, might be explained in several ways. A possible proximate factor could be that late-breeding females experience more favourable feeding conditions than those which breed early, since in the pine plantations of the Hoge Veluwe peaks in caterpillar abundance occur late in the season, on average about 20 days after fledging (van Balen, 1973). On the other hand, the tendency to have shorter intervals late in the breeding season could well be part of the general tendency to shorten the breeding cycle as the season progresses, an adaptation which ultimately favours late-reared young by allowing them to settle as early as possible.

Duration of interval and survival after fledging

The question arises whether the fledglings of a first brood suffer when the female parent, instead of looking after them, starts a second brood even while they are still in the nest. As a first step, this question was studied by comparing the recovery percentages of the fledglings from overlapping and non-overlapping first broods. Contrary to our initial expectation, those from overlapping first broods proved to have a significantly higher local survival until the next breeding season. The recovery percentages of overlapping and non-overlapping young were 13.1 and 6.0 per cent for the Hoge Veluwe and 13.0 and 7.4 per cent for Liesbosch, respectively. Since these values could be biased by differences in the contributions from different years, for example by a large contribution to the group of overlapping broods in years with high survival (such as 1964), it was decided to analyse the within-year variation, once again with a rank correlation method. For this, the duration of the interval of each brood was compared with the percentage survival of the fledglings, measured from its contribution to the next local breeding population.

Once more, no significant correlation was found in the Liesbosch data, but the results for the Hoge Veluwe strengthen the preliminary conclusion. In the latter area first broods with short intervals had a significantly higher local survival after fledging than broods with long intervals ($\tau_c = -0.141$, $P < 0.01$). In our view this finding indicates that there are differences between the pairs, especially between the females, in ability to raise nestlings and to produce a second clutch. It is also possible that the above-mentioned finding is partly the result of differences in food abundance within very small areas. Some females succeed in having their young fledged in a condition ensuring a high survival, and are also capable of producing a second clutch within a short interval, whereas other females have a lower performance in both the rearing of nestlings and the production of a second clutch.

So far we have looked primarily at the proximate side of the problem, and we may conclude that the shortening of the interval between first and second broods occurs when and where conditions permit. Now we come to the question of whether the shortening of the interval is important for the survival of the young of the second brood — especially after fledging — which is the ultimate aspect of the problem. A comparison of the recovery percentages (until the next breeding season) of overlapping and non-overlapping second broods showed that the young from overlapping second broods made a larger contribution to the next breeding population than those from non-overlapping second broods. The recovery percentages amounted to 4.7 and 2.0 per cent in the Hoge Veluwe and 2.2 and 1.3 per cent in Liesbosch, respectively. As explained above (p. 161) a more adequate method is provided by the analysis of the within-year variation with a rank correlation method. This analysis showed a negative correlation between the duration of the intervals between first and second broods in the Hoge Veluwe and the percentage of local recoveries of the young from the second brood, but the correlation was not significant ($P = 0.10$).

This finding does not mean, however, that a short interval has no advantages for the fledglings, because in the correlation analysis we did not take into account differences in timing of the broods within the season. As we have seen

Table 14.1 Local survival of young Great tits fledged in successive periods of the breeding season. Each breeding season is divided into 5-day periods starting from the earliest clutch of the season. Sum of data from 1955 to 1966.

Period	A Liesbosch			B Hoge Veluwe		
	Number fledged	Number recovered	Percentage recovered	Number fledged	Number recovered	Percentage recovered
1	902	84	9.3	968	83	8.6
2	1479	112	7.6	1240	114	9.2
3	1581	83	5.2	1404	64	4.6
4	770	30	3.9	667	39	5.8
5	136	1	0.7	347	29	8.4
6	106	2	1.9	138	12	8.7
7	99	0	0	133	3	2.3
8	132	5	3.8	98	2	2.0
9	214	4	1.9	328	20	6.1
10	334	7	2.1	349	11	3.2
11	95	1	1.1	496	17	3.4
12	52	0	0	206	4	1.9
13	21	0	0	111	2	1.8
>13	4	0	0	39	0	0

Note: This table can be compared with figure 2 in Kluyver (1971), but keeping in mind that A and B were erroneously interchanged there.

Table 14.2 Summary and statistical significance of the results shown in table 14.1.

Period	Percentage recovered		χ^2	P
	Liesbosch	Hoge Veluwe		
a (1–4)	6.5	7.0	0.825	0.364
b (5–9)	1.7	6.3	20.155	<0.001
c (10–>13)	1.6	2.8	2.318	0.128

Comparing $a-b$ 　$\chi^2 = 24.630$ 　$\chi^2 = 0.622$
　　　　　　　　$P < 0.001$ 　　$P = 0.430$

Comparing $b-c$ 　$\chi^2 = 0.049$ 　$\chi^2 = 15.992$
　　　　　　　　$P = 0.826$ 　　$P < 0.001$

earlier, short intervals occur more frequently late in the season than early, and, moreover, lateness as such is usually related to poor survival of the fledglings, as mentioned at the beginning. This is shown in table 14.1, which gives the percentages of local recoveries in the next breeding season for young fledged in successive periods of the breeding season. In spite of some irregularities, the general picture indicates decreasing chances for the young. As can be seen from table 14.2, this decrease in local survival to the next breeding season starts earlier in Liesbosch than in the Hoge Veluwe area.

Second-brood young from broods with short intervals should fledge a few days earlier and, hence, make a larger contribution to the next breeding population.

Discussion

As shown in *General Aspects*, the duration of the interval between first and second brood differs significantly between the two main study areas. This is corroborated by data collected in Oosterhout and Vlieland and by data from larch plantations near Lingen (Germany) (Winkel, 1975). From Winkel's data a mean interval of 39.0 days (13.5 days incubation and 25.5 days from hatching to the start of the second brood) can be calculated. Evidently, the interval between the first and second broods increases in the sequence from pure deciduous woods, via mixed woods, to pure coniferous plantations. This is also the sequence in which the breeding density and, most probably, the feeding conditions during the breeding season decrease (van Balen, 1973).

The general conclusion drawn in the preceding section was that the shortening of the interval between first and second brood occurs when and where the conditions permit. The factors affecting the duration of the interval (food abundance, number of first-brood young reared, date of laying) were summarised under two headings: general food abundance and the specific situation and abilities of the parents. The fact that these factors were only found to have effect in the marginal habitat is not surprising. The feeding conditions in the oak forest are evidently favourable to a degree such that food is in general not a limiting factor for the duration of the interval, and that differences in the condition or the abilities of the females are compensated for by an abundant food supply.

The factors affecting the *frequency* of second clutches in a population have not yet been discussed, but is of interest to discuss them here, and to examine their relationship with the factors influencing the length of the interval.

Berndt (1938) and Kluyver (1951) found that the percentages of second clutches differ among habitats, low values occurring in deciduous and high values in coniferous forests. This general tendency also holds for the habitats we studied, and an extremely high percentage of second clutches (90 per cent) was found for the larch forest studied by Winkel (1975).

The relationship with breeding density is also evident, years with high density being associated with low percentages of second clutches (figure 8 of Kluyver, 1951). Since it is known, furthermore, that breeding densities in deciduous woods are much higher than those in coniferous forests (van Balen, 1973), the effect of population density probably explains to a large extent the difference in the frequency of second clutches between different habitats.

The between-habitat differences in the length of the interval between first and second broods are probably not related to differences in breeding density, since no relationship was found between interval and breeding density within one habitat (between years). As is clear from the present results, food abundance seems to be the main factor underlying differences in mean interval between the habitats but is apparently relatively unimportant in determining whether or not a second clutch is produced.

Besides the breeding density, other factors also affect the percentage of second clutches in a population.

Kluyver (1951, p. 71) established a negative correlation between the mean initial date of the first clutches and the percentage of second clutches. Since the end of the breeding season is more or less fixed by the occurrence of competing events in the annual cycle, such as the moult, a late start means a shortening of the breeding season, with the result that some of the females are too late to form a second clutch. This makes Kluyver's finding plausible.

In this connection one would expect short intervals in late breeding seasons, because only females able to produce a second clutch without delay would be able to produce a second clutch at all. No such correlation could be found, however. Possibly the influence of food abundance on the timing of the second clutch is so dominant that the effects of other factors are masked.

However, the within-year analysis has shown that late-breeding females tend to have short intervals. This relationship was only found for the Hoge Veluwe data. If short intervals at the end of the breeding season result from a fixed closing of the season (that is, long intervals do not occur, because the laying date of the second clutch would fall outside the breeding season), one would expect the above-mentioned relationship to occur in both types of habitat. However, the explanation given earlier that late females in the Hoge Veluwe area experience relatively favourable feeding conditions, is more plausible.

Kluyver (1963, table 2) found that the percentage of second clutches in a population increases with increasing size of the first clutch, in combination with decreasing numbers of young reared in the first brood. In other words low success of the first clutch makes the occurrence of a second clutch more likely.

On the other hand, the interval between the first and second broods decreases with decreasing numbers of fledged young, but there is no correlation with the size of the first clutch. This implies that high nestling mortality in the first brood is often followed by a second clutch with a short interval.

One would expect young from broods with low nestling survival to fledge in a poor condition (van Balen, 1973, pp. 81–82) and, therefore show poor survival after fledging. However, as already mentioned short intervals occur when the nestling survival is low and also when the survival of the first-brood young after fledging is high, from which it should follow that low nestling survival is related to good survival after fledging. This is not consistent with the findings of van Balen (1973), possibly because the females that start a second clutch quickly do not form a random sample from the population. These females and their surviving young have clearly profited from the nestling mortality, which must have been high in relation to the feeding conditions. This mechanism can be demonstrated by the experiments done on Vlieland, where in years with an experimentally reduced number of nestlings the intervals were very short and the survival of the fledglings was high (Kluyver, 1971).

It seems odd, from the ultimate point of view, that in oak woods only a small proportion of the females start a second brood. This cannot be the result of a food shortage, because van Balen (1973) has shown that food is abundant in oak woods during the breeding season. Although the peak in caterpillar abundance decreases sharply at the end of the first breeding cycle, the abundance in the period of egg formation for the second clutch is still much higher in oak than in pine woods. Later on, the caterpillar abundance is more or less the same in both types of wood. Another argument for the view that food is not limiting in late oak-wood broods is provided by the observation that the parents spend a smaller part of the day in feeding their young in oak than in pine woods and that the nestling mortality in these broods does not differ appreciably from that in pine woods (van Balen 1973). These considerations make it unlikely, either proximately or ultimately, that the feeding conditions in oak woods prevent the tits from starting a second clutch. Nevertheless, second clutches are much more common in pine woods.

Local survival of the fledglings, measured up to the next breeding season, tends to decrease for young fledged in successive periods (table 14.1). This decrease seems to occur later in a habitat with pines predominating than in oak woods, as can be demonstrated with data from the Hoge Veluwe and Liesbosch. Table 14.2 shows that the decline in local survival in Liesbosch is already apparent for broods started after the twentieth day of the breeding season. This implies that fledglings of late first broods and repeat broods are already at a disadvantage compared with earlier fledglings. The decline in local survival of the fledglings in the Hoge Veluwe area occurs considerably later. This difference in local survival makes it less profitable to raise a second brood in Liesbosch than in the Hoge Veluwe, compared with the success of the first brood.

The local survival up to the next breeding season of the breeding birds is higher in Liesbosch than in the Hoge Veluwe (van Balen, unpublished). Since it is more difficult to raise young in the Hoge Veluwe than in Liesbosch, it is possible that this has a negative effect on the condition of the parents in the Hoge Veluwe, resulting in a higher mortality after the breeding season. On the other hand, the difference in survival between the two habitats may be due to a difference in feeding conditions in winter, which is reflected in the condition of the tits (van Balen, 1967). Whatever the cause, the higher local survival of the breeding females in Liesbosch gives them a better chance to take part in the reproduction in the next season.

If in both oak and pine woods the rearing of a second brood increases the risk of dying for the parents, it seems more worthwhile to take this risk in pine woods, where the survival of late-fledged young is relatively high and the survival of the parents relatively low. In oak woods, where the survival of late-fledged young is relatively low and parent survival is relatively high, it seems less profitable to take the risks involved in the production and rearing of a second brood. This difference in reproductive behaviour in the two types of habitat, which differ markedly in food availability and breeding density, gives the impression of different 'strategies'. In our opinion, the effect of breeding density on the occurrence of second clutches may be considered as a steering mechanism that maximises the number of offspring during the female's lifespan at different density levels.

Acknowledgements

We thank Dr J. A. L. Mertens for permission to use his detailed observations from 1961 and 1962, and Dr J. W. Woldendorp for his constructive remarks on the manuscript.

References

Balen, J. H. van (1967). The significance of variations in body weight and wing length in the Great Tit, *Parus major*. *Ardea*, **55**, 1 – 59.

Balen, J. H. van (1973). A comparative study of the breeding ecology of the Great Tit *Parus major* in different habitats. *Ardea*, **61**, 1 – 93.

Berndt, R. (1938). Über die Anzahl der Jahresbruten bei Meisen und ihre Abhängigkeit vom Lebensraum, mit Angaben über Gelegestärke und Brutzeit. *Dt. Vogelwelt*, **63**, 140 – 151, 174 – 181.

Deckert, G. (1968). *Der Feldsperling*. Neue Brehm Bücherei no.'398. A. Ziemsen Verlag, Wittenberg Lutherstadt.

Delmée, E. (1940). Dix années d'observations sur les moeurs de la Mésange charbonniére et de la Mésange bleue. *Le Gerfaut*, **30**, 97 – 129, 169 – 187.

Doom, D. (1965). Insektenplagen in bossen en andere houtopstanden in 1964. *Ned. Bosbouw Tijdschr*, **37**, 170 – 179.

Heinroth, O. (1909). Beobachtungen bei der Zucht des Ziegenmelkers. *J. f. Ornith.* **57**, 56 – 83.

Källander, H. (1974). Advancement of laying of Great Tits by the provision of food. *Ibis*, **116**, 365 – 367.

Kendall, M. G. (1955). *'Rank Correlation Methods'*. 2nd edn. Ch. Griffin & Company, London. 196 pp.

Kingsley Siddall, C. (1909). Notes on the breeding of the Great Tit. *Br. Birds*, **3**, 186 – 187.

Kluyver, H. N. (1951). The population ecology of the Great Tit, *Parus m. major* L. *Ardea*, **39**, 1 – 135.

Kluyver, H. N (1963). The determination of reproductive rates in Paridae. *Proc. 13th Int. Ornith. Congr*, 706 – 716.

Kluyver, H. N. (1966). Regulation of a bird population. *Ostrich, Suppl.* **6**, 389 – 396.

Kluyver, H. N. (1971). Regulation of numbers in populations of Great Tits (*Parus m. major*). *Proc. Adv. Study Inst. Dynamics Numbers Popul.* (*Oosterbeek, 1970*), 507 – 523.

Lack, D. (1930*a*). Double brooding of the Nightjar. *Br. Birds*, **23**, 242 – 244.

Lack, D. (1930*b*). A further note on double brooding of the Nightjar. *Br. Birds*, **24**, 130 – 131.

Lack, D. (1955). British Tits (*Parus spp*) in nesting boxes. *Ardea*, **43**, 50 – 84.

Lack, D. (1957). Notes on nesting Nightjars. *Br. Birds*, **50**, 273 – 277.

Löhrl, H. (1974). *Die Tannenmeise*. Neue Brehm Bücherei no. 472, A. Ziemsen Verlag, Wittenberg Lutherstadt.

Musselwhite, D. (1930). Great Titmouse laying a second clutch before departure of first brood. *Br. Birds*, **24**, 75 – 76.

Palmgren, P. (1932). Zur Biologie von *Regulus r. regulus* (L.) und *Parus atricapillus borealis*. Selys. *Acta zool. fenn.* **14**, 1 – 113.

Perrins, C. M. (1965). Population fluctuations and clutch-size in the Great Tit, *Parus major L. J. Anim. Ecol.,* **34**, 601 – 647.

Perrins, C. M. (1970). The timing of birds' breeding seasons. *Ibis,* **112**, 242 – 255.

Rheinwald, G. (1971). Schachtelbruten der Kohlmeise (*Parus major*). *Vogelwelt,* **92**, 231 – 232.

Stresemann, E. (1927 – 34). *Aves.* Walter de Gruyter & Co, Berlin.

Vollbrecht, K. (1938). Zur Balz der Nachtschwalbe. *Beitr. Fortpfl. Biol. Vögel,* **14**, 105.

Winkel, W. (1975). Vergleichend-brutbiologische Untersuchungen an fünf Meisen-Arten (*Parus* spp.) in einem niedersächsischen Aufforstungsgebiet mit Japanischer Lärche *Larix leptolepis. Vogelwelt,* **96**, 41 – 63, 104 – 114.

Zink, G. (1959). Zeitliche Faktoren im Brutablauf der Kohlmeise. *Vogelwarte,* **20**, 128 – 134.

15 Latitudinal gradients in clutch size: an extension of David Lack's theory

Dr D. F. Owen*

The Problem

In the European robin, *Erithacus rubecula*, there is a gradient in mean clutch size from the northern to the southern limits of the range of the species. Mean clutches of 6.3 occur in Scandinavia, 5.9 in central France, 4.9 in Spain, 4.2 in North Africa, and 3.5 in the Canary Islands (Lack, 1954). There is, in addition, a small longitudinal gradient within Europe, amounting to a decrease of about 0.4 from east to west, and in Britain clutches are slightly smaller than in areas immediately across the Channel in continental Europe. The gradient is, however, mainly latitudinal, and the pattern of geographical variation in clutch size in the robin is repeated in many species of birds, most of them passerines.

This variation is an example of a trend that occurs both within and between species. The lowest clutch sizes occur in the tropics where, in the passerines in particular, the most frequent clutch is two — counterparts in temperate latitudes lay between four and six. It is most impressive to find that very nearly all species of the common garden birds in the tropics produce a clutch of only two eggs. The trend of decreasing clutch size from the polar regions to the equator is so widespread, and there are so few exceptions, that a general explanation is required.

There are of course birds in which there is little or no latitudinal variation in clutch size. Most of them belong to groups that produce a clutch of only one or two eggs; well-known examples are pigeons and nightjars which throughout the world normally have a clutch of two. The groups that show conspicuous latitudinal variations are those that at least potentially can produce three or more eggs.

Lack's view (1947) that clutch size is adjusted by natural selection to correspond with the largest number of young for which the parents can, on average, provide food, has been widely accepted, extensively criticised, and repeatedly modified, but remains supported by the considerable body of direct evidence that has accumulated since the theory was first proposed. Lack's view was, I believe, initially stimulated by the attempts of Moreau (1944) to explain variations in clutch size in African birds. An alternative proposal, that clutch size is adjusted

*Denis F. Owen was David Lack's field assistant from 1951 to 1958. At Oxford he worked on the population biology of rooks, herons and tits and, in collaboration with Lack, on the food of swifts. He has subsequently taught and done research in ecology in the United States, Uganda, Sierra Leone and Sweden, and in 1974 was appointed Distinguished Visiting Professor at the University of Massachusetts. He is the author of 5 books and about 120 scientific papers.

by natural selection to balance the 'expected' mortality of the population, is no longer considered by itself as a satisfactory explanation, although some of the modifications to Lack's theory attempt to incorporate elements of this view. In most species of birds (there are exceptions) the upper limit on the number of eggs normally laid in a clutch does not appear to be set by what is physiologically possible: there are many examples of birds of a variety of species laying very large numbers of eggs if the eggs are removed, one by one, as they are laid.

In this chapter I accept the validity of Lack's interpretation of the significance of clutch size, and believe that he has gone a long way towards explaining the latitudinal gradients. But at the same time I think that a crucial factor in the feeding ecology of birds has been overlooked, both by Lack and by subsequent authors, and hence the theory I shall propose is not so much a new one as an extension of Lack's. To put things in perspective I will first summarise the contrasting theories that have been put forward to explain latitudinal gradients.

Tropical and Temperate Clutch Sizes: Contrasting Viewpoints

The vast majority of tropical passerines lay a clutch of two eggs; their temperate counterparts produce clutches that are more than twice as large. Lack (1947) correlated this difference with the amount of daylength available for the parents to gather food for their young. In spring and summer, when birds breeding at high latitudes are feeding young in the nest, the days are long and there is more time for gathering food than in the tropics where there is less daylight. This, Lack argued, could account for the difference in average clutch size between high- and low-latitude birds. One difficulty with this explanation is that it cannot easily be applied to nocturnal species like owls, which also have smaller clutches at low latitudes. For example, the Barn owl, *Tyto alba,* usually has a clutch of two or three eggs in the tropics and more than five in temperate areas. The trend ought to be the other way round as the nights are short at high latitudes when barn owls feed their young, although perhaps some allowance should be made for occasional daytime feeding in this species. Nor, incidentally, can the daylength theory be applied to nocturnal mammals whose litter sizes (in the few species for which there are adequate data) decrease with latitude. And if, as has been suggested, birds breeding at very high latitudes, where in summer there is continuous daylight, persist in the trend of increasing clutch size at even higher latitudes, the daylength theory is again weakened.

Another difficulty is that of defining exactly what is meant by food for the young. At high latitudes there are relatively few kinds (species) of food items available, but each tends to be abundant; at low latitudes, insectivorous birds in particular (but the same applies to fruit- and seed-eaters, as well as those that take vertebrate prey) are faced with an incredible diversity of potential food items, few of them abundant as species. Indeed it is likely that a bird collecting food for its young in a tropical forest will be forced to gather almost as many species as individuals within a given period of time. Hence low-latitude, tropical birds not only have less time for food gathering but are faced with a much wider range of possible food items.

Skutch (1949) argued that the small clutches typical of tropical birds have evolved as a means of reducing, to some extent, predation of the nestlings. With fewer nestlings to feed, the parents make fewer visits to the nest which then has less chance of being discovered by a predator. Many observers (myself included) have, however, noted that the nestlings of tropical birds are far more prone to the attacks of predators than those of temperate species. Skutch also observed that in certain tropical birds (he cites examples from Central America), parents with young often appear to be inactive, spending substantial periods in idleness, an observation that does not seem to apply to the more active temperate species. Skutch was prepared to accept, at least in part, Lack's theory in so far as it applies to temperate species, but maintained that it did not hold for tropical species. He argued also that reproductive rates are adjusted to balance death rates, a view Lack rejected, and that, at least for birds, has since found few serious proponents. There is no doubt that Skutch was extremely worried about Lack's views, even claiming to having been 'shocked' on reading them. Commenting on Skutch's response, Lack (1949) accepts many of his observations and, characteristically, uses them to support his own theory.

More recently it has been suggested that temperate populations are reduced (in winter and through losses on migration) to an extent below the carrying capacity of the environment. As a result, breeding populations tend to be low in relation to the availability of food in spring and summer, and there is thus less competition, which in turn favours larger clutches. Tropical populations, on the other hand, are probably much closer to the carrying capacity of the environment, and, with no significant food surplus, competition is intensified, and clutch sizes are correspondingly smaller. This viewpoint, which seems to originate from Ashmole (1961), has also been used to explain the slightly higher clutches in species inhabiting savanna than in forest species living at the same latitude (Lack and Moreau, 1965). Tropical forest is thought to provide a more stable and uniform environment with populations near the carrying capacity, whereas savanna, which is more seasonal, approaches some of the characteristics of high latitudes, including higher death rates and a tendency for populations to fall below the carrying capacity.

A similar position is taken by Cody (1966) who writes, 'It is known that in temperate regions, because periodic local catastrophes reduce and maintain populations below the carrying capacity of the habitat, natural selection is proceeding to maximise r, the reproductive rate. . . . In these regions, any phenotypic variation which enables parents to rear more offspring will be selected for. Any increase in clutch size, up to a limit determined by natural resources, would suffice to increase the reproductive rate. In the tropics, however, with a more climatically stable environment where the advent of such catastrophes is rare, populations will be at saturation density, and any adaptive variations that will increase the carrying capacity will usually be favoured by natural selection.' In current terminology Cody is postulating that r-selection operates at high latitudes and is replaced by K-selection at low latitudes. This means that birds breeding in 'unstable environments' will tend to have larger clutches than those breeding in 'stable environments', stability always being defined in terms of climate. The tropics in particular, but also certain other areas, such as the coast, are thought to be stable, while high latitude areas, especially those inland, are thought to be unstable. There appear to be uncertainties about islands. Some may be stable, others unstable; it is really a question of how stability is assessed and who is making the assessment.

In one respect Cody is following Skutch in believing that the tropics are some-how different from higher latitudes, and that so far as birds are concerned, the selection pressures for optimum clutch size are different. If Cody is right then there must be switches between r- and K-selection from place to place (and presumably from time to time), and in some areas and in certain seasons birds occur at densities close to the carrying capacity while in others they are below it. Cody's theory also depends on the proposition that there are indeed recognis-able stable and unstable environments.

Lack, Moreau, and Cody are not the first to believe that the tropics are more stable than higher latitudes; indeed the belief is widespread and is repeated in many ecology textbooks. But even if true there remains the difficulty of deter-mining exactly what effect stability has on the clutch size of birds. In the follow-ing section I suggest that once the nature of stability is explored it is not difficult to envisage its effect.

The Theory

I am convinced that Lack was basically correct and that in general clutch size is adjusted by natural selection to correspond with the largest number of young for which parents can provide and, of course, I accept the now overwhelming evi-dence that, in many birds, there is a decrease in clutch size from the poles to the equator.

The stability of tropical environments that is claimed to lead to K- rather than r-selection on the reproductive rates of birds is perhaps best defined in terms of complexity, which in turn arises because of the high diversity of species of plants and animals in the tropics. As is well known the tropics are characterised by high species diversity, even within small areas, and few of the species are common. Proceeding from the equator there is a gradient of decreasing species diversity and increasing species abundance. At high latitudes there are relatively few species, but each is likely to be more common than its tropical counterpart, and in some instances may be exceedingly common.

In no group is this trend more pronounced than in the plant-feeding insects. It is not easy (or even necessary) to offer precise figures, but all the evidence suggests that there are between ten and twenty times as many species in a tropical area than in an area of comparable size in the temperate region. For example, about fifteen species of butterflies may be expected in an English garden, while in a West African garden of comparable size there may be three hundred species, twenty times as many, but none as common as the most abundant species in the English garden. Defoliation is unusual in tropical forest but occurs regularly at high latitudes, usually caused by the caterpillars of just one or two exceedingly abundant species of Lepidoptera. Plant-feeding insects, and their insect predators and parasites, are the chief source of food for many species of birds.

The same trend of diversity applies to other potential food items, including fruits, seeds, fish, small mammals, and, of course, small birds themselves. Even within Europe a bird like the robin is faced with a much greater diversity of possible food in the south than in the north of its range.

Birds as a group may be described as 'conservative opportunists'. They seek and find food items in a way that depends heavily on previous experience with similar items, and are quick to form 'search images' so that once having successfully discovered a food they return again and again for more of the same. This behaviour is particularly evident in insectivorous birds (most of the passerines), but probably applies equally to predators of vertebrates and to fruit- and seed-eaters.

The role of search images in the feeding ecology of birds has been explored both theoretically and experimentally (Tinbergen, 1960; Gibb, 1962; Croze, 1970; Allen, 1972). If, as would seem reasonable, it is accepted that the formation of search images is an integral part of the feeding behaviour of birds, it follows that in a high latitude area where food diversity is low and where many food items are seasonally very abundant, it is relatively easier for parents to feed their young than in the tropics where the diversity of possible foods is high and where none is especially common. Thus an insectivorous bird feeding its young in a tropical forest is faced with a huge variety of possible food items, each species occupying a separate niche, and each differing in appearance and behaviour from other species. It must be assumed that such complexity makes food gathering more difficult. In a deciduous woodland a bird is able to utilise foods of the same species time and time again, and is much more readily able to form search images.

Moreover, in the tropics, there is a much greater proportion of inedible insects than in temperate areas; some of them are toxic, and it must be assumed that items which at first sight may seem suitable as food are in fact unsuitable. The same may be true of fruits, perhaps even of seeds: a tropical bird has to be more careful.

I suggest that the effect of high species diversity in the potential food of tropical birds results in food being less easy to obtain. Clutch size is therefore adjusted by natural selection to correspond with the largest number of young the parents can, on average, successfully feed.   For the majority of tropical passerines this has led to the evolution of a clutch of two. Conversely, the low diversity of potential food at high latitudes results in food being easier to obtain and clutch sizes are correspondingly higher. Even at very high latitudes where there is almost continuous daylight I would expect clutches to go on increasing with latitude.

My suggestion also accounts for the anomalous situation in owls which should, on Lack's daylength theory, show the reverse of the trend that occurs in diurnal species. That they do not suggests that other factors are involved, and I believe that it is increased prey diversity, which in turn results in greater difficulty in obtaining food, that accounts for the low clutch size in tropical owls. The same is probably true of the species of nocturnal mammals that have smaller litters at lower latitudes, although with mammals the main effect is likely to be a restriction in the availability of food to the parent. It is not difficult, for example, to imagine that a low latitude mammal might have more difficulty in obtaining seeds from a variety of different species of plants than a high latitude mammal feeding on the seeds of relatively few species.

The discovery by Lack and Moreau (1965) that savanna birds have slightly higher clutches than forest birds at the same latitude is also consistent with my suggestion. Although potential food items are more varied in tropical savanna than in a temperate area there is less diversity than in a tropical forest. Indeed tropical forest emerges as the environment where because of the extreme complexity, obtaining food is likely to be most difficult.

My suggestion, therefore, accepts Lack's daylength hypothesis as a partial explanation of latitudinal gradients in clutch size, and introduces as an additional factor latitudinal gradients in the diversity of potential food. The suggestion accounts for the hitherto difficult exceptions, like owls, and can be used to explain variations in clutch size between islands and the mainland and between coastal and inland areas. There is no need to invoke the existence of r- and K-selection, nor to postulate that birds be grouped into species that are r- or K-selected. This division, which seems to me to be untenable, relies on the assumption that stable and unstable environments exist and can be recognised as such, and that it is possible to state that bird populations are either at or below the carrying capacity of the environment.

Many tropical birds produce two or more consecutive clutches during a season and may raise as many young as temperate species. This, I believe, is an adaptation to the difficulty experienced by tropical birds in feeding large families. There are of course temperate species that produce several broods in a season but multiple broods are more frequent in the tropics. Three broods of two can be considered as the tropical equivalent to one brood of six in a temperate area.

Finally, the observation by Skutch (1949) that certain species of tropical birds show signs of inactivity while raising their families can now be incorporated into the present suggestion. Skutch believed that such inactivity is an indication that parents are not hard pressed to secure food and that the availability of food is not a crucial factor affecting clutch size. I suggest that inactivity can be interpreted in terms of uncertainty about where the next food item can be found. In some instances it might even be uncertainty over the palatability of food.

Testing the Theory

One obvious possibility is an analysis of the variety of food brought to nestlings of the same or similar species breeding at different latitudes. At high latitudes diversity might be expected to be low while at low latitudes it might be high. If information were available for a range of latitudes it would be possible to make use of a diversity index to estimate 'difficulty' in providing food for nestlings. Unfortunately, information on the food of birds is rarely presented in a way that enables the computation of a diversity index. Not all food items are separated into species or groups of related species and there is a marked tendency to lump rare items together as 'others'. If diversity is to be computed, then all items, rare and common, must be listed, preferably but not necessarily by species. Listings by genera and families are acceptable provided the same procedure is adopted at the different localities that are to be compared.

Suppose a bird feeding young in the nest obtains during a given period of time food items belonging to k different species, the number of each species obtained being $n_1, n_2 \ldots \ldots n_k$ ($\Sigma n = N$). The efficiency of the bird could be measured directly by N and the degree of difficulty experienced by k, if we assume that the greater the number of species the more difficult the process of collecting food. But each species may appear at a certain frequency and the overall diversity of food brought to the nestlings is better measured by the index λ, proposed by Simpson (1949) as an estimate of the probability of any two items chosen at random being the same, in this instance the same species. Simpson's index may be computed from

$$\lambda = \frac{\sum_{j=1}^{k} n_j (n_j - 1)}{N(N - 1)}$$

For the purpose of analysing food samples from nestling birds it is perhaps more appropriate to have an estimate of the probability of two items taken at random being different, and this is easily obtained from,

$$\beta = 1 - \lambda$$

The estimate of β has a minimum of zero (when only one species is taken) and a maximum of one (when every item is different). I think that β can be used as a measure of a bird's difficulty in securing food for its young, assuming of course that diversity is indicative of difficulty.

I would expect β to vary markedly between species. In swifts, for example, it would be high, but in owls, especially those breeding at high latitudes, it would be low. Comparisons should, therefore, be restricted to different populations of the same species or to closely similar species, such as those in the same genus. Latitudinal (or other) gradients in clutch size could then be correlated with β.

As an example of computing a diversity index for food items brought to nestlings table 15.1 gives the values of β for the food brought to nestling jays, *Garrulus glandarius*, at two localities in different seasons. The food items are not in this instance separated into species but into 'kinds' of food: oak-feeding caterpillars, beetles, acorns, and so on. Each locality and season are treated in the same way and so a direct comparison is possible. As shown there are marked differences between localities and seasons, all of them significant at the 1 or 5 per cent levels. In this example there is no question of relating diversity to variations in clutch size, and there is insufficient information to relate it to nestling survival; the

Table 15.1 Diversity of food items brought to nestling jays at two localities in different seasons (based on figures given in Owen, 1956).

	Index of diversity, β, ± s.e.*		
Wytham, Oxford			
	1953	0.044	±0.014
	1955	0.187	±0.017
Virginia Water, Surrey			
	1954	0.641	±0.021
	1955	0.398	±0.034
	1956	0.271	±0.030

*The s.e. of β is the same as for λ, and an equation for its estimation is given by Simpson (1949).

values are given simply to show that the computation of comparative diversities is a possibility that might be explored more fully.

The chief objection to the use of a diversity index for food items brought to nestlings is that the variety of food encountered by the parents may not be reflected in the food brought to the young. Indeed it is possible that a bird faced with finding food in an area where there is a high diversity of potential food items would find only those that are relatively frequent and miss many of the rare species. Finding food may be more difficult in high diversity areas but may or may not be reflected in the food actually brought to the young. If birds are as conservative as they appear to be, it is likely that they will be confused by the unexpected, unpredictable and novel stimuli presented by rare and unfamiliar food items to an extent that such items do not appear in food brought to the nestlings. My colleague Dr T. D. Sargent suggests that diversity is a perfect bane to birds and that if a food item has not been tried before (1) it does not exist, (2) it is frightening, and (3) it should be ignored. There does not seem to be any way of estimating diversity so as to take into account not only the food taken but also the food avoided or overlooked.

There are a few species of birds of wide latitudinal distribution for which it should be possible to attempt a correlation of food diversity with clutch size. One of these is the Cattle egret *Bubulcus ibis,* a widely distributed species whose breeding range is currently expanding. Mean clutch sizes of 2.6 are recorded in southern Ghana, 2.9 in Senegal, while in the southern USA, where the species has recently become established, mean clutches of 3.5 occur in Florida, 3.7 in Maryland, and 3.3 in South Carolina (Jenni, 1969). Higher mean clutches probably occur in southern Europe but precise figures are not available, and there is evidence (summarised by Jenni, 1969) of considerable variation in clutch size throughout the global range of the species. The food of nestling Cattle egrets has been studied in a few places. As in other Ardeidae, they regularly regurgitate food when alarmed, and compared with most species of birds it is not difficult to obtain quantitative and representative data on their diet. Other possible species include the Barn owl and the Long-eared owl *Asio otus,* which both show latitudinal variations in clutch size.

My view, then, is that environmental stability can to a large extent be equated with diversity, and that diversity determines the ease with which birds are able to feed their young. Tropical regions are the most difficult, and high latitudes the easiest. This view seems to explain all known geographical variations in clutch size and avoids the necessity of postulating that bird populations in some areas are maintained below the carrying capacity of the environment. And, of course, my suggestion immediately explains why so may birds are migratory and breed at high latitudes in the summer.

Acknowledgements

This chapter results from ideas developed in 1974 while I was teaching a graduate course in tropical ecology in the Department of Zoology, University of Massachusetts. I thank the Department for the invitation to teach the course, and Dr T. D. Sargent for some stimulating discussions. Mr D. O. Chanter computed the values of β given in table 15.1.

References

Allen, J. A. (1972). Evidence for stabilizing and apostatic selection by wild black-birds. *Nature, Lond.*, **237**, 348 - 349.

Ashmole, N. P. (1961). The biology of certain terns. D. Phil. thesis, University of Oxford. (Cited by Lack and Moreau, 1965).

Cody, M. L. (1966). A general theory of clutch size. *Evolution*, **20**, 174 - 184.

Croze, H. (1970). Searching image in carrion crows. *Z. Tierpsychol. Suppl.*, **5**, 1 - 85.

Gibb, J. A. (1962). L. Tinbergen's hypothesis of the role of specific search images. *Ibis*, **104**, 106 - 111.

Jenni, D. A. (1969). A study of the ecology of four species of herons during the breeding season at Lake Alice, Alachua County, Florida. *Ecol. Monogr.*, **39**, 245 - 270.

Lack, D. (1947). The significance of clutch-size. *Ibis*, **89**, 302 - 352.

Lack, D. (1949). Comments on Mr Skutch's paper on clutch-size. *Ibis*, **91**, 455 - 458.

Lack, D. (1954). *The Natural Regulation of Animal Numbers*. Clarendon Press, Oxford.

Lack, D. and Moreau, R. E. (1965). Clutch-size in tropical passerine birds of forest and savanna. *L'Oiseau*, **35**, 76 - 89.

Moreau, R. E. (1944). Clutch-size: a comparative study, with special reference to African birds. *Ibis*, **86**, 286 - 347.

Owen, D. F. (1956). The food of nestling jays and magpies. *Bird Study*, **3**, 257 - 265.

Simpson, E. H. (1949). Measurement of diversity. *Nature, Lond.*, **163**, 688.

Skutch, A. F. (1949). Do tropical birds rear as many young as they can nourish? *Ibis*, **91**, 430 - 455.

Tinbergen, L. (1960). The natural control of insects in pine woods. l. Factors influencing the intensity of predation by songbirds. *Arch. Néerl. Zool.*, **13**, 265 - 343.

16 The role of predation in the evolution of clutch size

Dr C. M. Perrins*

Introduction

There has been considerable discussion about the adaptive nature of clutch size and whether or not the average clutch can be related to the maximum number of young that the parents can successfully raise (Lack, 1954; Wynne-Edwards, 1962). In nidicolous species, those which feed their young in the nest, it is possible to see at least some of the advantages to a bird of being conservative in the number of eggs that it lays. Although parent birds tend to bring more food to larger than to smaller broods, the increase in the amount of food brought is not proportional to the increase in the number of mouths to be fed. As a result, the individual young in large broods receive less food than those in smaller broods and leave the nest lighter in weight. In at least a few species the probability of survival has been correlated with the weight at fledging (Perrins, 1965). Thus the lowered survival rates of the young in large broods may be sufficient to outweigh the initial advantage that the large brood had in terms of greater numbers of young. In such circumstances, parents with broods of average size may be at least as productive as, and sometimes more productive than those with larger broods (Perrins and Moss, 1975).

Such arguments apply most easily to nidicolous species; they are less easy to apply to species where the young are nidifugous and feed themselves after they have left the nest. In these circumstances the number of feeding birds increases at the same rate as the increase in the brood size. It has been suggested that in some of these species, such as the game birds and ducks, the condition of the female and her level of nutrition at the time of laying may be one of the factors governing the viability and size of the clutch laid. Not only may the female run out of reserves for forming eggs, but also the 'quality' of the eggs laid may differ and the hatching and survival rates of the young may vary in relation to such 'quality' (Theberge, 1971). If this is so the female might possibly improve the chances of survival of her offspring by laying a smaller number of good 'quality' eggs rather than a larger number of poorer quality.

In this chapter I consider the effect of predation on the size of the family, and demonstrate that predation on nests could, in certain species, have been a factor

*Dr C. M. Perrins has worked on the breeding biology of a number of birds round Oxford and on seabirds in Wales. At present he is at the Edward Grey Institute of Field Ornithology where his main research is centred on a long-term study of the population dynamics of the Great tit.

involved in the evolution of clutch size. Others (Cody 1966, Ricklefs, 1969) have also discussed some of the effects of predation in the evolution of clutch size. The basic argument is very simple, namely that the larger the clutch the longer the female takes to lay the eggs, and thus the longer that the clutch will be at risk. (Most female birds lay an egg per day, though some species may lay only every second day or even less frequently. In the species with which I am most concerned here, the birds do not normally start to incubate until the end of the laying period or close to it.)

Clutch size not only affects predation rates during the egg stage. Larger broods of nidicolous species may be taken by predators more frequently than smaller ones; possibly because they are more noisy than small broods (Perrins, 1965) but also possibly because the parents feeding large broods have to visit the nest more frequently thus revealing its position. Some of the arguments may be applied to nidifugous species also; a large brood will probably be more vulnerable to predation than a smaller one in certain circumstances.

This chapter is largely hypothetical; I am concerned to show the sort of selective effects that are theoretically possible rather than to show the exact quantitative effects of predation. The latter is impossible anyway since the necessary data do not exist for accurate quantification of some of the factors discussed below.

The Most Productive Clutch Size and Predation Rates

As an example of the argument presented, we may take a species such as a tit (*Parus*), where a large and variable clutch is laid, and where predation is low compared with many species nesting in the open. Some 18 per cent of first clutches at Wytham (our study area near Oxford) are taken by predators (mainly weasels, *Mustela nivalis*). On average a tit clutch will be available to predators for about 22 days (9 days to lay 10 eggs and 13 days for incubation). If we assume that the rate of predation is constant throughout the laying and incubation periods, a daily predation rate, m, may be calculated by equating survival rates, according to the formula

$$1 - m = (1 - M)^{1/(c+i)} \qquad\qquad (1)$$

where M is the proportion of clutches taken by predators, c is the clutch size, i is the length of the incubation period less one day. This formula takes no account of repeat clutches, assuming that M refers to first clutches only. Applying it to the tit gives a daily predation rate of 0.9 per cent.

However, tits that laid only 8 eggs would, on average, suffer fewer losses from predation whereas those that laid 12 would, on average, suffer more heavily; one would expect clutches of 12 to be taken by predators approximately 20 per cent more frequently than clutches of 8 (they are at risk for 24 days compared with 20). If this is true, it raises the first of many analytical problems for the field worker, since records of completed clutches will not be truly representative of the frequency of the clutches of different sizes, the larger ones being taken more frequently during the laying period than the smaller ones and hence being under-recorded.

From the evolutionary viewpoint, a bird should lay a clutch which will give rise to the most productive brood size (as used by Charnov and Krebs, 1974); in other words, a bird should go on laying eggs until the product of the clutch and its probability of survival reaches a maximum. Thus considering solely the effect of a constant daily predation rate, m, on the average productivity of clutch sizes, c, then by simple probability theory

$$P_c = c(1 - m)^{c+i} \qquad (2)$$

where P_c is the productivity of clutch size c, i is the incubation period less one day. The most productive clutch size is that clutch which maximises P_c.

Using such an argument, it can be shown that there comes a time when the increase of one in the clutch size produces fewer, not more, surviving young, since the increase in the production of young becomes outweighed by the disproportionate increase in the predation. Thus natural selection will have favoured birds which did not exceed clutches of this size. It may be shown, by simple calculus, that the most productive clutch size, c^*, in equation 2, is given by the formula

$$c^* = \frac{-1}{\log_e(1 - m)} \qquad (3)$$

so that the most productive clutch size is dependent on the level of the daily predation rate, m. An example of this relationship between the most productive clutch size and the daily predation rate is given in figure 16.1.

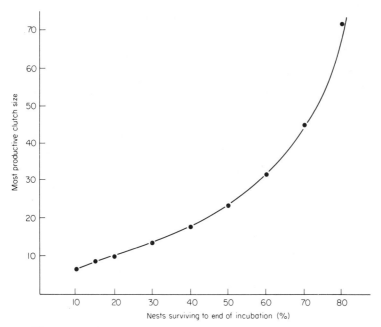

Figure 16.1 Most productive clutch size at different predation rates for a bird laying four eggs and incubating them for 13 days. For explanation see text.

In figure 16.1 the most productive clutch size has been calculated for a bird where the total number of days at which the clutch is at risk is 16 (as would be fairly typical for the Blackbird *Turdus merula*, which lays four eggs (over 3 days) and incubates them for about 13 days). The most productive clutch size has been calculated for a wide range of predation rates and shows that the average predation rate has a marked effect on the most productive clutch; this varies from 72 when only 20 per cent of nests are lost to predators, to about 7 when 90 per cent of the nests are lost.

The higher predation rates shown may seem unrealistic, but they may not be for all species. Lack (1954) listed a number of species and their losses during the egg stage. In 10 species of ducks, 33.5 per cent of the nests were lost during the egg stage; in 11 species of game birds this figure was 61 per cent. In nidicolous species nesting in the open 55 per cent of nests were lost during this stage, in nests in holes about 33 per cent; for various reasons these observed figures may be underestimates. In more recent detailed studies, as many as 86 per cent of nests started by blackbirds in woodland were lost though this figure covers the nestling stage as well (Snow, 1958) while Ant-birds *Gymnopithys bicolor*, lost 80 per cent of their nests before the end of incubation (Willis, 1973) and the Black-and-white manakin *Manacus manacus*, lost at least 60 per cent of its clutches to predators (Snow, 1962).

The information calculated from figure 16.1 is based on a bird such as the blackbird, where a clutch of four eggs is laid and incubated for 13 days. The most productive clutch size will vary with the length of time for which the eggs are at risk. With the same overall predation rate, clutches which are at risk for longer will have a larger most-productive size than those at risk for shorter periods of time. This results from the fact that, given a constant *overall* predation rate, a clutch which is at risk for a longer period must have a lower *daily* predation rate than one at risk for a shorter period. Hence the relative value of more eggs in the former is more advantageous than in the latter. This may seem an unreasonable way of looking at the problem; should one not compare birds having the same daily rate of loss? I have not done this here for two reasons. First, what few data there are provide information on the losses through the whole of the incubation period. Secondly, such information suggests that the proportion of nests of, say, the Partridge *Perdix perdix*, which are taken by predators is not higher than that

Table 16.1 Daily predation rates and most productive clutch size for different species suffering the same total predation rate

Species	Proportion of nests surviving incubation and laying periods	No. of eggs laid	Incubation period	Total days at risk	Daily survival rate	Most productive clutch size
Partridge						
Perdix perdix	0.7	15	21	35	0.9898	98
Mallard						
Anas platyrhynchos	0.7	7	23	29	0.9878	81
Blackbird						
Turdus merula	0.7	4	13	16	0.9779	45

for blackbirds, even though the nests of the former are at risk for a longer period. Thus, although one might suppose that both these species would be equally vulnerable to most of the common predators, this does not seem to be the case; the daily survival rates of the partridge are higher than those of the blackbird. Table 16.1 gives daily predation rates and most productive clutch size for three species with the same overall predation rate.

Strictly speaking, the variation in nesting losses arises through the variation in clutch size and the period of incubation is not involved since the latter is relatively constant with respect to clutch size. However, it seemed more realistic to calculate the losses for the whole of the period when eggs are in the nest since almost all the data available are for the whole of this period.

From figure 16.1 it can be seen that, theoretically at least, the average predation rates experienced by a species could affect the clutch size evolved by that species. However, the range of clutches predicted by such a simple calculation is for the most part outside the range observed in nature, often wildly so. Few birds lay clutches as large as 15 (the average for the common partridge in England). However, such figures show the qualitative effect of predation and without consideration of additional factors which might result in heavier predation of large clutches or broods. Some of these are discussed below.

Factors Affecting Predation Rates of Clutches

Several factors could result in an increase in predation rates as the clutch increases. Not all of these will apply to all species and for almost none of them are quantitative data available. However, theoretically at least, they could lead to selection favouring smaller broods. Some of these possible disadvantages are listed below.

(1) A larger clutch would be more visible to a predator than a smaller clutch. This is most likely to hold for birds that lay on the open ground, but may at times apply with other types of nest.

(2) The intensity of predation may increase with the season. This is the case in the Great tit *Parus major* (Perrins, 1965), perhaps because the predators' own young are hunting for food later in the season. Since a larger clutch takes longer to lay than a smaller one, if one considers clutches started on the same day, risk to the larger one will not only exist for longer; it will also be extended later into the season. Such small intervals of time may seem unimportant, but are not necessarily so.

(3) Birds laying large clutches may deplete their own reserves more severely than those laying smaller clutches. This might require them to spend some time between the end of laying and the start of incubation replenishing these reserves. The resulting gap between laying and incubation would increase the length of time that the eggs were unprotected in the nest and hence the risk of predation. Gibb (1950) showed that large clutches laid by Great tits took longer to hatch than smaller ones, apparently as a result of an extension of the interval between completion of the clutch and the start of incubation. However, in this case, the situation is complicated by the fact that the large clutches are started early in the season and hence the effect could be related to the time of year rather than to clutch size.

(4) The increased risk of predation for a large clutch carries with it a risk that

the laying or incubating female may also be taken by the predator; any increase in female mortality as a result of such an extension of the laying period would further reduce the optimum clutch size. Since the extension of the egg period results from an extension of the laying period rather than of the incubation period, one might suppose that the female laying a large clutch is only at risk (compared with the female of a smaller clutch) for the extra nights of the laying period (when she normally roosts on the nest). However, it is also possible that since laying females must collect more food than non-laying birds (since they need the extra food to form the eggs) they may put themselves at greater risk from predation by doing so, either because they are less cautious or merely because they need to spend more time in exposed areas in order to get the food.

Factors Affecting Predation Rates on Broods

Once the young have hatched, different factors prevail. In nidicolous species, a number of authors have shown that as the size of the brood increases, the parents' feeding rates do not increase proportionately. The result is that the larger the brood, the less well nourished the young. Hungry young may often be noisier than young in smaller, better-nourished broods; at least among titmice the larger broods are not only noisier, but also attract more predators than smaller broods (Perrins, 1965).

To achieve full size, the young in large broods may need to spend longer in the nest than the young in smaller broods. O'Connor (1973) has shown such an effect in the Blue tit, *Parus caeruleus*, as has Dawson (1972) for the House sparrow, *Passer domesticus*. Exceptionally, poorly-nourished broods may be in the nest for considerably longer than the normal period (Gibb and Betts, 1963). Clearly, any such delay in fledging increases the chance of predation.

Additional losses may arise in larger broods if, as a result of working harder, the parents are less alert with the result that they reveal the nest to predators. Skutch (1949) has used such an argument to explain the low family size of some tropical birds; a small brood needs only infrequent visits and this may enable the parents to be particularly cautious. In the pine plantations of the Breckland, Gibb and Betts (1963) noted high predation of parent tits when feeding young, the predator being the sparrowhawk, *Accipiter nisus*. The tits were finding it hard to obtain sufficient food for their young and were leaving the pines to feed in deciduous trees nearby. As a result of their long journey or lack of attention they fell victim to the sparrowhawks. Even if only one parent is killed the brood is not likely to survive.

Some of the factors listed above for nidicolous species may not apply to nidifugous species. Once hatched, many nidifugous species feed for themselves, only using their mother for shelter and as a guard against predators. Nevertheless in some game birds the female not only leads the young to food, but may also scratch up the ground to help the young in their search for food. Larger broods will need more such care from their mother, and may (as with nidicolous species) be less well nourished and hence noisier and so more in danger from predation. Even where young in broods of all sizes are equally well nourished as a result of their own foraging powers, the larger broods may still be more vulnerable to predation than broods of smaller size. To obtain their food such broods will either have to spread out over larger areas, or, as happens with flocks of different sizes

(Morse, 1970), they will have to travel faster in order to come into contact with more food. Whichever is the case, the larger broods will be in greater danger than smaller ones of coming into contact with a predator.

It might be thought that, in contrast with nidicolous species, a complete nidifugous brood is unlikely to be taken by a predator since, when danger threatens, the young will quickly spread out and hide. Nevertheless, the predator after finding one young, will still probably cover the adjacent area particularly thoroughly and so several chicks may be taken. This has, in fact, recently been shown to be true for a wader (Safriel, 1975). A sample of natural broods was increased at hatching from four young to five. Compared with control broods of four young, the broods of five were markedly less successful. The reason for this appears to be that the predators searched harder in the area where they had just found a prey. The number of broods from which no young were raised was higher than expected.

Further, if in addition the predator takes the parent as a result of finding a chick, then the whole brood will be lost since the chances of the young (at least small young) surviving on their own are very low. Although these latter arguments apply particularly to nidifugous species they are equally applicable to many nidicolous species after leaving the nest, since the birds travel around as a family party for a period which is usually at least several days and is often longer. At times they may be very noisy and conspicuous.

The Effect of Other Mortality Factors on the Optimum Brood

The factors suggested above are likely to affect the chances of survival of large broods deleteriously compared with smaller broods. As mentioned previously, the effects of most of these factors have not been quantified for any species let alone a range of different species. Notwithstanding this, it is possible to see what effect varying the extra disadvantages would have on the optimum clutch size.

Let us assume that the disadvantages decrease linearly the survival rates for broods derived from clutches of size c, by an amount m_1 per egg laid. Then by an extension of the argument used in deriving equation 2, the average productivity of clutches of size c becomes

$$P_c = c(1 - m)^{c+i}(1 - cm_1) \tag{4}$$

Equation 4 assumes that the disadvantages caused by large brood size act sequentially and independently of the clutches being taken by predators during laying and incubation. As previously, the most productive clutch size is that which maximises P_c. The effect of varying levels of m_1, the disadvantage per egg, on the relationship between optimal clutch size and daily predation rate during laying and incubation (m) is shown in figure 16.2, for a bird with an incubation period of 13 days.

Although additional disadvantages as high as 3 per cent per additional egg (shown in figure 16.2 by the 0.97 line) may be unnaturally strong, one can see that, even with relatively small selective disadvantages of 1.5 per cent and overall losses of about 50 per cent, the most productive clutch is already reduced to about 15.

It is, perhaps, worth noticing at this point that, the higher the overall predation rate, the greater the effect of a small increase in predation. For example, a change from 10 per cent predation to 15 per cent only reduces survival by about

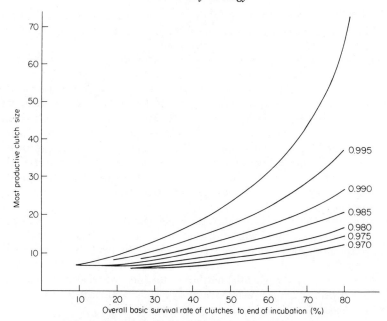

Figure 16.2 The most productive clutch size in relation to production
rates. The upper curve is the same as in figure 16.1. The lower
curves show the effect of additional mortality (m_1), resulting from
increased clutch size, on survival. For explanation see text.

5.5 per cent (from 0.90 to 0.85) but a change from 85 per cent predation to
90 per cent reduces survival by 33 per cent (from 0.15 to 0.10). Hence at high
levels of predation, different predation rates for different brood sizes have a
greater effect on optimum brood size than is the case with low levels of predation.

The Advantages of Asynchronous Hatching

Lack argued that by hatching its young asynchronously, parent birds obtained a
brood whose size was relatively easily adjusted to the prevailing food conditions;
if food was scarce the smallest young swiftly died with minimum harm to the
larger ones, which could remain well fed throughout. Lack suggested that the
habit of asynchronous hatching had evolved in species with a very variable, and
unpredictable, food supply.

In the absence of predation, asynchronous hatching is probably a sound strategy
from the evolutionary viewpoint. However, in some species such as the tits, evidence cited above suggests that at least some of the predation results from the
predators locating the brood by sound. Hence hungry young are at greater risk
than well-fed young since they beg more noisily or more persistently for food.
Even a single hungry young is all that is necessary to put the whole brood at risk.
One might therefore expect that species in which there are high rates of predation
would not normally be able to 'risk' using the strategy of asynchrony.

There are a few broad generalisations which can be made about hatching
synchrony. Nidifugous species hatch synchronously (see, for example, Vince,
1964) as do many of the smaller species of passerines which, in general, suffer
high levels of predation. Those species which tend to have the most highly

asynchronous hatching include the diurnal birds of prey, the owls, the swifts and certain species of seabirds, species which on the whole are not heavily preyed on, either because of their size or because of the inaccessibility of their nests.

Many small passerines have young that do not hatch all together, but the spread of hatching is very variable between species and, in some species, at different times in the season. For example, in the House sparrow the spread of hatching is much more extended than in most of the tits. O'Connor (1973) has discussed what he called runting in this species whereby the birds adjust their brood size during the nestling period; losses from predation are very low in the House sparrow (Dawson, 1972). Hence, even for birds of similar size the tendency to hatch synchronously or not may have been modified through the selective pressures of predation.

It should perhaps be stressed that a bird could, theoretically, evolve two separate strategies. It could vary its clutch size and also vary the point at which it starts to incubate its eggs in relation to the termination of laying. Hence both small or large clutches could be associated with either asynchronous or synchronous hatching. Probably it would be advantageous to either hatch synchronously (especially if predation is likely) or very asynchronously (so that the smallest young will quickly die if the larger ones are still hungry).

The relevance of the strategy of asynchronous hatching to clutch size is that, if the brood has a high risk of predation, it is probably dangerous to hatch the young asynchronously. Hence the best strategy for such a species might well be to be conservative and lay a clutch related to the number of young it can raise in a normal year. Although occasionally a larger clutch would result in more young surviving, usually hatching asynchronously will greatly increase the numbers of nests lost to predators. Hence the best strategy for such birds is probably to be conservative in their clutch size.

Discussion

As stressed before, it is not suggested that all the clutch sizes observed in nature have evolved solely as a result of predation. Clearly, a wide variety of other factors are involved. In particular, effects of nutrition have been clearly demonstrated in a number of nidicolous species. Obviously, these would come into play most strongly in the situation where large broods might be raised because of little predation.

However, in certain circumstances predation may have had such an effect on nesting success that it has contributed to the selection of the clutch size laid by birds. From the simple calculations made above we can draw a number of conclusions.

First, clutches are likely to be smaller when predation rates are higher. There is a widespread tendency for hole-nesting birds to have higher nesting success than birds in open nests and also for them to have higher clutches.

Secondly, if, as has been suggested, predation rates are particularly high in the tropics compared with temperate regions, one would expect to find that many species laid smaller clutches in the tropics than those laid by closely related species elsewhere; this is indeed the case (Moreau, 1944).

One explanation put forward to explain low clutch size in the tropics compared with temperate regions does so in terms of the length of the day available for feeding; this argument founders on the fact that nocturnal species should not encounter the same problems as the diurnal species, but yet they too have smaller clutches in the tropics than in temperate regions (Lack, 1947 – 48). Further, in at least one species, the Snow bunting, *Plectrophenax nivalis*, the clutch size continues to increase as one goes northwards from the arctic circle, even though effective daylength does not increase (Salomonsen, 1972).

One would not expect the following groups of birds to have modified their clutches as a result of predation.

(1) In birds laying very small clutches the incubation period is so long compared with the laying period that the laying of an extra egg results in only a little extra disadvantage in terms of predation. Consider, for example, a bird that lays one egg and incubates it for 20 days; if it were to lay a second egg, it would only increase its risk of predation (assuming that the extra young could be adequately fed and cared for) by some 5 per cent, yet it would be doubling the number of young raised. Hence it is difficult to accept that birds laying very small clutches, such as many tropical species of passerines, have evolved such a clutch solely as a result of predation pressure as has been suggested (Skutch, 1949; Ricklefs, 1969). The only thing that would militate against laying a second egg would be if the bird took a very long time to collect the food to form the second egg. For example, some shearwaters may take as long as two weeks to form an egg and the Mallee fowl, *Leipoa ocellata* may take a week (Serventy, 1958; Frith, 1962). If the interval between the eggs were as long as a week the predation rate in our example would increase by some 35 per cent and so relatively more of the potential advantage would be lost.

(2) Birds with very long incubation periods would not have modified their clutch size for basically the same reason as the one above, namely that the incubation period is long in relation to the laying period and so the delay caused by laying a further egg is small in relation to the advantage to be gained.

In contrast to the above, one would expect those species with large clutches or with short incubation periods to be those where the effects of predation on clutch size would be most marked. Once again, I stress that it is unlikely that the clutch sizes of such species have been evolved solely because of the pressures of predation. However, when predation was high, clutch size may well have been modified as a result of such pressures. In addition, other aspects of nesting (for example, nest type, dispersion of nests, size of eggs in relation to length of incubation or fledging periods and the point at which the young birds leave the nest) are likely to have been influenced by the dangers from predation. These, however, are beyond the scope of this chapter. For these very small clutches to have been evolved in response to predation, the differences in rates of predation between different brood sizes must have been higher than those used in this chapter. Information to show whether or not this is so does not exist. I conclude that the very small clutches laid by many tropical species seem, on the basis of this discussion, to be adapted to something other than predation, possibly the complications of finding food for the young, as suggested by Owen in chapter 15.

Acknowledgements

I thank the late Dr M. I. Webber for advice concerning the formulae presented in this paper. Drs D. G. Dawson, J. R. Krebs and P. J. Bacon kindly read and criticised earlier versions of the manuscript.

References

Charnov, E. L. and Krebs, J. H. (1974). On clutch-size and fitness. *Ibis*, **116**, 217-219.

Cody, M. L. (1966). A general theory of clutch-size. *Evolution*, **20**, 174-184.

Dawson, D. G. (1972). The breeding ecology of House Sparrows. D.Phil. thesis, Oxford University.

Frith, H. J. (1962). *The Mallee Fowl.* Angus and Robertson, London.

Gibb, J. A. (1950). The breeding biology of the Great and Blue Titmice. *Ibis*, **92**, 507-539.

Gibb, J. A. and Betts, M. M. (1963). Food and food supply of nestling tits (Paridae) in Breckland Pine. *J. Anim. Ecol.*, **32**, 489-533.

Lack, D. (1947-48). The significance of clutch-size. *Ibis*, **89**, 302-352, **90**, 25-45.

Lack, D. (1954). *The Natural Regulation of Animal Numbers.* Oxford University Press, Oxford.

Moreau, R. E. (1944). Clutch-size: a comparative study, with special reference to African birds. *Ibis*, **86**, 286-347.

Morse, D. H. (1970). Ecological aspects of some mixed species foraging flocks of birds. *Ecol. Monogr.*, **40**, 119-168.

O'Connor, R. J. (1973). Growth and metabolism in some insectivorous birds compared with a granivorous species. D.Phil. thesis, Oxford University.

Perrins, C. M. (1965). Population fluctuations and clutch-size in the Great Tit, *Parus major* L. *J. Anim. Ecol.*, **34**, 601-647.

Perrins, C. M. and Moss, D. (1975). Reproductive rates in the Great Tit. *J. Anim. Ecol.*, **44**, 695-706.

Ricklefs, R. E. (1969). An analysis of nesting mortality in birds. *Smithsonian Contr. Zool. Nr.*, **9**, 1-48.

Safriel, U. N. (1975). On the significance of clutch-size in nidifugous birds. *Ecology*, **56**, 703-708.

Salomonsen, F. (1972). Zoogeographical and ecological problems in arctic birds. *Proc. 15th Int. Ornith. Congr.*, 25-77.

Serventy, D. L. (1958). Recent studies on the Tasmanian Mutton-bird. *Aust. Museum Mag.*, **12**, 327-332.

Skutch, A. F. (1949). Do tropical birds raise as many young as they can nourish? *Ibis*, **91**, 430-455.

Snow, D. W. (1958). The breeding of the Blackbird, *Turdus merula* at Oxford. *Ibis*, **100**, 1-30.

Snow, D. W. (1962). A field study of the Black and White Manakin, *Manacus manacus* in Trinidad. *Zoologica*, **47**, 65-104.

Theberge, J. B. (1971). Population fluctuation and changes in the quality of Rock Ptarmigan in Alaska. Ph.D. Thesis, University of British Columbia.

Vince, M. A. (1964). Synchronization of hatching in American Bobwhite Quail. (*Colinus virginianus*). *Nature*, **203**, 1192-1193.

Willis, E. O. (1973). Survival rates for visited and unvisited nests of Bi-coloured Antbirds. *Auk*, **90**, 263-267.

Wynne-Edwards, V. C. (1962). *Animal Dispersion in Relation to Social Behaviour.* Oliver and Boyd, Edinburgh.

17 A note on the evolution of clutch size in altricial birds

Dr R. E. Ricklefs*

Introduction

David Lack (1947; 1954) suggested that clutch size in birds is adapted to correspond to the maximum number of young that parents can nourish. The studies that Lack's pioneering work stimulated have revealed both supporting and contrary evidence (see Lack, 1954, 1966; Wynne-Edwards, 1962; Skutch, 1949; 1967; Klomp, 1970; Hussell, 1972). Lack's hypothesis also has been criticised on theoretical grounds. Skutch (1949) suggested that predation could determine optimum clutch size if the loss of nests to predators were to increase as brood size increases, hence as feeding visits, begging and general activity about the nest increase. As Lack (1949) pointed out in response, whether predation or starvation is the principal cause of death, the optimum brood size is none the less that which produces the most young on average.

Skutch (1949; 1967) and, later, Wynne-Edwards (1962; 1963) have championed the hypothesis that brood size is adjusted so recruitment of young in the population balances the mortality of adults. In this case, optimum brood size would be determined indirectly by factors affecting adult mortality rather than by the food resources available for rearing young. Skutch and Wynne-Edwards' hypothesis has been discounted by most ecologists inasmuch as it lacks a sound evolutionary mechanism (Williams, 1966; Wiens, 1966; Lack, 1966). None the less, indications that some species normally rear fewer young than they can feed adequately have not been accounted for properly. For example, adults do not often forage constantly during the nestling period, and some species are able to rear larger than normal broods; there are also species in which females alone rear broods as large as those of other species in which both parents feed the young. Numerous explanations have been offered for these phenomena, but compelling evidence for a single explanation has not been found.

Brood sizes below those that can be reared by parents have been predicted from mathematical treatments of optimum reproductive effort (Williams, 1966; Gadgil and Bossert, 1970; Charnov and Krebs, 1973; Goodman, 1974; Pianka and Parker, 1975). These models are based on the trade-off between fecundity and adult

*Dr Robert E. Ricklefs was born in San Fransisco, California. He received his A.B. in Biology from Stanford University in 1963 and his Ph.D. in Biology from the University of Pennsylvania in 1967. After a year of postdoctoral study at the Smithsonian Tropical Research Institute in the Panama Canal Zone, he returned to the faculty of the University of Pennsylvania, where he is now Associate Professor. This chapter was written while Dr Ricklefs was a Guggenheim Fellow on leave at Stanford University.

mortality resulting from increments of reproductive effort. According to the theory, adults optimally should rear a brood smaller than the maximum possible so as to increase their own chances of survival and thus extend their further opportunities to breed – that is, they limit the size of the present brood for the sake of future reproduction. The models do not predict how much smaller than the maximum possible broods should be (Ricklefs, 1977) and they do not, therefore, increase the weight of evidence either supporting or refuting Lack's original hypothesis.

Lastly, in an environment having unpredictable fluctuations in food supply, clutch size may be adapted to the level of food resources available in the best season, resulting in some young starving during most years (Lack, 1954; Ricklefs, 1965). If young are starved selectively in accordance with the order of hatching (and hence of the size of young) within the brood, nestlings in excess of the maximum number that can be reared *during any particular year* can be starved with little ill effect on the survivors.

The factors that can reasonably be expected to influence the optimum clutch size of a species have almost certainly in the main been identified (see reviews by Klomp, 1970; von Hartman, 1971 and Cody, 1971; see also Ricklefs, 1970; Snow, 1970 and Foster, 1974a; 1974b). It remains for ecologists to determine the relative contribution of each factor to the variation in clutch size between species. That the importance of each factor has not been evaluated adequately is due, in part, to a lack of hypotheses or models with testable predictions, particularly models that predict expected levels of predation and starvation of nestlings.

In this chapter, I develop a simple model relating optimum clutch size to mortality of the eggs and young, particularly that due to starvation and predation. Predictions concerning the relationship between mortality and clutch size are tested to the extent made possible by published studies of nesting success. I am not concerned with the compromise between fecundity and adult survival arrived at by optimising reproductive effort. I assume here that an optimum level of reproductive effort is achieved, by which the ability of adults to nourish their young is determined. Given this constraint, I ask how predation, competition for food among nestlings, and variability in the environment influence the optimum size of the clutch.

Structure of the Model

Components of survival

The number of young to be expected from a nesting attempt, which I shall call production, is the product of the clutch size and the probability of survival of the young. This relationship may be expressed mathematically as $P = CS$ where P is the expected production, C the clutch size, and S the probability of survival. The probability of survival may be thought of as having four components, each influenced differently by clutch size and duration of the nest period.

S_1 – independent of both clutch size and duration of nest period. Hatching success, to the extent determined by infertility of eggs, falls into this category.

S_2 – influenced by duration, but not clutch size. Most predation and death due to weather conditions, which happen with finite probability during small time intervals of the nest period, determine survival of this type.

S_3 – dependent on clutch size only. This component of survival probability includes hatching failure due to difficulties of incubating large clutches, and starvation of young due to chronic food shortage. Starvation caused by brief food shortages brought on by bad weather, for example, would be properly placed in the second category if entire broods died, and in the last category if some young were not affected.

S_4 – dependent on both clutch size and duration. Predation that is influenced by the number of eggs or young in the nest belongs here.

The expression for production can now be written

$$P = CS_1 S_2 S_3 S_4 \tag{1}$$

We may further specify S_2, S_3, and S_4 in terms of the duration of the nest period and clutch size. If the probability of predation or death due to bad weather were constant throughout the nest period, the number of young surviving would decline exponentially during the nest period, according to the expression

$$S_2 = \exp(-m_2 t) \tag{2}$$

where m_2 is the expected mortality rate from predation and storms and t is the length of the time interval.

Mortality due to starvation

Because the relationship of starvation to brood size is not well known and may differ from one species to another, a general, flexible form of expression relating S_3 to C is best suited to our purposes. The function should equal 1 when clutch size is 0, remain near 1 until a clutch size is reached corresponding to the largest number of young that can be nourished, and decrease rapidly as clutch size increases above that point. The relationship

$$S_3 = \exp(-aC^X) \tag{3}$$

where a and X are constants, satisfies these requirements and allows considerable flexibility of form.

Mortality due to predation

An equation similar to that for S_2 can be written when the influence of predation on survival depends on clutch size. The flexibility of the expression for S_4 should match our ignorance of this influence in natural populations. Beginning with

$$S_4 = \exp(-m_4 t) \tag{4}$$

whose form is identical to the equation for S_2, we designate m_4 to be a function of clutch size in accordance with the relationship

$$m_4 = bC^Y \tag{5}$$

where b and Y are constants. The expression for S_4 thus becomes

$$S_4 = \exp(-tbC^Y) \tag{6}$$

When C is equal to 0, S_4 equals 1. The decrease in S_4 as clutch size increases is made greater by increasing b and Y. When Y is 1, mortality increases in direct proportion to clutch size; when Y is 2, mortality increases in proportion to the square of clutch size, and so on.

Optimisation of clutch size

Having defined the components of survival probability in terms of clutch size and duration of the nest period, the expression for productivity can now be rewritten as

$$P = CS_1 \exp(-m_2 t)\exp(-aC^X)\exp(-tbC^Y) \tag{7}$$

We may determine the optimum clutch size, the value of C that maximises P, by differentiating the equation for P with respect to C and evaluating the resulting derivative at the point $P = 0$. The derivative of P with respect to C is found most readily for the logarithmic form of equation (7),

$$\log P = \log C + \log S_1 - m_2 t - aC^X - tbC^Y \tag{8}$$

whose derivative with respect to C is

$$\frac{\text{d}\log P}{\text{d}C} = \frac{1}{C} - Xac^{X-1} - YtbC^{Y-1} \tag{9}$$

The logarithm of P, and thus P itself, is greatest at that value of C for which $\text{d}\log P/\text{d}C$ equals 0, or

$$\frac{1}{C} = Xac^{X-1} + YtbC^{Y-1} \tag{10}$$

Multiplying through by C, we obtain

$$1 = Xa\hat{C}^X + Ytb\hat{C}^Y \tag{11}$$

(the hat over the C denotes that it is an optimum value). S_1 and m_2, neither of which varies with respect to clutch size, do not appear in the solution for optimum clutch size. Clutch size-dependent predation or clutch size-dependent starvation alone can determine optimum clutch size, or the two may act in combination. As long as either X or Y is positive, an optimum clutch size is defined by equation (11). If X and Y were both equal to 0, mortality would be independent of clutch size and the optimum clutch size would be infinitely large. We may be certain that X and Y are positive in nature, but their values remain to be determined.

Skutch's Predation Hypothesis

Skutch (1949) suggested that if predation were to increase as brood size increased the optimum clutch size could be less than the maximum number of young the parents could nourish. Nest concealment, reduced number of feeding trips to the nest and lack of begging cries in young have been cited as evidence of the importance of predation to the lives of some species of birds. Indeed, Skutch's hypothesis has been invoked to explain the small clutches of some tropical species that are known to suffer great nest mortality (Skutch, 1949; Snow, 1970).

If predation acted alone to determine optimum clutch size, the expected level of predation on nests could be evaluated from the model presented in this chapter. Set X equal to 0 because starvation is presumed, for the time being, to have no influence on optimum clutch size; then equation (11) becomes

$$1 = Ytb\hat{C}^Y \qquad (12)$$

which may be rearranged to give

$$\hat{C} = \left(\frac{1}{Ytb}\right)^{1/Y} \qquad (13)$$

The relationship between C, Y, and b, setting t equal to 15 days, is shown in figure 17.1. Clutch size-dependent predation probably is important only during the nestling period when activity around the nest could be expected to vary in proportion to the number of young in the brood. Fifteen days is an average nestling period. Any given optimum clutch size may be obtained by any number of combinations of b and Y, with the general restriction that as Y increases, b must decrease in a particular manner for each value of $\hat{C}$.

We may evaluate the level of mortality caused by clutch size-dependent predation on clutches of optimum size by rearranging equation (4) into the form

$$C^Y = \frac{1}{Ytb} \qquad (14)$$

From (5), we have $C^Y = m_4/b$, and thus

$$m_4 = \frac{1}{Yt} \qquad (15)$$

The level of mortality depends only on Y and t and is not influenced by $\hat{C}$ or b. The relationship of m_4 to Y, assuming $t = 15$, is shown in the lower portion of figure 17.1. We can only guess at the value of Y in natural populations. If predators located nests by activity, particularly feeding trips by adults to and from the nest, and if activity increased in direct proportion to brood size, we could reasonably estimate Y as 1 and thus predict a mortality rate, due to clutch size-dependent predation alone, of almost 7 per cent per day.

If the young of large broods begged more loudly than the young of small broods because they were underfed, the commotion emanating from a nest might increase in proportion to the square or cube of clutch size, but it is difficult to imagine an effect of higher order. For $Y = 3$, we would expect mortality from predation to exceed 2 per cent per day. Daily mortality rates for whole clutches and broods, calculated from Skutch's data for lowland species of birds in the humid neotropics, average about 4 per cent per day (Ricklefs, 1969). Most of this mortality is, indeed, caused by predators but we would be mistaken to assume that it is of the type that varies in accordance with clutch size.

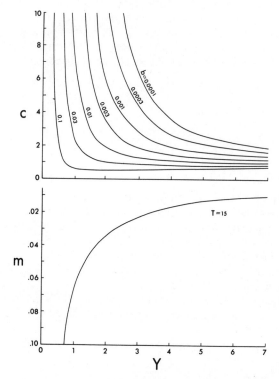

Figure 17.1 Above: optimum clutch size ($\hat{c}$) as a function of Y and b when clutch size-dependent mortality determines $\hat{C}$ (see text). Below: daily mortality (m) resulting from clutch size-dependent predation as a function of Y; m is independent of optimum clutch size. Nestling period (t) is 15 days.

The magnitude of clutch size-dependent predation can be estimated by comparing mortality rates during the nestling period to those during the incubation period, when activity around the nest is minimal and there are no young to beg. The amount by which mortality during the nestling period exceeds that during the incubation period provides a generous estimate of clutch size-dependent mortality at the optimum clutch size. In the species studied by Skutch, mortality during the nestling period is 1.2 per cent per day *less* than during the incubation period (Ricklefs, 1969). This differential predation undoubtedly occurs because the most readily locatable and accessible nests are preyed on first, mostly during the incubation period. None the less, clutch size-dependent predation must on average be a minor component of mortality and of little importance to the evolution of clutch size.

Clutch size-dependent mortality is more plausible for arctic species, in which daily mortality during the nestling period exceeds that during the incubation period by an average of 1.7 per cent (Ricklefs, 1969). This excess may be due partly to death from exposure, to which nestlings are probably more susceptible

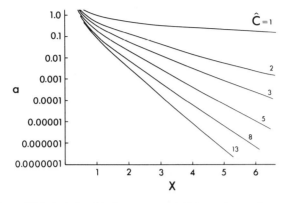

Figure 17.2 Relationship between a and X for optimum clutch size of 1, 2, 3, 5, 8, and 13, when clutch size-dependent starvation determines optimum clutch size (see text).

than brooded eggs, and partly to starvation. Yet in open arctic habitats, where the overall level of predation is low compared to temperate and tropical habitats, the locations of nest sites are not protected by cryptic adult and nestling behaviour and predators may cue on activity around the nest. None the less, even if clutch size-dependent mortality were as high as 1.5 per cent per day, Y would still have to be about 5 if predation were the sole determinant of optimum clutch size. If Y were 5, and optimum clutch size were 5, clutch size-dependent mortality would increase by 2.5 times from clutch size 5 to 6, and by more than 5 times from clutch size 5 to 7. Changes in mortality of this magnitude, and the activity at the nest required for such changes, should be detectable in experiments involving manipulation of brood sizes.

Even if large values of Y were to occur, they would probably result only from intensified begging for food by young in large broods. Inasmuch as begging noises from a brood are linked to the availability of food and thus reflect the state of nutrition of the young, predation may be thought of as a factor that intensifies the selective force of starvation on optimum clutch size rather than influencing optimum clutch size directly.

Food Resources and Optimum Fecundity

If we assume clutch size-dependent predation to be negligible and set the value of Y in equation (11) at 0, optimum fecundity may be expressed by

$$\hat{C} = \left(\frac{1}{aX}\right)^{1/X} \tag{16}$$

The relationship of $\hat{C}$ to a and X is shown in figure 17.2. The constant X determines how sharply survival declines as clutch-size increases; a determines the range of clutch sizes over which survival declines most rapidly. If X is small, and thus

Evolutionary ecology

Table 17.1 The constants X and a of the equation $S_3 = \exp(-aC^X)$ fitted to the relationship between survival and clutch or brood size. The correlation coefficient of the regression (r) and the optimum clutch size ($\hat{C}$) predicted by X and a (equation (16)) are also tabulated.

Species	Period	Clutch or brood size		r	a	X	$\hat{C}$	Source
		Mode	Range*					
Black-footed albatross† *Diomedea nigripes*	Nestling	1	1– 2	1.00	0.400	2.62	1.0	Rice and Kenyon (1962)
Manx shearwater† *Puffinus puffinus*	Nestling	1	1– 2	1.00	0.0513	5.21	1.3	Harris (1966)
Heron *Ardea cinerea*	Nestling		2– 5	0.98	0.000698	4.20	4.0	Owen (1960)
	Fledging to September 1		2– 4	0.93	0.000757	5.24	2.9	Owen (1960)
Buzzard *Buteo buteo*	Nestling	3	3– 4	1.00	0.0435	1.69	4.7	Mebs (1964)
Kestrel *Falco tinnunculus*	Nestling	6	5– 7	0.99	0.000864	2.99	7.3	Cavé (1968)
Western gull† *Larus occidentalis*	Nestling, 1971	3	1– 6	0.96	0.0132	1.72	9.0	Coulter (unpublished)
	1972	3	1– 6	0.96	0.0144	1.89	6.8	Coulter (unpublished)
Forsters tern *Sterna forsteri*	Nestling	2	1– 3	0.97	0.148	1.82	2.1	Coulter (unpublished)
Arctic tern *Sterna paradisea*	Nestling	2	2– 3	1.00	0.0812	2.00	2.5	Lemmetyinen (1973)
Woodpigeon† *Columba palumbus*	Nestling	2	2– 3	1.00	0.000507	5.32	3.0	Murton, Isaacson and Westwood (1974)
	Fledging to 1 month	2	2– 3	1.00	0.000741	8.51	2.4	Murton, Isaacson and Westwood (1974)

Species	Period	Mode	Clutch or brood size Range*	r	a	X	$\hat{C}$	Source
Swift *Apus apus*	Nestling	2	2 – 4	0.99	0.00145	4.49	3.1	Perrins (1964)
Pied flycatcher *Fidecula hypoleuca*	Fledging to breeding	7	6 – 9	0.99	0.000294	3.43	7.5	Haartman (1967)
Great tit *Parus major*	Nestling	10	3 – 12	0.99	0.00692	1.47	22.8	Perrins (1965)
	Fledging to 3 months‡		9 – 13	0.96	2.17×10^{-14}	12.81	9.6	Lack, Gibb and Owen (1957)
Blue tit *Parus caeruleus*	Fledging to 3 months	11	3 – 14	0.95	0.000390	3.22	8.0	Lack, Gibb and Owen (1957)
Blackbird *Turdus merula*	Nestling∮	4	2 – 5	0.85	0.0156	1.58	18.6	Snow (1958)
Starling *Sturnus vulgaris*	Nestling†	5	3 – 7	0.88	0.00148	2.68	7.9	Ricklefs and Hussell (unpublished)
	Fledging to 3 months	5	4 – 8	0.98	0.000410	3.76	5.6	Lack (1948)
Common grackle *Quiscalus quiscula*	Nestling	5	4 – 6	0.92	5.33×10^{-7}	8.55	4.2	Willson *et al.* (1971)
Boat-tailed grackle *Cassidix mexicanus*	Nestling	3	3 – 5	0.99	0.000565	5.21	4.8	Tutor (1962)

*Range of brood sizes used to calculate constants *a* and *X*.
† Larger than normal clutches were made by the investigators.
‡ Broods of 1949 – 50, 1952 – 55 only; no clutch size-dependence in 1947 – 48.
∮ Botanic Garden only, no clutch size-dependence in Wytham Wood.

survival declines greatly as clutch size increases, small changes in *a* cause large changes in optimum clutch size. If *X* is large, optimum clutch size is more resistant to change in *a*.

The effect of *X* on the shape of the curve relating survival to clutch size is shown in figure 17.3 for several combinations of *X* and *a* resulting in an optimum clutch size of 5. As *X* increases, the curve becomes steeper. By multiplying survival and clutch size, we can see that 5 is always the most productive clutch size for the combinations of *a* and *X* depicted in figure 17.3, but production is more evenly distributed among clutch sizes for small values of *X*. As *a* increases, the peak of the production curve shifts to the left.

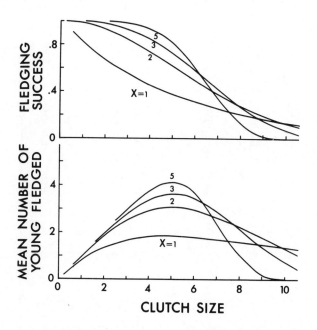

Figure 17.3   Relationship between survival and clutch size for different values of *X* when the optimum clutch size is 5. Lower portion of graph shows number of young produced (survival × clutch size).

X and a in natural populations

The equation for S_3 (equation 3) may be transformed into an equation relating S_3 to a linear combination of terms in *a* and *X* by twice taking its logarithm. The resulting equation

$$\log(-\log S_3) = \log a + X \log C \tag{17}$$

can be fitted to data for S_3 and *C* by linear regression.

Values of *X* and *a* so obtained are presented in table 17.1. The fit of equation (17) to the data varies from species to species (see figure 17.4), but the correlation

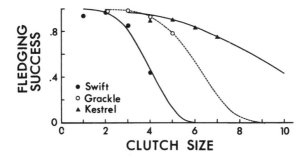

Figure 17.4 Relationship between survival and clutch size in the Swift, Boat-tailed grackle and Kestrel. Curves are fitted equations of the form $S_3 = \exp(-aC^X)$ for each species (see table 17.1 for values of constants a and X). Data for smallest clutch sizes of each species were not included in calculating the equations for the curves.

coefficient of most of the regressions was greater than 0.98. In several cases, including those of the Swift and Kestrel, shown in figure 17.4, small clutches had poorer survivorship than medium-sized clutches. In such cases, the date for the smaller clutches were not used to fit the curve. Because of the small sample size in some species, data for broods of more than one size were sometimes lumped.

A more critical problem with the data was that not all the mortality is caused by clutch size-dependent starvation. Indeed, starvation and other death related to undernourishment are rarely identified as important factors in field studies (Ricklefs, 1969). Much irrelevant mortality is eliminated by considering survivorship only during the nestling period. In some cases, when the percentage fledging from broods of the most successful size was below 90 per cent, the fledging success of all brood sizes was increased by a common factor to bring the most successful to 99 per cent. This is justifiable to the extent that small broods are free of the effects of starvation. Most of the species listed in table 17.1 nest in holes or other inaccessible sites and therefore do not suffer heavy predation. To what degree differences in survivorship between clutches of different size are due to starvation is not known, but the survivorship data none the less demonstrate selection for optimum clutch size.

Values of X and a were used to calculate optimum clutch sizes ($\hat{C}$) from equation (16) – see table 17.1. In many cases, the predicted optimum clutch size corresponded closely to the observed modal clutch size; in other cases, it was much greater (for example, Western gull, Great tit, English Blackbird, and Woodpigeon). In the Heron, Woodpigeon, Great tit, and Starling, $\hat{C}$ calculated from survival after leaving the nest was both lower and closer to the modal clutch size than that calculated from survival during the nestling period. The deleterious effects of large clutch size on survival evidently are not fully expressed until the young have left the nest. Estimates of $\hat{C}$ based on survival of Starlings and Pied flycatchers banded as nestlings and recovered several months after fledging agree reasonably with modal clutch size. Only in the English Blackbird do survival data fail to provide agreement between $\hat{C}$ and the observed mode. Post-fledging data are not, however, available for this species.

The survival function

In table 17.1, X varies between 1.5 and 12.8, with an average of about 4. The significance of this variation is not readily apparent. Variation in X is not obviously associated with particular patterns of reproductive or feeding behaviour. A simple model relating survival to the nutrition of young reveals at least one source of variation in X. Assume that a particular species can rear no more young than some maximum number ($\ddot{C}$) determined by resource availability. Furthermore, all young hatching from clutches that are less than the maximum rearable size fledge successfully. Larger clutches produce exactly the maximum number that can be nourished. For example, if 3 were the maximum rearable clutch, clutches of 1, 2, and 3 would be 100 per cent successful; clutches of 4, 75 per cent successful (3/4); clutches of 5, 60 per cent successful (3/5); and so on. The relationship of survival to clutch size calculated in accordance with this model for maximum rearable clutches of 1, 2, 3, 4, 5, and 8 is shown in figure 17.5. When these curves are fitted by equation (3), relating survivorship (S_3) to X, a, and C, values of X are found to vary between 3.7 and 4.6 (table 17.2). Of course, equation (3) does not describe the curves in figure 17.5 perfectly and predicted optimum clutch sizes ($\hat{C}$) are somewhat larger than the maximum rearable clutch. None the less, values of X close to 4.0 in natural populations probably represent situations in which the number of young reared reaches a plateau as clutch size increases above the maximum rearable clutch. In this model, no penalty is imposed on the survival of the young for laying a clutch larger than the maximum rearable size, but neither does the productivity of large clutches increase above that point.

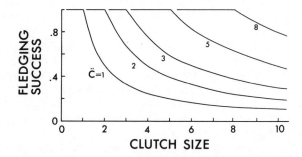

Figure 17.5 Relationship between survival and clutch size in accordance with the maximum rearable clutch model for maximum rearable clutches ($\ddot{C}$) of 1, 2, 3, 5, and 8.

According to the model of a maximum rearable clutch, X decreases and the predicted optimum clutch size increases as variation about the mean size of the maximum rearable clutch increases, either from one year to the next or from one territory or breeding site to another (ignoring variation among individuals in this model). Assuming the maximum rearable clutch to be a variable having a normal distribution with mean $\bar{C}$ and standard deviation s_c, we can calculate the expected relationship between survival and clutch size averaged over all years (figure 17.6).

Table 17.2   Constants X and a in the equation $S_3 = \exp(-aC^X)$ fitted to values of survival calculated in accordance with the maximum rearable clutch size model. Survival of young in clutches of the maximum rearable size was arbitrarily set at 0.95.

Clutch of maximum rearable size	Range of clutch size*	r	a	X	$\hat{C}$
1	1 – 2	1.000	0.0513	3.756	1.55
2	2 – 4	0.974	0.00414	3.844	2.94
3	3 – 6	0.963	0.00108	3.747	4.35
5	5 – 9	0.967	0.0000965	4.071	6.86
8	8 – 13	0.977	0.00000458	4.590	10.45

*Range used to calculate a and X.

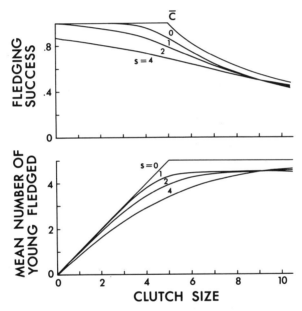

Figure 17.6   Relationship between survival and clutch size (above) and between production and clutch size (below) for a mean maximum rearable clutch of 5, with standard deviations of 0, 1, 2, and 4. To calculate survival, the percentage of years with each maximum rearable clutch is determined from a normal distribution. Survival for any clutch size is then the weighted mean of survival values for that clutch size for the various maximum rearable clutch sizes.

As variation increases, the average fledging success of small clutches decreases because the proportion of years having smaller maximum rearable clutches increases. As a result, the relationship between success and clutch size flattens out, X decreases, and $\hat{C}$ increases (table 17.3).

The relationship between success and clutch size can also be influenced by adding a penalty or premium to survival of young in broods above the maximum rearable size. The relationship represented by the dashed line in figure 17.7 was obtained by imposing a factor reducing survival by one-tenth for each egg laid over the maximum rearable. Thus if $\hat{C}$ were 5, the survival of young raised from clutches of 8 would be multiplied by 0.7. The addition of a penalty increases X and reduces $\hat{C}$ (table 17.4). By similar argument, placing a premium on large clutches decreases X and increases a for any maximum clutch specified.

Table 17.3 Constants X and a in the equation $S_3 = \exp(-aC^X)$ fitted to values of survival calculated in accordance with the maximum rearable clutch size model for a mean maximum rearable clutch size of 5 and standard deviations between 0 and 4.

Standard deviation of maximum rearable clutch size	r	a	X	$\hat{C}$
0	0.967	0.0000965	4.071	6.86
1	0.985	0.00243	2.621	6.88
2	0.997	0.0137	1.792	7.92
4	0.999	0.0756	1.005	12.99

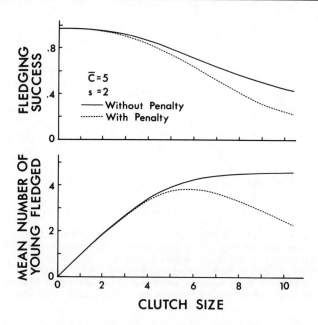

Figure 17.7 Relationship between survival and clutch size (above) for an average maximum rearable clutch size of 5 ± 2 (s.d.). Dashed line represents the addition of a penalty reducing survival by a factor of 0.1 for each egg in excess of the maximum rearable clutch size (see text). Lower portion of graph shows the relationship between productivity and clutch size with and without having a penalty applied.

Table 17.4 Constants X and a in the equation $S_3 = \exp(-aC^X)$ fitted to values of survival calculated in accordance with the maximum rearable clutch hypothesis when a penalty is applied to the survival of clutches larger than the maximum rearable clutch size is 5.

Standard deviation of maximum rearable clutch size	Penalty applied*	r	a	X	$\hat{C}$
0	No	0.967	0.0000965	4.071	6.86
0	Yes	0.957	0.0000232	5.035	6.04
2	No	0.997	0.0137	1.792	7.92
2	Yes	0.999	0.00772	2.280	5.87

*Penalty reduces survival by a factor of 0.1 for each egg in excess of the maximum rearable clutch size.

Discussion

Any factor that causes nesting success to decrease as clutch size increases, potentially could determine optimum clutch size, the optimum being defined as the clutch size that yields the greatest number of young. Only factors whose effect on survival varies with clutch size can influence the optimum. Other agents, acting independently of clutch size, are inconsequential, regardless of the mortality they cause. Hence the factor that determines optimum clutch size does not necessarily cause the greatest mortality.

According to the model developed in this chapter, optimum clutch size ($\hat{C}$) is determined by an equation of the form

$$1 = Xa\hat{C}^X + Yb\hat{C}^Y$$

where the constants X and a pertain to clutch size-dependent starvation or other mortality related to undernourishment, and Y and b pertain to clutch size-dependent predation (Skutch's hypothesis). (In equation (18), b absorbs both b and t (length of nest period) of equations (6) and (11).) The component of survival influenced by clutch size-dependent starvation (S_3) is equal to $\exp(-aC^X)$; the component of survival influenced by clutch size-dependent predation (S_4) is equal to $\exp(-bC^Y)$. Therefore, equation (18) may be rearranged to give

$$X \log S_3 + Y \log S_4 = -1$$

at optimum clutch size ($\hat{C}$). If optimum clutch size were not influenced by starvation ($S_3 = 1$ and $X = 0$), the proportion of young in broods of optimum size that escaped predation (S_4) would be $\exp(-1/Y)$. As we have seen above, either the level of mortality, the value of Y, or both must be unreasonably high to result in an optimum clutch size within the range observed in natural populations.

Although predation does not have the major role in determining optimum clutch size, clutch size-dependent predation will have some influence on optimum clutch size as long as $Y > 0$. For example, if Y were to equal 1 (predation increases in direct proportion to clutch size), S_4 could reasonably equal

0.90 for the optimum clutch size, and predation would exert about 10.5 per cent of the selective force acting on clutch size (equation (18)) the remainder being due to undernourishment.

It is difficult to attribute a greater role to predation in the light of data on nesting mortality and our understanding of predatory behaviour. If predators were attracted to nests by activity, clutch size-dependent predation would be nil during the incubation period and its magnitude would be revealed by the difference in mortality rate between nestling and incubation periods. In a survey of nesting mortality among passerines (Ricklefs, 1969), this difference was found to average 1.7 per cent for arctic species, -0.1 per cent for temperate species, -1.22 per cent for humid tropical species, and 1.4 per cent for arid tropical species. A portion of the excess nestling mortality could be due to starvation of young in the arid tropical sample and to death from exposure in the arctic sample. In any event, mortality due to clutch size-dependent predation probably does not amount to more than 1 per cent per day in any locality and may be nonexistent in the humid tropics, for which region Skutch (1949) proposed a major role for predation in the optimisation of clutch size,

If clutch size-dependent predation on the optimum clutch size does not exceed 1 per cent per day, predation could assume a major role in determining optimum clutch size only if Y were much larger than 1. The response of predators to the number of trips made by the parents to feed their young, which would increase in direct proportion to brood size if food were not a limiting factor, would not be likely to result in a value of Y greater than 1. Larger values of Y could be caused only by an increase in the frequency and intensity of begging as brood size increased. By virtue of the increased number of young begging, predation might be expected to increase in direct proportion to brood size. Therefore, parental feeding trips and begging cries together might enhance the ability of a predator to find a nest in proportion to the square of clutch size. For Y equal to 2, clutch size-dependent predation (m_4) equal to 0.01 per day and a nestling period (t) of 15 d, S_4 would be 0.86, the term YbC^Y in equation (18) would be 0.30, and hence predation would account for 30 per cent of the selective force on optimum clutch size.

Values of Y could exceed 2 if begging intensity increased out of proportion to the number of young in the brood. To be sure, undernourishment and competition for food among the brood would increase the commotion disproportionately in nests with large broods, but availability of food and the nutritional level of the young would then become part of the predation component of mortality. This component would more properly be placed with S_3 than with S_4. Predation probably constitutes less than 10 per cent, and certainly less than 30 per cent, of the selective force determining optimum clutch size.

With predation relegated to a minor role, food availability remains the principal factor determining optimum clutch size and causing clutch size to vary from one species to another. Optimum clutch sizes, calculated from the relationship of survival to brood size (table 17.1), are consistent with observed clutch sizes in most cases, although the selective force of undernourishment may not be felt fully until after the young have left the nest (for example, Great tit and Woodpigeon). Studies on the Herring gull (Paynter, 1949), Yellow-headed blackbird (Willson, 1966), and the English Blackbird in Wytham Wood (Snow, 1958) revealed no dependence of survival on clutch size. In these cases, selection on brood size may have been expressed later in the life cycle or the samples may not have been representative of the population as a whole. Negative evidence is not sufficient to reject the hypothesis that clutch size is determined primarily by the availability of food.

The relationship between survival and clutch size reveals the competitive relationship between sibs in a brood. As we saw earlier, if there is a maximum rearable clutch size below which no starvation occurs and above which productivity of the clutch levels off, X will be close to 4 for maximum rearable clutch sizes between 1 and 8 (table 17.2). Values of X and a calculated from field studies are compared to the maximum rearable clutch model in figure 17.8. In several studies (Western gull, terns, Woodpigeon, Great tit (nestling period), and European Blackbird), X lies far to the left of values predicted (from the model; in six studies (Manx shearwater, Heron (fledglings), Woodpigeon (fledglings), Great tit (fledglings), Common grackle and Boat-tailed grackle, X lies far to the right of predicted values.

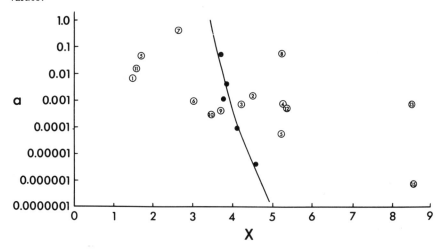

Figure 17.8 Relationship between a and X calculated from field observations on several species of birds (see table 17.1) compared with the relationship expected from the maximum rearable clutch model (solid line; see table 17.2). 1, Great tit; 2, Swift; 3, Heron (nestling); 4, Heron (fledgling); 5, Boat-tailed grackle; 6, Kestrel; 7, Black-footed albatross; 8, Manx shearwater; 9, Starling; 10, Pied flycatcher; 11, Blackbird; 12, Woodpigeon (nestling); 13, Woodpigeon (fledgling); 15, Common grackle.

Factors reducing X below that predicted from the maximum rearable clutch model include variability in the environment and premiums placed on large clutch size. Compared to other species listed in table 17.1, those with small values of X possibly experience relatively little variation in their environments. Large broods undoubtedly carry an energetic premium because of the increased insulation afforded by a large number of nestlings (Royama, 1966; Mertens, 1969). Even in doves, broods of two have less than twice the energy requirement of single nestlings (Brisbin, 1969). This effect occurs in precocious species as well (Osbaldiston,

1966, Kleiber and Dougherty, 1934). But because the effect should apply to all species, it cannot explain the occurrence of small values of X in a few species.

In all species having low values of X, except terns, predicted optimum clutch size is much greater than the observed modal clutch size. This suggests that low values of X are obtained when the effect of undernourishment has not yet been expressed in survival. Indeed, when optimum clutch size is calculated from survival through the fledgling stage, it corresponds closely to modal clutch size in the Great tit and the Woodpigeon. During the fledgling stage, X greatly exceeds values predicted by the maximum rearable clutch model.

If not expressed in survival, undernourishment presumably would be reflected in the weights of young produced by broods of different size. Fledging weights decrease on average with increasing brood size in most species that have been studied (for example, Lack, 1948; Perrins, 1965), and survival of young after they leave the nest has been found to be positively correlated with fledging weight in the Great tit (Lack, Gibb and Owen, 1957).

Values of X lying to the right of the maximum rearable clutch model probably represent the enforcement of a penalty on survival of young from large clutches. If young in excess of the maximum rearable clutch were not selectively starved, their competition with sibs for food would reduce the number of young eventually surviving to less than the maximum rearable clutch (see figure 17.7). Large values of X indicate the failure of selective starvation to reduce broods to the maximum rearable size at an early age.

Values of X correspond closely to the maximum rearable clutch model in four species (Swift, Kestrel, Pied flycatcher and Starling (fledgling)), in which the predicted and modal clutch sizes are also similar. The only common factor among these species is their hole-nesting habit, but the relevance of this trait to X is unclear, particularly as another hole nester, the Great tit, exhibits small X.

Regardless of how X is determined, the mortality caused by the factor or factors that determine clutch size can be calculated from

$$S_3{}^X S_4{}^Y = 1/e \tag{20}$$

derived from equation (19). If we assume Y to be 0 $(S_4{}^Y = 1)$ for a moment, we see that the level of mortality from undernourishment $(1 - S_3)$ decreases as X increases, from 63 per cent at $X = 1$ to 39 per cent at $X = 2$, 22 per cent at $X = 4$, and 11 per cent at $X = 9$. Mortality in broods larger than the optimum size would be greater for any given X. Even when clutch size-dependent predation accounts for half the selective force responsible for determining equilibrium clutch size $(S_4{}^Y \approx 0.6)$, starvation still must be an important source of martality, being 39 per cent at $X = 1$, 22 per cent at $X = 2$, 12 per cent at $X = 4$, and 5 per cent at $X = 9$.

Although retarded growth of young is frequently cited as evidence for undernourishment in field studies, starvation and other causes of death related to undernourishment are rarely cited in studies of nesting success. In six species of passerines for which nesting losses were tabulated by cause, starvation accounted for the deaths of only 2.2 per cent of all young (6.3 per cent of all losses) (Ricklefs, 1969). Evidently, death due to starvation is either not recognised or happens after fledging, or the model presented here does not adequately explain the observations. Considering how difficult it is to measure the effects of undernourishment, it does not seem unreasonable to attribute a mortality level of 20 per cent to this factor.

Because a few studies have not revealed any dependence of either survival or fledging weight on brood size, we should find it illuminating to estimate the degree to which optimum clutch size could be reduced below the maximum rearable size by a component of clutch size-dependent predation. If predation is responsible for proportion Z of the selective force on optimum clutch size, equation (18) may be written

$$XaC^X = 1 - Z \tag{21}$$

and

$$\hat{C} = \left(\frac{1 - Z}{aX}\right)^{1/X} \tag{22}$$

Predation would reduce optimum clutch size to a factor $(1 - Z)^{1/X}$ of its predicted size in the absence of clutch size-dependent predation. For X equal to 4, this factor is 0.97 for $Z = 0.1$, 0.95 for $Z = 0.2$, and 0.91 for $Z = 0.3$. It can be shown that clutch size-dependent predation will increase survival (S_3) by a factor, $\exp(Z/X)$, over its level in the absence of predation, hardly enough to affect our ability to detect any relationship between starvation and clutch size in field studies.

Two factors that could reduce the dependence of survival on clutch-size have not yet been considered. First, individual variation in clutch size may correspond to the ability of the individual to rear young (Perrins and Moss, 1975). If the correlation were perfect, all individuals would lay a clutch of optimum size, and starvation would attain similar levels in clutches of all sizes. The correlation between the size of a clutch and its optimum cannot, of course, be perfect, if only because the male's ability to gather food is not a part of the female's genotype. But the tendency of clutch size to increase with age and, presumably, with experience (Coulson and White, 1961; Kluijver, 1935; Richdale, 1957) and the ability of birds to adjust clutch size in accordance with prevailing feeding conditions (see Lack, 1954; Klomp, 1970), suggest that the correlation between the size of a clutch and the optimum may be sufficiently great as to obscure the dependence of mortality on clutch size. This problem can be circumvented in field studies by artificially adjusting the sizes of completed clutches (Hussell, 1972; Murton, Isaacson and Westwood, 1974).

A second factor that could obscure the relationship of mortality to clutch size would be the ability of adults to adjust their feeding rate to the size of their brood when the optimum foraging effort for the population is generally less than the maximum possible. In this case, variation in clutch size would be expressed in the survival of the adults rather than that of their young and it would thus be difficult to detect.

The complete elucidation of factors governing optimum clutch size may require measuring the relationship to clutch size of the rate of recruitment of young into the adult population and the survival of adults in studies in which the number of eggs in nests is artifically manipulated. Such studies would be practical in few species although weights of fledglings and adults may provide reasonable indices of probability of survival. In many of the studies referred to in this chapter, the dependence of mortality on clutch size is sufficient to account for modal clutch size, and undernourishment is shown to be the primary factor in the optimisation of clutch size. It would appear, then, that Lack's hypothesis pertains to most variation in clutch size among species of birds. But the factors governing the availability of food in the habitat and the optimum effort adults should expend to gather food remain a challenging problem for evolutionary ecologists.

Acknowledgements

I am grateful for helpful comments and criticism from Malcolm Coulter, Mercedes Foster, and Christopher M. Perrins. This study was supported in part by the US National Science Foundation (grant no. GB 42661) and a fellowship from the John Simon Guggenheim Foundation.

References

Brisbin, I. L., Jr (1960). Bioenergetics of the breeding cycle of the Ring Dove. *Auk*, 86, 54 – 74.

Cavé, A. (1968). The breeding of the Kestrel, *Falco tinnunculus* L., in the reclaimed area Oostelijk Flevoland. *Neth. J. Zool.*, 18, 313 – 407.

Charnov, E. L. and Krebs, J. R. (1974). On clutch-size and fitness. *Ibis*, 116, 217 – 219.

Cody, M. L. (1971). Ecological aspects of reproduction, in *Avian Biology* (eds. D. S. Farner and J. R. King), Academic Press, New York and London, 461 – 512.

Coulson, J. C. and White, E. (1961). An analysis of the factors influencing the clutch size of the Kittiwake. *Proc. Zool. Soc. Lond.*, 136, 207 – 217.

Foster, M. S. (1974a). A model to explain molt – breeding overlap and clutch size in some tropical birds. *Evolution*, 28, 182 – 190.

Foster, M. S. (1974b). Rain, feeding behavior, and clutch size in tropical birds. *Auk*, 91, 722 – 726.

Gadgil, M. and Bossert, W. H. (1970). Life historical consequences of natural selection. *Am. Nat.*, 104, 1 – 24.

Goodman, D. (1974). Natural selection and a cost ceiling on reproductive effort. *Am. Nat.*, 108, 247 – 268.

Haartman, L. von (1967). Clutch-size in the Pied Flycatcher. *Proc. 14th Int. Ornith. Congr. Oxford 1966*: 155 – 164.

Haartman, L. von (1971). Population dynamics, in *Avian Biology* (eds. D. S. Farner and J. R. King), Academic Press, New York and London, 391 – 459.

Harris, M. P. (1966). Breeding biology of the Manx Shearwater *Puffinus puffinus*. *Ibis*, 108, 17 – 33.

Hussell, D. J. T. (1972). Factors affecting clutch size in arctic passerines. *Ecol. Monogr.*, 42, 317 – 364.

Kleiber, M. and Dougherty, J. E. (1934). The influence of environmental temperature on the utilization of food energy in baby chicks. *J. gen. Physiol.*, 17, 701 – 726.

Klomp, H. (1970). The determination of clutch-size in birds. A review. *Ardea*, **58**, 1-124.

Kluijver, H. N. (1935). Waarnemingen over de levenswijze van den Spreeuw (*Sturnus v. vulgaris*) met behulp van geringde individuen. *Ardea*, **24**, 133-166.

Lack, D. (1947). The significance of clutch-size. I and II. *Ibis*, **89**, 302-352.

Lack, D. (1948). Natural selection and family size in the Starling. *Evolution*, **2**, 95-110.

Lack, D. (1949). Comments on Mr. Skutch's paper on clutch size. *Ibis*, **91**, 455-458.

Lack, D. (1954). *The Natural Regulation of Animal Numbers*, Clarendon Press, Oxford. 343 pp.

Lack, D. (1966). *Population Studies of Birds*, Clarendon Press, Oxford. 341 pp.

Lack, D., Gibb, J. and Owen, D. F. (1957). Survival in relation to brood-size in tits. *Proc. Zool. Soc. Lond.*, **128**, 313-326.

Lemmetyinen, R. (1973). Breeding success in *Sterna paradisea* Pontopp. and *S. hirundo* L. in southern Finland. *Ann. Zool. Fenn.*, **10**, 526-535.

Mebs, T. (1964). Zur Biologie und Populationsdynamik des Mäusebussards (*Buteo buteo*). *J. Ornith.*, **105**, 247-306.

Murton, R. K., Westwood, N. J. and Isacson, A. J. (1974). Factors affecting egg-weight, body-weight and moult of the Woodpigeon *Columba palumbus*. *Ibis*, **116**, 52-73.

Owen, D. F. (1960). The nesting success of the Heron *Ardea cinerea* in relation to the availability of food. *Proc. Zool. Soc. Lond.*, **133**, 597-617.

Osbaldiston, G. W. (1966). The response of the immature chicken to ambient temperature, in *Physiology of the Domestic Fowl* (eds. C. Horton-Smith and E. C. Amoroso) Oliver and Boyd, Edinburgh and London, 228-234.

Paynter, R. A., Jr (1949). Clutch-size and the egg and chick mortality of Kent Island Herring Gulls. *Ecology*, **30**, 146-166.

Perrins, C. M. (1964). Survival of young Swifts in relation to brood-size. *Nature*, **201**, 1147-1148.

Perrins, C. M. (1965). Population fluctuations and clutch-size in the Great Tit, *Parus major. J. Anim. Ecol.*, **34**, 601-647.

Perrins, C. M. and Moss, D. (1975). Reproductive rates in the Great tit. *J. Anim. Ecol.*, **44**, 695-706.

Pianka, E. R. and Parker, W. S. (1975). Age-specific reproductive tactics. *Am. Nat.*, **109**, 453-464.

Rice, D. W. and Kenyon, K. W. (1962). Breeding cycles and behavior of Laysan and Black-footed Albatrosses. *Auk*, **79**, 517-567.

Richdale, L. E. (1957). *A Population Study of Penguins*. Oxford University Press, London and New York, 195 pp.

Ricklefs, R. E. (1965). Brood reduction in the Curve-billed Thrasher. *Condor*, **67**, 505-510.

Ricklefs, R. E. (1969). An analysis of nesting mortality in birds. *Smithson. Contrib. Zool.*, **9**, 1-48.

Ricklefs, R. E. (1970). Clutch-size in birds: Outcome of opposing predator and prey adaptations. *Science*, **168**, 599-600.

Ricklefs, R. E. (1977). On the evolution of reproductive strategies in birds: reproductive effort. *Am. Nat.*, in press.

Royama, T. (1966). Factors governing feeding rate, food requirements and brood size of nestling Great Tits *Parus major*. *Ibis*, **108**, 313 - 347.

Skutch, A. F. (1949). Do tropical birds rear as many young as they can nourish. *Ibis*, **91**, 430 - 455.

Skutch, A. F. (1967). Adaptive limitation of the reproductive rate of birds. *Ibis*, **109**, 579 - 599.

Snow, D. W. (1958). The breeding of the Blackbird *Turdus merula* at Oxford. *Ibis*, **100**, 1 - 30.

Snow, B. K. (1970). A field study of the Bearded Bellbird in Trinidad. *Ibis*, **112**, 299 - 329.

Tutor, B. M. (1962). Nesting studies of the Boat-tailed Grackle. *Auk*, **79**, 77 - 84.

Wiens, J. A. (1966). On group selection and Wynne-Edwards' hypothesis. *Am. Scient.*, **54**, 273 - 287.

Williams, G. C. (1966a). Natural selection, the costs of reproduction, and a refinement of Lack's principle. *Am. Nat.*, **100**, 687 - 692.

Williams, G. C. (1966b). *Adaptation and Natural Selection.* Princeton University Press, Princeton, New Jersey.

Willson, M. F. (1966). Breeding ecology of the Yellow-headed Blackbird. *Ecol. Monogr.*, **36**, 51 - 77.

Willson, M. F., St John, R. D., Lederer, R. J. and Muzas, S. J. (1971). Clutch size in grackles. *Bird Banding*, **42**, 28 - 35.

Wynne-Edwards, V. C. (1962). *Animal Dispersion in Relation to Social Behaviour*. Oliver and Boyd, Edinburgh and London, 653 pp.

Wynne-Edwards, V. C. (1963). Intergroup selection in the evolution of social systems. *Nature*, **200**, 623 - 626.

18   Clutch size in the Compositae

Professors D. A. Levin and B. L. Turner*

The number and sizes of seeds produced by plants of different species have
exercised the minds of plant evolutionists and ecologists for decades. Seed size
is relatively constant within species but may vary greatly between species
(Salisbury, 1942; Harper, Lovell and Moore 1970). Seed number is subject to
very great phenotypic modification, but mean seed number may vary substantially
between species. The importance of seed size and numbers in relation to seedling
survival and to the ability of populations to replace themselves in time and space
was first realised by Salisbury (1942) and amplified by Harper et al., (1970),
Janzen (1971), Baker (1972) and Levin (1974).

Little attention has been given to the number of ovules per flower, in spite
of the fact that this number is relatively constant and often used by systematists
as a diagnostic character. The number of ovules per flower or per head in the
Compositae may be referred to as clutch size in plants, in that the development of
a seed cluster from these ovules represents a discrete reproductive episode and
energy commitment in time and space (Johnson and Cook, 1968). Clutch size,
like seed size and number per plant and other reproductive components of a life
history strategy, will be shaped by natural selection to produce the highest individual fitness. Clutch size reflects a compromise between many selection pressures
acting on the reproductive and vegetative features of plants (Harper and White,
1974). It is a manifestation of how plants package energy allocated to sexual
reproduction. However, it provides no information about how the plant is budgeting its total reproductive energy allocation or how many flowers a plant is producing throughout the growing season. The latter is a function of resource availability
or habitat hospitality. The number of ovules per flower (that is, clutch size) is well
buffered against these variables. There is a relatively constant allocation of resources per flower or head.

The tendency for clutch sizes in birds (Lack, 1947, 1948; Lack and Moreau,
1965; Moreau, 1944; Cody, 1966; Skutch, 1967), lizards (Tinkle, Wilbur and
Tilley, 1970), and mammals (Lord, 1960) to decrease with latitude is well
documented. In temperate mountainous areas some groups of bird species at
high altitudes have larger clutches than those in adjacent lowlands (Cody, 1971).

*Professor D. A. Levin received his Ph.D. from the University of Illinois in 1964. After
staff appointments at Yale University and the University of Illinois at Chicago Circle, he
was appointed Professor of Botany at the University of Texas in 1972. His current research
interests include evolutionary and ecological plant genetics, adaptive strategies in plants,
and plant – animal co-evolution.

*Professor B. L. Turner received his Ph.D. from Washington State University in 1953; on
the faculty of the University of Texas since that date, he was appointed Professor of Botany
in 1960. His research interests include biochemical systematics, chromosome numbers in
the Compositae and Leguminosae, vegetation of the world in relation to climate (especially
that of deserts), gypsum plants in North America and numerical systematics.

It is of interest to know whether similar patterns occur in plants, and whether clutch size in plants is a function of plant growth form or size.

The existence of these relationships were sought in the tribe Heliantheae of the family Compositae. The tribe is composed of perhaps 2500 species distributed among approximately 150 genera (Turner, 1977). It is a New World tribe with centres of diversity in the lower montane tropical and subtropical regions. It contains many tropical genera, some of which extend into alpine regions on the one hand, and remote Pacific Islands on the other. The tribe contains annual and perennial herbs and shrubs. The number of ovulate flowers per head varies widely within and among genera. Each ovulate flower contains one ovule.

Materials and Methods

Our source of material was the herbarium at the University of Texas, Austin. Several herbarium specimens of 1007 species in the tribe Heliantheae were studied. Each of the species was scored for the average number of ovules per head, geographical distribution and growth form. The numbers of genera and species considered in each of the subtribes of the Heliantheae are as follows: Lagascinae 2,11; Millerinae, 10,35; Melampodinae, 21, 162; Ambrosinae, 4,21; Petrobinae, 1,2; Zinninae, 5,41; Verbesininae, 37,420; Coreopsidinae, 13,165; Galinsoginae, 6,84; Madiinae, 8,66.

To determine whether differences between clutch sizes of tropical, temperate, and alpine plants, and between annual herbs, perennial herbs, and shrubs were statistically significant, and whether geographical area and growth form were independent variables, we used a model II analysis of variance for unequal sample sizes using the method of fitting constants as described by Steel and Torrie (1960, p. 257).

Mature seeds were obtained from the ray flowers of 87 species. Mean weight was determined for each species. In view of the small sample size, our analysis is confined to pairwise comparisons of the mean seed weights of shrubs, perennial herbs and annual herbs using a t-test.

Results

Clutch size among species of Heliantheae covers a broad spectrum, varying from 1 to over 500. Heterogeneity resides between genera as well as within them. In several genera clutch size is constant, the diagnostic number being 1 in *Clibadium, Lagascea* and *Icthyothere*, 5 in *Parthenium*, 8 in *Engelmannia* and 10 in *Berlandiera*. In genera where there is variation the number may be small or large. For example, in *Iva* clutch size varies from 1 to 7, in *Zinnia* from 10 to 60, and in *Espeletia* from 17 to 300.

If we treat all species of the Heliantheae collectively, it is evident that clutch size displays a pattern in space as well as with respect to growth form (table 18.1). The mean clutch size of tropical species is 51.2 versus 82.7 and 89.6 in temperate and alpine species, respectively. The differences between the tropical assemblage and the other two are statistically significant ($F_s = 23.01; P < 0.01$). The mean clutch size of perennial herbs is 78.14 as compared with 48.62 in annual herbs and 51.89 in shrubs and trees. The differences between the *perennial herbs* versus *other growth* forms is statistically significant ($F_s = 17.70$;

Table 18.1 Mean clutch size as a function of geographical region and growth form. Sample sizes in parentheses.

Growth form	Region			
	Tropics	Temperate	Alpine	Form mean
Annual herb	39.39 (163)	62.25 (106)	110.0 (1)	48.63
Perennial herb	61.31 (251)	99.03 (161)	103.53 (34)	78.14
Woody	49.88 (249)	69.42 (33)	44.91 (11)	51.89
Region mean	51.63	82.77	89.65	62.62

$P < 0.01$). The effect of region is independent of the effect of growth form on clutch size, that is, there is no interaction between these variables ($F_s = 1.8$).

The divergent clutch sizes of *herbaceous* versus *woody* plants evident at the tribal level also are evident within genera. For example, in *Espeletia* the mean clutch size in shrubs is 37.4 versus 108.0 in herbs ($t_s = 3.31; P < 0.01$). In *Bidens,* the mean size in shrubs is 29.4 versus 71.7 in herbs ($t_s = 3.69; P < 0.01$); and in *Verbesina* the value for shrubs is 53.2 versus 81.7 in herbs ($t_s = 3.88; P < 0.001$). We do not mean to infer that all genera with both forms show this disparity. However, it is notable that in none of the genera studied did the clutch size of shrubs exceed that of perennial herbs.

In broadly distributed genera, we found no evidence of significant clutch size differentials between tropical and temperate species.

Since each ovule may develop into a seed subsequent to fertilisation, it is important to consider the weight of single seeds and the total seed weight per head. There is considerable heterogeneity in seed weight among species of the Heliantheae, the values ranging from 0.2 to 24 mg. In general seed weights vary as a function of growth form but not of geography. The mean weight is 1.68 mg in annuals, 4.45 mg in perennial herbs, and 11.7 mg in shrubs. The differences between the shrubs and perennial herbs ($t_s = 3.17; P < 0.01$) and annuals ($t_s = 9.72; P < 0.01$) are significant, as is the difference between the two herb classes ($t_s = 2.71; P < 0.01$). The mean weight for tropical species is 5.7 mg versus 4.8 mg in temperate species. The difference is not statistically significant.

We had expected an inverse correlation between seed weight and clutch size, but a statistically significant one was not forthcoming, neither among all species collectively nor among species of each growth form. The zones in which the seed weights and clutch sizes of species fall within a two-dimensional plot are shown in a somewhat idealised form in figure 18.1. Although the reproductive spectra overlap, each growth form does have a distinctive position in the diagram. Shrubs tend to have relatively heavy but few seeds, compared with the herbs.

For each growth form, the representative amount of energy invested per head exclusive of accessory parts (e.g., bracts) may be determined by multiplying the mean clutch size by the mean seed weight. The head weight for average species of shrubs, perennial herbs and annual herbs are 560, 350 and 90 mg, respectively. The representative investment per head in temperate and tropical species also may be computed from clutch size and seed weight data. The head weight of an average tropical species is approximately 300 versus 400 mg for the average temperate species.

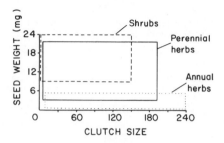

Figure 18.1 The zones of distribution of clutch size and seed weight in species of annual herbs, perennial herbs and shrubs in a somewhat idealised form.

Discussion

The diversity in the number and kinds of reproductive structures in the Compositae is well understood. Almost invariably, adaptations relating to successful pollination have been viewed as independent of those relating to successful seed development. However, as noted by Burtt (1976), the architecture of the head must meet the demands of the flowering and fruiting phases of the reproductive cycle. The features of flowering and fruiting phases must be the product of co-evolution.

The quality of flowers and seeds as resources depends in part on their proximity or density, since the more energy the pollinator or herbivore must devote to obtaining a given calorific intake, the less will be the net energy gain (Schoener, 1971; Wilbur, 1977). It is not surprising then that the rate of pollination (Free, 1970; Heinrich and Raven, 1972; Willson and Ratheke, 1975; Heinrich, 1975) and of seed infestation (Janzen, 1970; Elliott, 1974; Vandermeer, 1975) are functions of the proximity of flowers and seeds. The relationship ostensibly holds for flowers and seeds clustered in the heads of composites. The heads of most species of Compositae are exploited by many classes of pollinators including beetles, flies, bees, butterflies, moths and wasps (Leppik, 1970). The pollinators respond to the nutritional reward of the head as a whole, as each flower offers little reward (Heinrich and Raven, 1972). The heads of many species also are exploited by various species of phytophagous insects, whose larvae feed on the rich resource afforded by pollen and developing seeds (Burtt, 1961; 1977). Diptera and Coleoptera are relatively inactive, tending to eat some or all developing seeds within a head, but do not move between heads. From the foregoing, we may conclude that clutch size in the Compositae, in large measure is a compromise between selection for optimising pollinator service and seed-set, and selection for minimising predation pressure by herbivores.

There is anecdoctal evidence that inactive seed predators (insect larvae) are richer in numbers and species, and are more specialised in the tropics than in temperate regions (Ehrlich and Raven, 1964; Schoener, 1971; Janzen, 1973a; 1973b). Thus the potential damage to ovules and developing seeds of species with large clutch sizes may be much greater in evolutionary and ecological time in the tropics, all other factors being equal. This suggests that the difference in mean clutch sizes of temperate and tropical Heliantheae may be due to greater pest pressure in the tropics. Reducing the number of ovules and their developing seed per head, and distributing these resources more uniformly in space (that is, on the plant) would afford a means of escape. This may be accomplished by reducing the total number of flowers in the head, or by converting hermaphroditic flowers into flowers which produce pollen but not ovules. The latter avenue of escape may have no impact on the quality of heads as resources for pollinators. The former would probably require increasing the nutritional reward of single flowers (Heinrich and Raven, 1972; Heinrich, 1975). Both forms of clutch size reduction have occurred in the evolutionary history of tribes and genera in the Compositae (Cronquist, 1955; Stebbins, 1967; Burtt, 1977).

Greater predator pressure also may explain why the clutch size of shrubs is smaller than that of perennial herbs. It seems likely that shrubs are more predictable resources for phytophagous insects than are perennial herbs, since individuals are longer lived and populations occur in more stable habitats (Harper and White, 1974). In shrubs selection for small energy commitment per clutch has been confounded by selection for heavier seeds which are necessary for seedling establishment in environments with low illumination and high competition (Carlquist, 1966; Harper *et al.*, 1970; Baker, 1972). The seed weight per head in the average shrub is 560 mg compared with 350 mg in the average perennial herb. Although clutch size in annuals may be influenced by biotic pressures, it seems likely that the unpredictable energy status and survivorship of annuals relative to perennials favour reduced clutch size. The rationale for our hypothesis is as follows. The energy which can be mobilised to fill developing seeds during a specific timespan in a specific part of the plant probably is much less predictable in annuals than in perennials, since annuals lack an energy reserve from which to draw (Kozlowski and Keller, 1966; Mooney, 1972). Secondly, in plants with a short life expectancy, there is a selective premium on reproducing when reserves have reached a low threshold rather than on storing reserves for larger reproductive commitments later in the growing season. Also in plants with a short life expectancy, rapid maturation and dispersal of seeds is selectively advantageous. Johnson and Cook (1968) also argue that time and the availability of resources determine clutch size in buttercups.

Not only is the clutch size of annuals smaller than that of perennials, but their seeds tend to be smaller. The mean of annual seed is 1.68 versus 4.45 mg in perennials. The smaller seeds of annuals probably tend to increase dispersability (Salisbury, 1942; Harper *et al.*, 1970; Baker, 1972). It is important to recognise that seed weight per head in the average annual is only 90 mg whereas in the average perennial herbs it is 350 mg.

Although seed weight per head in annuals could be the product of independent selection for clutch size and for seed size, both characters may be the byproduct of selection for heads which encumber little energy. As noted by Stebbins (1967), the weight of the seed-bearing head must be congruent with the architecture of the plant as a whole. The slender branches and relatively weak vascular system of many small annuals could not support, in a structural and physiological sense, anything but small clutches.

We do not wish to leave the impression that small clutch size is associated with small reproductive effort, that is the proportion of the total energy budget of a plant that is devoted to reproductive processes. Although we have not measured reproductive effort, there is sufficient information in the literature to permit some general statements. Harper *et al.* (1970) points out that annual herbs, perennial herbs, and woody plants allocate a progressively smaller fraction of their resources to reproduction. Reproductive effort in annual species of Compositae (Harper and Ogden, 1970; Gaines *et al.*, 1974; Powell, personal communication) average roughly 25 per cent as compared to 15 per cent in perennial herbs (Gaines *et al.*, 1974; Ogden, 1974; Abrahamson and Gadgil, 1973). Estimates for woody composites are not available; reproductive effort probably is less than 10 per cent (Mooney, 1972). These general patterns suggest that small clutch size notwithstanding, annual species of the Heliantheae will have, on average, larger reproductive efforts than perennial herbs. Small clutch size in shrubs will be associated with relatively small reproductive risks in time.

Heliantheae and other composite tribes evolved in semi-arid conditions in warm-temperate regions, and gave rise to derivatives that invaded the tropics and evolved small clutch size (Bentham, 1873; Cronquist, 1955; Stebbins, 1967; Leppick, 1970; Turner, 1976; Burtt, 1977). However, some genera of the moist tropics have spread into montane regions of the tropics or into temperate regions, and in doing so some have increased their clutch size. A prime example is the genus *Espeletia*. The genus is composed of 100 species, all of which are endemic to the northern Andes of South America. The lowland tropical species, *E. nerifolia*, is believed to be the most primitive member of the genus. This species, or something similar to it, is believed to have given rise to more specialised species which occupy essentially alpine regions of the tropical Andes (Smith and Koch, 1935). *E. nerifolia* has female sterile disc flowers and an average of 17 fertile ray flowers. Its derivatives have more conspicuous inflorescences and greater seed production by virtue of more ray flowers. Some bizarre species have 250 or more ray flowers. The disc flowers have remained sterile. Another method of increasing clutch size has been used in genera where there are one or fewer flowers per head. This is to aggregate heads into compound units, each of which has the superficial appearance of being a single head. Compound heads occur in one genus within the Heliantheae, and in several genera that are not members of this tribe (Cronquist, 1955; Burtt, 1961; Leppik, 1970).

In this study we have demonstrated clutch size differences that are related to plant geography and growth form. These differences ostensibly are shaped by biotic selective agents, but also may be by-products of general physiological and morphological adaptations. There is no reason to assume that the patterns observed in the Heliantheae are unique to that tribe or the Compositae as a whole. We hope that our study will stimulate analyses of clutch size in other families within ecogeographical contexts, as clutch size is a central reproductive adaptation which has been ignored by plant evolutionists.

Acknowledgements

We are indebted to R. E. Ricklefs and B. L. Burtt for their comments on an earlier version of this paper and to A. Temperton for statistical assistance. This study was supported in part by US National Science Foundation grant BMS – 73 06687, to D. A. Levin.

References

Abrahamson, W. G. and Gadgil, M. (1973). Growth form and reproductive effort in goldenrods (Solidago, Compositae). *Am. Nat.,* **107**, 651 - 661.

Baker, H. G. (1972). Seed weight in relation to environmental conditions in California. *Ecology,* **53** 997 - 1110.

Bentham, G. (1873). Notes on the classification, history and geographical distribution of Compositae. *J. Linnean Soc. Lond.,* **13**, 335 - 577.

Burtt, B. L. (1961). Compositae and the study of functional evolution. *Trans. Bot. Soc. Edinb.,* **39**, 216 - 232.

Burtt, B. L. (1977). Aspects of diversification in the capitulum, in *Biology and Chemistry of the Compositae* (ed. V. H. Heywood) Academic Press, New York, in press.

Carlquist, S. (1966). The biota of long-distance dispersal. II. Loss of dispersability in the Pacific Compositae. *Evolution,* **20**, 30 - 48.

Cody, M. L. (1966). A general theory of clutch size. *Evolution,* **20**, 174 - 184.

Cody, M. L. (1971). Ecological aspects of reproduction, in *Avian Biology,* (ed. D. S. Farner, J. R. King and K. C. Parkes) Academic Press, New York.

Conquist, A. (1955). Phylogeny and taxonomy of the Compositae. *Am. midl. Nat.,* **53**, 478 - 511.

Ehrlich, P. R., and Raven, P. H. (1964). Butterflies and plants: a study in coevolution. *Evolution,* **18**, 586 - 608.

Elliott, P. F. (1974). Evolutionary responses of plants to seed-eaters: pinesquirrel predation on lodgepole pine. *Evolution,* **28**, 221 - 231.

Free, J. B. (1970). Insect Pollination of Crops. Academic Press, New York.

Gaines, M. S., Vogt, K. J. Hamrick, J. L. and Caldwell, J. (1974). Reproductive strategies and growth patterns in sunflowers (Helianthus). *Am. Nat.* **108**, 889 - 894.

Harper, J. L., Lovell, P. H. and K. G. Moore. (1970). The shapes and sizes of seeds. *A. Rev. Ecol. Syst.,* **2**, 465 - 492.

Harper, J. L. and Ogden, J. (1970). The reproductive strategy of higher plants. I. The concept of strategy with reference to *Senecio vulgaris* L. *J. Ecol,* **58**, 681 - 698.

Harper, J. L. and White, J. (1974). The demography of plants. *A. Rev. Ecol. Syst.* **5**, 419 - 463.

Heinrich, B. (1975). Energetics of pollination. *A. Rev. Ecol. Syst.,* **6**, 139 - 170.

Heinrich, B. and Raven, P. H. (1972). Energetics and pollination. *Science,* **176**, 597 - 602.

Janzen, D. H. (1970). Herbivores and the number of tree species in tropical forests. *Am. Nat.,* **104**, 501 - 528.

Janzen, D. H. (1971). Seed predation by animals. *An. Rev. Ecol. Syst.,* **2**, 465 - 492.

Janzen, D. H. (1973a). Comments on host-specificity of tropical herbivores and its relevance to species richness, in *Taxonomy and Ecology* (ed. V. H. Heywood) Academic Press, New York.

Janzen, D. H. (1973b). Tropical agroecosystems. *Science,* **182**, 1212 - 1219.

Johnson, M. P., and Cook, S. A. (1968). 'Clutch size' in buttercups. *Am. Nat.,* **102**, 405 - 411.

Kozlowski, T. T. and Keller, T. (1966). Food relations of woody plants. *Bot. Rev.* **32**, 293 – 382.

Lack, D. (1947). The significance of clutch size, Parts. I and II. *Ibis,* **89**, 302 – 352.

Lack, D. (1948). The significance of clutch size, Part III. *Ibis*, **90**, 25 – 43.

Lack, D. (1966). *Population Studies of Birds.* Clarendon Press, Oxford.

Lack, D. and Moreau R. E. (1965). Clutch size in tropical passerine birds of forest and savanna. *Oiseau Suppl.*, **35**, 75 – 89.

Leppik, E. E. (1970). Evolutionary differentiation of the flower head of the Compositae. II. *Ann. Bot. Fenn.*, **7**, 325 – 352.

Levin, D. A. (1974). The oil content of seeds: an ecological perspective. *Am. Nat.*, **108**, 193 – 206.

Lord, R. D. (1960). Litter size and latitude in North American mammals. *Am. Midl. Nat.*, **64**, 488 – 499.

Mooney, H. A. (1972). The carbon balance of plants. *A. Rev. Ecol. Syst.* **3**, 315 – 346.

Moreau, R. E. (1944). Clutch size: a comparative study with reference to African birds. *Ibis*, **86**, 282 – 347.

Ogden, J. (1974). The reproductive strategy of higher plants. II. The reproductive strategy of *Tussilago farfara* L. *J. Ecol.*, **62**, 219 – 324.

Salisbury, E. J. (1942). *The Reproductive Capacity of Plants.* Bell. London.

Schoener, T. W. (1971). Theory of feeding strategies. *A. Rev. Ecol. Syst.*, **2** 369 – 404.

Schoener, T. W. (1974). The compression hypothesis and temporal resource partitioning. *Proc. natn. Acad. Sci. U.S.A.,* **71**, 4169 – 4172.

Skutch, A. F. (1967). Adaptive limitation of the reproductive rate of birds. *Ibis*, **109**, 1 – 16.

Smith, A. C., and Koch, M. K. (1935). The genus Espeletia: a study in phylogenetic taxonomy. *Brittonia,* **1**, 479 – 543.

Stebbins, G. L. (1967). Adaptive radiation and trends of evolution in higher plants, in *Evolutionary Biology Vol. 1.* (ed. Th. Dobzhansky, M. K. Hecht and W. C. Steere), Appleton – Century – Croft, New York.

Steel, R. G. D. and Torrie J. H. (1960). Principles and Procedures of Statistics. McGraw-Hill, New York.

Tinkle, D. W., Wilbur, H. M. and Tilley, S. G. (1970). Evolutionary strategies in lizard reproduction. *Evolution*, **24**, 55 – 74.

Turner, B. L. (1977). Fossil history and geography, in *Biology and Chemistry of the Compositae,* (ed. V. H. Heywood) Academic Press, New York.

Vandermeer, J. H. (1975). A graphical model of insect seed predation. *Am. Nat.* **109**, 147 – 160.

Wilbur, H. M. (in press). Propagule size, number and dispersion pattern in Ambystoma and Asclepias. *Am. Nat.*

Wilson, M. F. and Rathcke, B. J. (1974). Adaptive design of the floral display in *Asclepias syriaca* L. *Am. Midl. Nat.*, **92**, 47 – 57.

Section 4 Behaviour, adaptation and taxonomic relationships

This section draws together four chapters covering various aspects of evolutionary adaptation. Lord Medway and Professor J. David Pye take up the problem of taxonomic relationships among the Indo-Pacific swiftlets; where Lack drew attention to nest building and nesting behaviour as useful taxonomic indicators in this group, these authors investigate both the ecological and the taxonomic significance of echolocation. Dr David Snow, whose research was for many years based at the Edward Grey Institute, examines the courtship patterns of manakins of the genus *Chiroxiphia* and discusses the evolution of their communal displays. In Chapter 21 Dr Amotz Zahavi tackles the paradox of sexual adornments, which are derived and perpetuated by sexual selection but confer obvious disadvantages in the day-to-day life of their wearers. Finally Dr William and Valentine Schaffer discuss the evolution of semelparity or monocarpy – the strategy of 'big-bang' reproduction – with special reference to desert yuccas and agaves. Their acknowledgement would have given particular pleasure to David Lack, who was a keen amateur botanist and also enjoyed seeing the spread of his basically ornithological ideas into other branches of natural history.

19 Echolocation and the systematics of swiftlets

Lord Medway and Professor J. D. Pye*

During his period of interest in swifts, David Lack drew attention to the impor-
tance of nesting behaviour and nest type as guides to the systematics of the
Apodidae (Lack, 1956). This approach has been particularly fruitful among the
Indo-Pacific assemblage known colloquially as 'swiftlets', recognised as a tribe
Collocaliini by Brooke (1970) and as a subfamily Collocaliinae by Condon (1975).
Among these swifts, in several instances, the type of nest built has proved to be
critical in specific determination. Another characteristic exhibited only by the
living bird that has claimed attention is the capacity of some (but not all) swiftlets
to orientate in darkness by echolocation. Because the component frequencies of
the click-like orientation sound fall largely within the range audible to man, it is
detectable without instruments in field conditions. As a consequence, the distribution of this capacity among the swiftlets is now reasonably well known. Together
with those included in this chapter, sound spectrograms have been published of
the orientation clicks of six taxa, representing four species. Tests of the ability to
detect and avoid obstacles in darkness have been made on three of these taxa. Data
are thus available for evaluating the importance of echolocation in the ecology of
these birds, and for considering its significance as a taxonomic character.

The recordings described for the first time here were obtained by Medway as
follows: those in Java were made on an EMI RE 321 tape recorder at $7\frac{1}{2}$ inches
(19 cm) per s with an STC 4032 G microphone (a combination found in preliminary tests to have a satisfactorily even frequency response between 0.15 and
10.0 kHz), in a recording chamber described by Medway (1967); those in Borneo
were made on an EMI L 2B tape recorder at $7\frac{1}{2}$ inches (19 cm) per s with a
Philips 9564/10 microphone, in the dark-room of the Sarawak Museum and in a

*After graduating from Cambridge, Lord Medway joined the Sarawak Museum (1956–58);
archaeological duties (especially at Niah cave) provided the initial opportunity to study the
cave-dwelling swiftlets. Subsequent research was encouraged under the supervision of Lord
Zuckerman at Birmingham, and later by a fellowship from the Yayasan Siswa Lokantara,
Indonesia (1960–61). Service with the University of Malaya (1961–70) provided further
opportunities. Now settled in Britain, he has still been able to keep field contact with this
fascinating group of birds on sporadic expeditions to the Indo–Pacific region. Since 1973
he has also been editor of *Ibis*.

*David Pye studied Zoology at Aberystwyth and Bedford College, London. He is
especially interested in the acoustic mechanisms of echolocation and in ultrasonic communication by animals. In his current research he uses low power microwave radar to measure the
motions of flying bats while recording them in the field. He and Lord Medway first met as
students on a field course. He is now Professor of Zoology at Queen Mary College, London.

Table 19.1 Provisional classification of the species of swiftlets, incorporating details of the type of nest and the ability to echolocate, as verified by experimentation or by the utterance of the distinctive click-like orientation sounds.

Species	Nest type[1]	Representative taxa tested	Ability	Echolocation Confirmatory reference
Giant swiftlet *Hydrochous gigas* (Hartert and Butler)	Externally supported; vegetable materials; sparse, moist nest cement[11]	*gigas*	No	Medway and Wells (1969)
Glossy swiftlet *Collocalia esculenta* (Linn.)	Self-supporting, bracket-shaped; vegetable materials; sparse, firm nest cement	*esculenta* *linchi* Horsfield and Moore *cyanoptila* Oberholser *uropygialis* Gray	No No No[4] No	Fenton (1975) Medway (1967) Cranbrook and Medway (1965) Medway (unpublished)
Pygmy swiftlet (*C.*) *troglodytes* Gray[5]	Self-supporting, bracket-shaped; vegetable materials; sparse, firm nest cement	—	Unknown	—
Mascarene swiftlet *Aerodramus francicus* (Gmelin)	Self-supporting, bracket-shaped; vegetable materials; sparse, firm nest cement	*francicus*	Yes	S. Temple; A. S. Cheke; J. F. M. Horne (all *in litt.*)
Seychelles swiftlet *A.* (*francicus*) *elaphrus* (Oberholser)	Self-supporting, bracket-shaped; vegetable materials; sparse, firm nest cement[6]	*elaphrus*	Yes	Penny (1975)
Indian edible-nest swiftlet *A.* *unicolor* (Jerdon)	Self-supporting, bracket-shaped; vegetable materials; sparse to moderate, firm nest cement	*unicolor*	Yes	Novick (1959)[7]
White-rumped swiftlet *A. spodiopygius* (Peale)[3]	Self-supporting, bracket-shaped; vegetable materials, copious, firm nest cement	*terraereginae* Ramsay	Yes	Pecotich (1974)
Himalayan swiftlet *A. brevirostris* (McClelland)[9]	Self-supporting, bracket-shaped; vegetable materials; sparse, firm nest cement	*vulcanorum* Stresemann	Yes	Medway (1962b)
Mossy-nest (Grey) swiftlet *A. vanikorensis* (Quoy and Gaimard)[10]	Externally supported; vegetable materials; sparse, moist nest cement	*vanikorensis* *granti* Mayr *natunae* Stresemann *salangana* Streubel[8]	Yes Yes Yes Yes	Medway (1975) Griffin and Suthers (1970) This chapter This chapter
Mountain swiftlet *A. hirundinaceus* (Stresemann)	Externally supported; vegetable materials; little or no nest cement[11]	*hirundinaceus*	Yes	Fenton (1975)

Species	Nest type[1]	Representative taxa tested	Ability	Confirmatory reference
			Echolocation	
Atiu swiftlet A. (leucophaeus) sawtelli[12] (Holyoak)	Externally supported; vegetable materials; sparse, sticky nest cement	*sawtelli*	Yes	Holyoak (1974)
Polynesian swiftlet A. leucophaeus (Peale)	Self-supporting or externally supported; vegetable materials; sparse, sticky nest cement[13]	*ocistus* (Oberholser)	Yes	Holyoak and J. C. Thibault, MS (correcting Holyoak, 1974)
Edible-nest swiftlet A. fuciphagus (Thunberg)	Self-supporting, bracket-shaped; pure, firm nest cement	*fuciphagus* *vestita* Lesson *amechana* Oberholser *germani* Oustalet *perplexa* Riley	Yes Yes Yes Yes Yes	This chapter Medway (1966) Medway (1966) Medway (1966) Medway (1966)
Black-nest swiftlet A. maximus (McClelland)	Self-supporting; bracket-shaped; feathers and copious firm nest cement	*maximus* *lowi* Sharpe	Yes Yes	Medway (unpublished) This chapter
Three-toed swiftlet (A.) papuensis (Rand)[14]	Unknown	—	Unknown	—

1 Unless indicated otherwise, nest descriptions are summarised from Medway (1966).

2 See Becking (1971).

3 Dr S. Somadikarta has drawn our attention to the fact that *unalashkensis* (*Hirundo unalashkensis* Gmelin, 1789) is a senior subjective synonym of *Collocalia spodiopygia townsendi* Oberholser, 1906, and has priority as specific name (see Stresemann, 1950, *Auk* **67**, 74). However, this name has not been used for 50 years or more and its revival is not recommended (*cf.* Lysaght, 1959, *Bull. Br. Mus.* (*nat. Hist.*) Ser. 1 (6), 339).

4 Cranbrook and Medway (1965) also established that there are no inaudible ultrasonic sounds emitted in flight in darkness.

5 The generic position of *troglodytes* follows Brooke (1970; 1972), but must be regarded as tentative until its capacity to echolocate is investigated.

6 See Procter (1972).

7 Under the name *Collocalia brevirostris unicolor*.

8 *Aerodramus* is masculine in gender, and adjectival specific names take the appropriate termination. The species name *salangana*, however, is a noun in opposition and therefore remains unaltered.

9 *A. brevirostris* is taken to include the taxa *whiteheadi*, *nuditarsus* and *orientalis*, each considered to be a separate species by Somadikarta (1967).

10 Including the *inquieta* group, placed as a separate species by Medway (1966).

11 Published descriptions are modified by a report by R. Schodde, I. S. Mason and W. Boles (MS), who describe a single nest made of vegetable materials 'applied thinly with saliva paste about the rim and down the hinge where the nest is attached to the rock' (R. Schodde, *in litt.* 1976).

12 Taxonomic position proposed by D. T. Holyoak and J. C. Thibault (MS).

13 Published descriptions are modified by Holyoak and Thibault (MS).

14 See Somadikarta (1967).

cave mouth in Bau district, Sarawak (Medway, 1959). Sonagrams of both sets of recordings were made with a Kay Electric Co. type 675 Sonagraph displaying 0 – 15 kHz over 0.7 s. Filter bandwidth at −3 dB was 600 Hz wideband and 60 Hz narrowband. Low frequency noise was suppressed by a 1 kHz high pass filter in most analyses. During the analysis, signal levels were monitored at all stages by oscilloscopes to prevent overloading and the consequent production of spurious harmonics.

Because our findings support the classification proposed by Brooke (1970; 1972), his generic nomenclature is used throughout the paper. Table 19.1 lists the species of swiftlets, as recognised by Medway (1966) and amended by Somadikarta (1967), Holyoak (1974) and Medway (1975), and further modified by the evidence assembled here; table 19.1 also suggests English vernacular names.

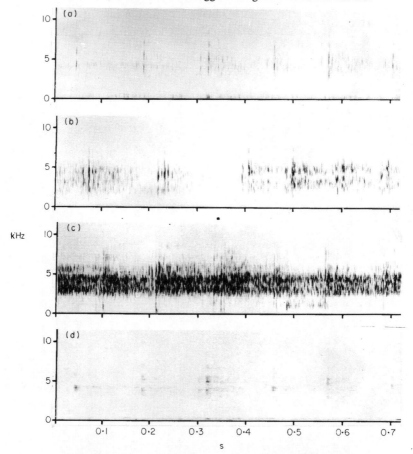

Figure 19.1 Sonagrams of Edible-nest Swiftlets, *Aerodramus f. fuciphagus.* (*a*), (*b*) and (*c*) show wideband displays of the orientation sounds of three different birds recorded in a flight chamber in Java, while in steady flight in total darkness. The double nature of the sound is clearly seen in (*a*); in (*b*) and (*c*) echoes of both calls are also prominent; (*d*) is a narrow band display of (*a*).

The Nature and Function of the Orientation Sounds

Representative sonagraphs and oscillograph displays of the orientation clicks of four taxa, prepared for this chapter, are presented as figures 19.1 – 19.4. Salient characteristics of these and of the clicks of two additional taxa investigated by other authors are summarised in table 19.2. In three species the clicks are normally double, that is they each consist of two successive sound impulses. In our recordings of *Aerodramus fuciphagus fuciphagus*, *A. vanikorensis salangana* and *A.*

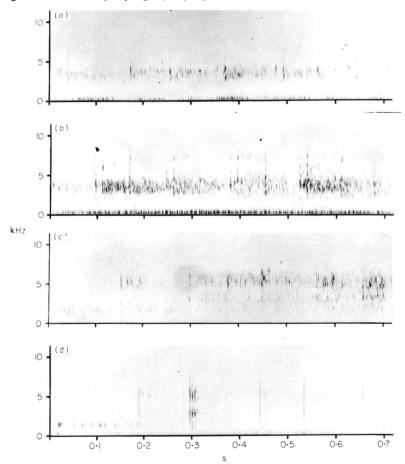

Figure 19.2 Wideband sonagrams of Mossy-nest Swiftlets, *Aerodramus vanikorensis*. (*a*) and (*b*) are two displays of a Javan bird *A. v. salangana*, recorded in the same conditions as *A. fuciphagus* in figure 19.1; the double impulses, together with their echoes are clearly shown. (*c*) and (*d*) are two displays of a Bornean bird, *A. v. natunae*, recorded in the darkroom of the Sarawak Museum with different equipment; in (*c*) the level of echoes largely obscures the clicks, but in (*d*) the high amplitude component of each doublet is clearly distinguishable and the weaker impulse preceding it is seen in several instances.

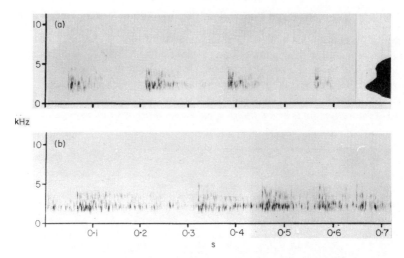

Figure 19.3 Sonagrams of the Black-nest swiftlet, *Aerodramus maximus lowi.* In (*a*), recorded in the darkroom of the Sarawak Museum, the sound pulse is not clearly separable from its echoes; in (*b*), recorded as a single bird flew in a cave mouth at night, the pulse is more clearly seen. The inset to the right of (*a*) shows the sound spectrum of a large number of birds recorded simultaneously in the cave.

vanikorensis natunae each impulse lasts about 2 ms: the second impulse of each doublet is invariably of higher amplitude, although there is variation in the degree of difference. This paired pattern is strongly reminiscent of the fruit bat, *Rousettus*, in which paired impulses (of equal amplitude and much higher sound frequency) are separated by a similar interval (figure 19.4*d*, left). The intervals between the two impulses of swiftlets are approximately constant at 15 - 16 ms, and in all impulses energy is distributed over a similar range of frequencies from about 2 to 10 kHz, in a broadly consistent pattern which, however, is not identical even in short series of pulses emitted by a single individual (figures 19.1, 19.2 and 19.4). The intervals between successive clicks in steady flight in darkness (measured from one major amplitude component to the next) in the recordings of *A. fuciphagus fuciphagus* are in the range 60 - 178 ms, in those of *A. vanikorensis salangana* 50 - 138 ms, and in *A. vanikorensis natunae* 62 - 200 ms. Instantaneous repetition rates calculated from these values are given in table 19.2.

In these characteristics, there is evidently close similarity between the orientation sounds of the species *A. fuciphagus, A. vanikorensis* and *A. hirundinaceus.* Differences in frequency range apparent in table 19.2 are probably attributable chiefly to differences in the sensitivity (in particular to higher harmonics) of the instruments used. Interspecific differences between the ranges of high amplitude frequences (that is, the principal components of the sound pulses) are less marked.

These overall similarities emphasise the distinctiveness of the orientation click of *A. maximus* (figures 19.3 and 19.4). At about 2.5 ms the sound pulse is 25 per

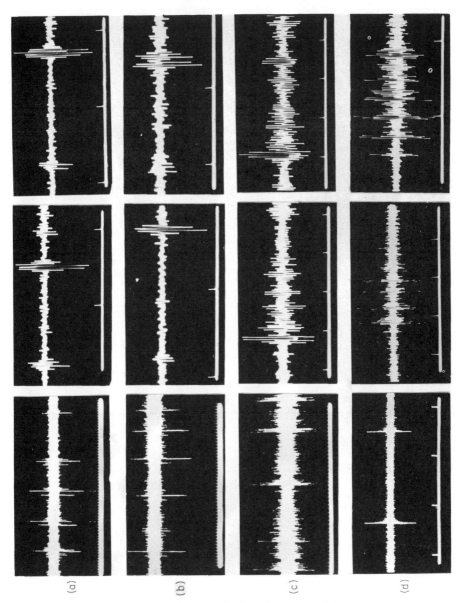

Figure 19.4 Oscillograph traces of orientation sounds of three
swiftlets and two other forms. (*a*) *Aerodramus f. fuciphagus*: train
of five clicks and two clicks at a faster sweep speed. (*b*) *Aerodramus
vanikorensis salangana:* train of six clicks and two clicks at a faster
sweep speed. (*c*) *Aerodramus maximus lowi*: train of five clicks and
two clicks at a faster sweep speed. (*d*) left: a double click of the
fruit bat, *Rousettus aegyptiacus*; centre and right: two compound
clicks of the oil bird, *Steatornis caripensis,* at different sweep speeds.
All time-markers show 10 ms intervals.

Table 19.2 Characteristics of the orientation clicks of six taxa of swiftlets, *Aerodramus* spp.

Taxon	A. f. fuciphagus	A. vanikorensis salangana	A. v. natunae	A. v. granti	A. h. hirundinaceus	A. maximus lowi
Locality	Java	Java	Sarawak	New Guinea	New Guinea	Sarawak
Reference	This chapter	This chapter	This chapter	Griffin and Suthers (1970)	Fenton (1975)	This chapter and Medway (1959)
Nature of click	Double	Double	Double	Double, ocasionally treble	Double	Single
Frequency range (kHz)	2 – 9+	2 – 9+	2 – 10	2 – 16	2 – 15+	1.5 – 7+
Range of instantaneous repetition rates (clicks/second)	5.6 – 17	7.3 – 20	5 – 16	3 – 20	4 – 21	3 – 11

Table 19.3 Calculated rates of wing beat of swiftlets, from the formula of Greenewalt (1960; 1962).

Taxon	Wing length (L) range (mm)	Source	$f = 3540/L^{1.15}$ (Hz)
A. f. fuciphagus	111 – 118	Medway (1962b)	15.7 – 14.7
A. v. salangana	115 – 123	Medway (1962b)	15.1 – 14.0
A. h. hirundinaceus	113 – 122	Rand and Gilliard (1967)	15.4 – 14.1
A. m. lowi	125 – 140	Smythies (1968)	13.7 – 12.0

cent longer than, for example, the single pulse of *A. fuciphagus* or *A. vanikorensis salangana*. In no instance is there any indication on the sonagrams of a double emission. Conceivably, this could be masked by echoes if the relative intensity of the two pulses was reversed, that is if the second pulse was of lower intensity than the first. Such a reversal would be an equally distinct character. The clearest of our records suggest alternatively that the click of *A. maximus* is a brief burst of sound impulses similar to that of the Venezuelan oil bird, *Steatornis* (compare figure 19.4c and d, centre and right). In the present sample of clicks of *A. maximus* the least interval between pulses, 90 ms, is less than previously reported (Medway, 1959), but still indicates a maximum instantaneous repetition rate lower than that found in the three other species of *Aerodramus*. The overall frequency range is narrower and between lower limits than for any other taxon (see table 19.2), although this difference is again less pronounced if high amplitude components only are compared.

Despite these contrasting features, the orientation clicks of the four species are sufficiently alike to allow the conclusion that all are produced in a similar manner. The nature of the sound-producing mechanism remains unknown. That it is not syringeal is suggested by the characteristics of the sound, in particular the variability in component frequencies of successive impulses, even when emitted by one bird over a brief period of time. Yet Medway (1962a) observed that *A. vanikorensis natunae* and other echolocating species which build nests of vegetable substances carry nest material in the feet rather than in the bill, as do non-echolocating swifts including the swiftlets *Collocalia esculenta* and *Hydrochous gigas* (Becking, 1971), and concluded that the orientation sound must therefore be uttered through the open mouth. Mixed sounds of this nature could probably be produced by some procedure independent of the respiratory system, for example by the action of the tongue against the palate as has been demonstrated in the fruit bats *Rousettus* spp. (Kulzer, 1958; 1960; Roberts, 1975). Although no investigation has yet been made, it is to be expected that the click-producing mechanism involves modification of this region and its nerve supply.

Harrisson (1966) observed that unfledged nestlings and restrained adults of *A. maximus lowi* and *A. vanikorensis natunae* uttered clicks only when the wings were moved, and postulated 'direct nerve and muscle linkage' between the wing and the emitting structure. Field observations, however, show that there is no obligatory connection between flight and the emission of clicks. It is, for instance, a commonplace observation that the orientation clicks commence abruptly as a flying bird approaches or enters a cave (Medway, 1959; Fenton, 1975; J. F. M. Horne, *in litt.*), and free-flying birds in large cave mouths click intermittently without observed concomitant changes in flight speed (Medway, 1959).

The rate of wing beat of swiftlets has not been ascertained. However, Greenewalt (1960, 1962) has shown that during steady flight the wing beats of birds do not normally vary more than 5 per cent from a value which, among a wide variety of species (excluding only humming birds), shows an approximately constant relationship to wing length expressed by the equation $f = 3540/L^{1.15}$, where f is in Hz and L in mm. Table 19.3 shows that only for *A. vanikorensis salangana* and *A. hirundinaceus hirundinaceus* do the calculated rates of wing beat fall within the range of observed instantaneous repetition rates of clicks. It is moreover unlikely that the rate of wing beat in steady flight could vary so continuously or between such wide limits as does the rate of clicking. A direct connection between the two systems is therefore improbable.

Tests of the acuity with which swiftlets can detect obstacles in darkness have produced somewhat inconsistent results (Medway, 1967; Griffin and Suthers, 1970; Fenton, 1975) doubtless attributable, at least in part, to the differences in materials used and the experimental situations. In theory, the resemblance between the clicks of the three taxa involved (*Aerodramus f. fuciphagus, A. vanikorensis granti* and *A. h. hirundinaceus*) indicates that the limits of sensitivity would be similar in all, that is in the detection of vertical cylindrical rods somewhere between 6.3 mm and 2 mm in diameter, as found by Griffin and Suthers (1970). Features of the clicks of *A. maximus lowi*, in particular the 25 per cent greater length of the pulse and the lower total range of frequencies involved, suggest that the echolocating system of this species must be relatively less sensitive to small targets and give poorer range discrimination (see, for example, Sales and Pye, 1974).

The diets of swiftlets are not known in detail. If reported gut contents of *Aerodramus* spp. (Chasen, 1931; Abdulali, 1942; Harrisson, 1976) are typical, at least a proportion of the prey would on theoretical grounds be below sizes detectable by an echolocating mechanism using sound pulses of the duration and frequency described above. Field observations of cave colonies have invariably demonstrated a diel periodicity of morning egress and evening ingress (for example, Medway, 1962a), only compatible with the conclusion that all or most food is obtained by daylight. Yet all swiftlets probably possess acute vison in dim light. For instance, its behaviour suggests that *Hydrochous gigas* is normally a crepuscular feeder (Medway and Wells, 1969), and *Aerodramus unicolor* has been observed feeding by night with the assistance of artificial illumination (Kershaw, in Ali and Ripley, 1970). Since *H. gigas* does not echolocate but *A. unicolor* does (table 19.1), it appears that in swiftlets the ability to hunt in poor light is unconnected with this capacity.

Both Medway (1962a) and Fenton (1975) have noted that echolocating species return to roost in the evening later than non-echolocating swiftlets sharing the same cave. Both authors concluded that the capacity to echolocate confers the freedom to fly after nightfall, and hence to travel further from the roost-cave and thus during the diurnal feeding period to exploit a greater range than is available to non-echolocating swiftlets, and that this factor may confer critical selective advantage. A more direct advantage is undoubtedly the ability to utilise as nesting (and roosting) sites the rock faces in the total darkness of cave interiors, a habitat which in the Indo-Pacific region is exploited by no other avian group. The success of this strategy is demonstrated by the enormous numbers of birds that occupy large caves in the region, for example, in Borneo (Medway and Smythies, in Smythies, 1968 ch. 5). The darkness of the cave provides protection from all but a few specialised predators – an egg-eating orthopteran and a predatory snake were noted by Medway (1962a). As Fenton (1975) has noted, cave swiftlets do not lay larger clutches than sympatric epigean swifts, but other features of their breeding biology are interpretable as responses to the increased security of the nest site. For instance, recorded incubation and fledging periods are relatively long: average values for *Aerodramus maximus lowi* (clutch of 1 egg) are 28 days and 65 days, respectively, and for *A. vanikorensis natunae* (2 eggs) 23 days and 45 days (Medway, 1962a), compared with 22 days and 40 days for *Apus affinis*

(2 – 4 eggs) (Ali and Ripley, 1970), a non-cavernicolous swift intermediate in size between the two swiftlets.

Rates of breeding success over 2 years (calculated from figures given by Medway, 1962*a*) were: for *Aerodramus maximus lowi*, hatching success 29.3 per cent of all eggs laid, and fledging success 57.3 per cent of eggs hatched; for *A. vanikorensis natunae*, hatching success 51.5 per cent and fledging success 74.1 per cent. By combining these figures and taking into account the species' normal clutch sizes, it can be estimated that among *A. maximus lowi* each pair produces on average 0.17 young per breeding attempt, and among *A. vanikorensis natunae* 0.76 young. Since in both species, in addition to replacements of lost clutches or broods, a proportion of successful pairs will lay again during the season (Medway, 1962*a*), these are minimal estimates of the total annual production of fledged young. Values of the same general order have been found among birds of the local epigean habitat: 0.34 juveniles per pair per year in lowland forest in Sarawak (Fogden, 1972), , and 0.4 in lowland forest in Peninsular Malaysia (Wells, in Medway and Wells, 1976, ch. 3). Fogden (1972), however, noted a very high rate of nest predation and, since the normal clutch of about half the birds in these samples exceeds two eggs (Wells, *op. cit.*), in percentile terms their breeding success is undoubtedly very low. Thus the safe nest site of cave swiftlets does apparently allow greater economy of reproductive effort, even though it does not lead to the production of strikingly larger numbers of fledged offspring, than among sympatric non-cavernicolous birds. In both communities it is necessary to assume that the low annual production of viable young is compensated by high adult survival rates. In this aspect, the security of the cave (which serves as roost site all year round for non-migratory swiftlets) probably again confers a significant advantage.

Phylogenetic Conclusions

Echolocating swiftlets do not invariably nest in total darkness; there are many records of nests situated in dim illumination in buildings, rock shelters or shallow caves. Although at such nest sites visual means of orientation may be adequate, the ability to echolocate still confers other advantages noted above. As argued previously by Medway and Wells (1969), it is improbable that the ability to echolocate, once acquired, would become unadaptive and be lost secondarily. We conclude that, while the echolocating taxa clearly evolved from non-echo-locating ancestors, selection in the reverse direction is highly unlikely.

Among the swiftlets, members of the species *Collocalia esculenta* and *Hydrochous gigas* have been shown not to utter click-like orientation sounds in flight in darkness, and to be unable to orientate in the absence of visual clues (table 19.2). *C. esculenta* has been shown by Orr (1963) to share aspects of palatine anatomy with the echolocating taxa '*inexpectata*' (a synonym of *A. fuciphagus*) and *A. spodiopygius*. The palatine anatomy of *H. gigas* has not been studied in detail. The species has been associated with other Collocaliini by features of external morphology and plumage, in which it shows no particularly close affinity to *C. esculenta*. The nests of these swiftlets are of dissimilar types, which do not suggest affinity between *gigas* and *esculenta* (table 19.1). Both types of nest can be matched among the echolocating swiftlets. Those of *C. esculenta* are similar to nests of *A. f. francicus* and *A. f. elaphrus,* and differ from nests of *A. unicolor* and *A. spodiopygius* only in the relative proportions of salivary nest cement (table 19.1). The nests of *H. gigas* are of a type similar to those of *A. vanikorensis,* to which the nests of *A. hirundinaceus, A. sawtelli* and *A. leucophaeus* are clearly related. There are thus *prima facie* grounds to support the proposal of either *C. esculenta* or *H. gigas* as ancestor of some or all of the echolocating swiftlets.

The close resemblance between the sonagrams of clicks of *A. fuciphagus,* of the three subspecies of *A. vanikorensis* and of *A. hirundinaceus,* is unlikely to have arisen by convergence, and therefore suggests a common phylogeny. Yet the nest of *A. fuciphagus,* self-supporting, bracket-shaped and composed solely of firm salivary cement (with the accidental addition of feathers or extraneous vegetable material in some cases), is not derivable from the externally supported vegetable nest, bound with moist cement (*vanikorensis*) or no cement (*hirundinaceus*). Both types of nest could, however, evolve by divergence from the bracket-shaped vegetable nest, bound with firm cement, which characterises *C. esculenta.* By this hypothesis, the taxa *francicus, elaphrus* and *unicolor* are primitive in the genus *Aerodramus.* Spectrographic analysis of the orientation clicks of these swiftlets would provide important evidence.

The position of *A. maximus* is uncertain. The distinctive features of its click suggest anatomical modifications of the bucco-pharyngeal region that differ from those of swiftlets which emit double clicks, but no study has been made. Its nest, composed of firm salivary cement and plumage apparently derived from the building birds, is distinctive and not clearly related to any other type of nest among the swiftlets. It is possible that the echolocating group is diphyletic, but again this hypothesis requires further knowledge of the diversity of the orientation click and its originating mechanism before it can be discussed authoritatively. Since they are linked by general similarities of size and plumage, for the time being it is preferable to place all echolocating swiftlets in a single genus, together with the morphologically similar species *A. papuensis,* of which the echolocating capacity is unknown.

The evidence summarised above thus points to the species *Collocalia esculenta* (including the taxon *marginata,* see duPont, 1971) as the nearest non-echolocating relative to the echolocating group. Information of the capacity of the Pygmy swiftlet *C. troglodytes,* to orientate in darkness will be critical in assessing its position. At present, its glossy plumage and small size associate it with *C. esculenta* (Brooke, 1972).

The type of the genus *Collocalia* Gray is *esculenta.* For the echolocating swiftlets, Condon (1975) used the genus-group name *Salangana* Streubel. By our reading of the original text, however, Streubel's usage (1848, p. 368) of the name 'Salanganae' (sic) did not in fact constitute a proposal of a genus-group name. Even if validly proposed, moreover, this name is a junior homonym of *Salangana* Lesson (itself a junior synonym of *Collocalia*). Although the criterion on which it was based is of no phylogenetic significance, and in fact divides species as well as the genus as at present understood, the prior generic name for the echolocating swiftlets appears to be *Aerodramus* Oberholser, type *innominata* Hume (Brooke, 1970).

Acknowledgements

We are grateful to Drs C. M. Perrins and B. Stonehouse for their comments on a first draft of this chapter, and to David T. Holyoak and Richard Schodde for permission to publish information supplied in correspondence.

References

Abdulali, H. (1942). The terns and edible-nest swiftlets at Vengurla, west coast, India. *J. Bombay nat. Hist. Soc.*, **43**, 446 – 451.

Ali, S. and Ripley, S. D. (1970). *Handbook of the Birds of India and Pakistan*. vol. 4. Oxford University Press, Bombay.

Becking, J. H. (1971). The breeding of *Collocalia gigas*. *Ibis*, **113**, 330 – 334.

Brooke, R. K. (1970). Taxonomic and evolutionary notes on the subfamilies, tribes, genera and subgenera of swifts (Aves: Apodidae). *Durban Mus. Novit.*, **9** (2), 13 – 24.

Brooke, R. K. (1972). Generic limits in old world Apodidae and Hirundinidae. *Bull. Br. ornith. Club*, **92**, 52 – 57.

Chasen, F. N. (1931). *Report on the Birds' Nest Industry of North Borneo*. Government Printer, Jesselton.

Condon, H. T. (1975). *Checklist of the Birds of Australia, 1. Non-passerines*. RAOU, Melbourne.

Cranbrook, Earl of, and Medway, Lord (1965). Lack of ultrasonic frequencies in the calls of swiftlets. *Ibis*, **107**, 258.

Dupont, J. E. (1971). *Philippine Birds*. Monograph series 2. Delaware Museum of Natural History.

Fenton, M. B. (1975). Acuity of Echolocation in *Collocalia hirundinacea* (Aves: Apodidae), with comments on the distribution of echolocating swiftlets and molossid bats. *Biotropica*, **7**, 1 – 7.

Fogden, M. P. L. (1972). The seasonality and population dynamics of equatorial forest birds in Sarawak. *Ibis*, **114**, 307 – 343.

Greenewalt, C. H. (1960). *Humming Birds*. Am. Mus. nat. Hist., Doubleday.

Greenewalt, C. H. (1962). Dimensional relationships for flying animals. *Smithson. misc. Collns.*, **144** (2), 1 – 46.

Griffin, D. R. and Suthers, R. A. (1970). Sensitivity of echolocation in cave swiftlets. *Biol. Bull.*, **139**, 495 – 501.

Harrisson, T. (1966). Onset of echo-location clicking in *Collocalia* swiftlets. *Nature, Lond.*, **212**, 530 – 531.

Harrisson, T. (1976). The food of *Collocalia* swiftlets (Aves, Apodidae) at Niah Great Cave in Borneo. *J. Bombay nat. Hist. Soc.*, 71, 376 – 393.

Holyoak, D. T. (1974). Undescribed land birds from the Cook Islands, Pacific Ocean. *Bull. Br. ornith. Club*, **94**, 145 – 150.

Kulzer, E. (1958). Untersuchungen über die Biologie von Flughunden der Gattung *Rousettus* Gray. *Z. Morph. Ökol. Tiere*, **47**, 374 – 402.

Kulzer, E. (1960). Physiologische und Morphologische Untersuchungen über die Erzeugung der Orientierungslaute von Flughunden der Gattung *Rousettus. Z. vergl. Physiol.*, **43**, 231 – 268.

Lack, D. (1956). A review of the genera and nesting habits of swifts. *Auk*, 73, 1–32.

Medway, Lord (1959). Echo-location among *Collocalia. Nature, Lond.*, 184, 1352–1353.

Medway, Lord (1962*a*). The swiftlets (*Collocalia*) of Niah cave, Sarawak. *Ibis*, 104, 45–66, 228–245.

Medway, Lord (1962*b*). The swiftlets (*Collocalia*) of Java, and their relationships. *J. Bombay nat. Hist. Soc.*, 59, 146–153.

Medway, Lord (1966). Field characters as a guide to the specific relations of swiftlets. *Proc. Linn. Soc. Lond.*, 177, 151–172.

Medway, Lord (1967). The function of echonavigation among swiftlets. *Anim. Behav.*, 15, 416–420.

Medway, Lord (1975). The nest of *Collocalia v. vanikorensis*, and taxonomic implications. *Emu*, 75, 154–155.

Medway, Lord and Wells, D. R. (1969). Dark orientation by the Giant Swiftlet *Collocalia gigas. Ibis*, 111, 609–611.

Medway, Lord and Wells, D. R. (1976). *The Birds of the Malay Peninsula*, vol. 5, Witherby, London.

Novick, A. (1959). Acoustic orientation in the cave swiftlet. *Biol. Bull.*, 117, 497–503.

Orr, R. T. (1963). Comments on the classification of swifts of the subfamily Chaeturinae. *Proc. 13th int. ornith. Congr.*, 126–134.

Pecotich, L. (1974). Grey swiftlets in the Tully River gorge and Chillagoe caves. *Sunbird*, 5, 16–21.

Penny, M. (1975). *The Birds of Seychelles and the Outlying Islands*, Collins, London.

Procter, J. (1972). The nest and identity of the Seychelles swiftlet *Collocalia. Ibis*, 114, 272–273.

Rand, A. L. and Gilliard, E. T. (1967). *Handbook of New Guinea Birds*. Weidenfeld and Nicolson, London.

Roberts, L. H. (1975). Confirmation of the echolocation pulse production mechanism of *Rousettus. J. Mammal.*, 56, 218–220.

Sales, G. D. and Pye, J. D. (1974). *Ultrasonic Communication by Animals*. Chapman and Hall, London, 281 pp.

Smythies, B. E. (1968). *The Birds of Borneo*. 2nd edn. Oliver and Boyd, London and Edinburgh.

Somadikarta, S. (1967). A recharacterization of *Collocalia papuensis* Rand, the three-toed swiftlet. *Proc. U.S. natn. Mus.*, 124 (3629), 1–8.

Streubel, A. V. (1848). Die Cypseliden des Berliner Museums. *Isis,* 41, 348–373.

20 Duetting and other synchronised displays of the blue-backed manakins, Chiroxiphia *Spp*

Dr. D. W. Snow*

The blue-backed manakins of the genus *Chiroxiphia* are apparently unique among birds in the organisation of their courtship displays. They belong to a family (Pipridae) in which many species show a high degree of sexual dimorphism and lek displays are highly developed. In the other manakins that have been studied, as in lek birds of other familes, each male defends and displays on his own court or perch within the general area of the lek; but blue-backed manakin males live in groups which jointly display on a number of perches within their territory. One of the most remarkable aspects of their organisation is that the advertising calls which are a prelude to the courtship display, and the first phase of the courtship display itself, are performed in perfect co-ordination by two males from the group (or, in one species, three or more birds), each of which appears to play an equal part. The highly synchronised duets that are important in the courtship sequence are thus quite different from duets described in other birds, the main function of which is to maintain contact between the members of permanently mated pairs (Thorpe, 1972).

Chiroxiphia is a well-marked genus, rather distinct from other manakin genera and clearly monophyletic. Adult males are black with a blue patch on the back (in *C. caudata* the rest of the body is also mainly blue) and a red crown patch. In three species the central pair of tail-feathers are elongated. Females are olive green. Like other manakins, they feed largely on berries, supplemented by insects. They are confined to forested areas, but are birds of drier kinds of woodland, secondary growth and forest edges and are not generally found in undisturbed primary forest. The genus is usually treated as consisting of four species, but they are closely related and allopatric and the recognition of species limits is somewhat arbitrary in the present state of knowledge. Figure 20.1 shows the distribution of the various forms. The Tobago population of *C. pareola* has received the most detailed study (Snow, 1963*a*), and something has been recorded of the displays of the three other species (references in Snow, 1963*b*).

*David W. Snow was born in Windermere, England. After three years' wartime and post-war service with the RNVR he read zoology at Oxford, and became a postgraduate student at the Edward Grey Institute. He worked on the comparative ecology and geographical variation of Palaearctic titmice and on blackbirds, and in 1957 became Resident Naturalist at the New York Zoological Society's tropical field station in Trinidad. After nearly five years there he became director of the Charles Darwin Research Station in the Galapagos Islands, returning to England in 1964 as Director of Research for the British Trust for Ornithology. He was appointed to his present post in 1968.

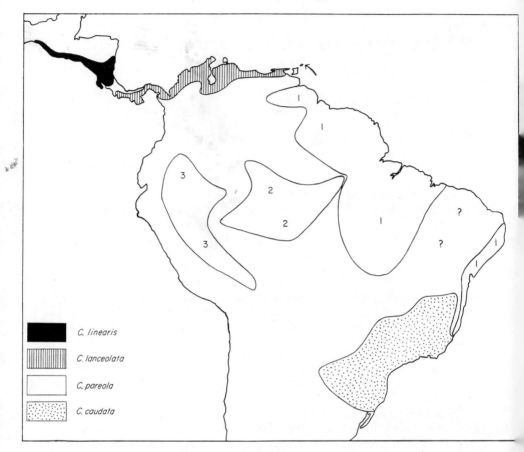

Figure 20.1 Distribution of *Chiroxiphia* species. The general areas
within which the species occur are indicated; distributions are not
continuous within these areas. *C. pareola* is divided into several
geographical races, as follows: (1) *pareola,* with very similar *atlantica*
in Tobago (indicated by arrow); (2) *regina,* with yellow instead of
red crown patch; (3) *napensis* and *boliviana,* similar to *pareola,* but
differing slightly in coloration and more markedly in mensural
characters. In the east of its range, *regina* extends to the west bank
of the lower Tapajós, being replaced on the right bank by *pareola.*
Regina, napensis and *boliviana* show at least as much divergence
from *pareola* as *linearis* does from *lanceolata* in Central America,
and they may have reached specific status.

This chapter describes and compares the duetting calls of all four species,
presents new information on the display of the least known and most distinct
species, *C. caudata,* and discusses the evolutionary implications of the system of
courtship found in this genus. The data are incomplete, and a subsidiary aim of
this chapter is to stimulate further field research into the Blue-backed manakins,
which are not difficult to trap and to study at close quarters.

Social Organisation and the Main Types of Display

We have only a rough idea of the social organisation of blue-backed manakin populations. Behaviour at the nest has not been studied, but there is every reason to believe that, as in other manakin genera, only the cryptically coloured female attends the nest. The sexes appear to keep mainly apart, except when the females visit the males' display ground. The males live in groups consisting of adult, sub-adult and immature birds, and spend a great part of their time in the vicinity of their display perches. Study of a group of males of *C. pareola* in Guyana, some of which were colour-ringed, showed that the group owned a number of display perches within a small area, and strongly suggested that there was a hierarchy within the group and that the dominant male called up subordinates to display with him (Snow, 1971). There is no reason to suppose that the social organisation is different in the other species. It is not known whether males belonging to the same group tend to be related to one another, but since they appear to be sedentary birds and are often patchily distributed it is likely that this is often the case.

In all four species, the displays can be divided into three main phases. (1) Two males come together and utter a repeated series of duetting calls in the trees above a display perch. Bouts of calling are initiated by one male (possibly the dominant bird of the group); a special call of this bird acts as a summons inviting another male of the group to join him. (2) After calling for some time the males fly down to the display perch and, side by side on the perch, alternately jump up fluttering their wings at the same time. In one species, *C. caudata,* one or more additional male may join in the display at this stage. Each jump is accompanied by a special call. If a female comes to the perch, the males face her and jump up alternately in front of her in a cartwheel or 'Catherine wheel' dance, each bird moving back as he hovers while the next bird shuffles up to take his place. When a female is present, this dance usually gets more and more excited and the jumping quicker as it proceeds, until it is suddenly brought to an end when one bird, probably the dominant one, utters one or more loud sharp notes. The males then fly off from the perch. (3) One male only, probably the dominant one, remains in the immediate vicinity of the perch and executes a mainly silent display round the female, flying back and forth across the perch with a floating flight that prominently displays the blue back. During brief landings between flights the red crown is displayed to the female. This display sequence culminates in copulation.

Duetting

The duetting calls of pairs of males are the most striking of the blue-backed manakins' vocalisations. They are loud and long, and since they are often uttered from high in trees they carry a long way through the forest. Though the term 'duetting' is used for convenience, in fact the calls range from duets in the common meaning of the term, in which each bird utters a different phrase, to pairs of identical calls uttered almost simultaneously. When the calls are nearly simultaneous it is not possible for the human ear to separate the contributions of the two birds, but the slight difference in timing gives the combined call a distinctive echoing or ringing quality.

Evolutionary ecology

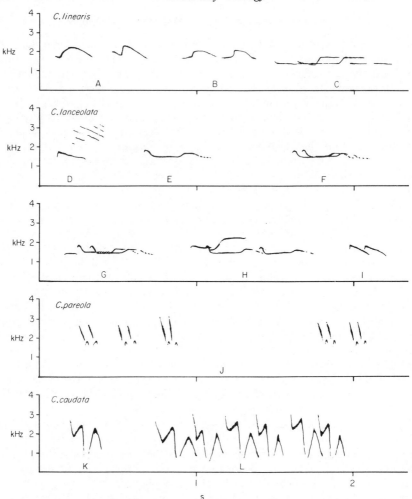

Figure 20.2 Sonagrams of duets and some other calls of *Chiroxiphia*
species. Recordings as follows: *C. linearis,* Monte Verde, Costa Rica
(B. K. Snow); *C. lanceolata,* Aragua, Venezuela (P. Schwartz);
C. pareola, Kanuku Mountains, Guyana (J. Lindblad); *C. caudata,*
Boraceia Forest Reserve, São Paulo, Brazil (D. W. Snow). For further
explanation, see the text.

There are striking differences in the duetting calls of the four species. There
may in addition be significant local differences within each species, but this is a
matter for future research. The following descriptions are based on available
recordings.

Chiroxiphia linearis

The main duetting call is a relatively low-pitched musical whistle of very pure
quality, described by many authors and giving the bird its local name, *toledo.*
In fully developed form (figure 20.2*C*) it consists of three pure notes; the first at

1.4 kHz and lasting about 0.25 s, is connected by an upward slur to the second note at 1.7 kHz and lasting about 0.13 s, followed after a very brief gap by the third note at 1.4 kHz. The whole call lasts about 0.6 s. In what appears to be typical timing, the second male begins to utter the call about 0.14 s after the first (figure 20.2*C*), with the result that the two higher-pitched middle notes do not overlap but follow one another with hardly a break. In the series analysed, most of the time intervals between successive duet phrases are about 2 s, with a few up to 4 s.

The calls uttered by one bird before the start of a duet (figure 20.2*A*), and the duetting calls at the beginning of the sequence (figure 20.2*B*), are shorter, higher-pitched and not clearly divided into sections of even pitch. They resemble the fully developed duetting call in beginning at a lower pitch (1.7 – 2.0 kHz), then rising to a higher pitch (2.1 – 2.3 kHz) and ending on a lower pitch about the same as that of the opening note. In the sequence of duetting calls analysed, there is less synchronisation early in the sequence than towards the end: figure 20.2*B* shows a phrase from early in the sequence, in which the interval between the two birds' calls is about 0.25 s.

Chiroxiphia lanceolata

This species too has acquired its local name in Panama from the *toledo* call uttered in the duet (Wetmore, 1972). Slud (1964) has described the call as heard in the extreme southwest of Costa Rica, where it comes closest to the range of *C. linearis*. 'Its "toledo", sounding to my ears more like "pericor", differs significantly from that of *linearis*. Whereas the latter species pauses momentarily after the first note, distinctly accents the second, and produces a clear short-syllable "toe-laydo", the present species does not pause after the first syllable, and prolongs the final syllable into a trough-shaped inquiring "dor" or "cor".' The calls recorded in Venezuela (figure 20.2*D – I*) are difficult to match with this description, suggesting that there is local variation. This seems the more likely in that the Venezuelan calls are themselves variable, and Slud also mentions a number of other calls in the repertoires of the Costa Rican birds. Nevertheless, it is clear that the main duetting call is of the same quality in this species as in *C. linearis,* and the sonagrams suggest that there is a basic similarity in structure.

The main element in the duet call of *C. lanceolata* is shown in figure 20. 2*E*. Measurement of sonagrams of 20 calls shows great uniformity: the initial frequency is 1.8 – 1.9 kHz; This is at once followed by a sharp downward slur and then a pure note at 1.4 – 1.5 kHz, lasting 0.16 – 0.18 s, followed by an upward slur to 1.8 – 1.9 kHz and then a downward slur to 1.4 – 1.5 kHz, at which pitch the call fades out in a series of pulses. The total duration of the call is about 0.40 s, but is difficult to measure accurately because of the gradual fading of the final note. If this is omitted, the duration is rather invariable in the series available, nearly all lasting 0.27 – 0.28 s, with extremes of 0.26 and 0.29 s.

Although to the human ear they sound much alike, the duet phrases in the series analysed are of four distinct types. The intervals between successive phrases are 1 – 3 s. (1) In the simplest type (figure 20.2*F*), each bird utters the main element only, the second bird beginning 0.05 – 0.06 s (in three instances) or 0.09 s (one instance) after the first. Some of these phrases are immediately preceded by a harsher, nasal note of quite different quality (figure 20.2*D*). (In other duet sequences, each duet phrase is immediately followed by the harsher note.) (2) The first bird gives a short note of 1.4 kHz, lasting 0.08 – 0.12 s; the second bird follows after a variable interval (0.04 – 0.09 s in three cases) with the main element, followed in turn after 0.08 – 0.19 s by the first bird, who also utters the main element (figure 20.2G). (3) The first bird utters a slightly descending note of 0.12 s duration, beginning at 1.8 and dropping to 1.7 kHz, followed by an upward slur and a pure note of 2.2 kHz lasting about 0.13 s. The second bird comes in with the main element, beginning about 0.09 s after the first, and the duet phrase ends with the first bird repeating the main element as the second bird ends. This complex duet phrase (figure 20.2H) lasts about 0.7 s. (4) The series ends with a single very simple duet phrase consisting of a sharp *chook* uttered by each bird, with an interval of 0.09 s between them (figure 20.2I).

In some of the sonagrams, the long overlapping middle part of the two calls produces a beat frequency resulting in amplitude modulation (figure 20.2G). The beat frequency of about 70 Hz provides a direct measurement of the difference in frequency between the two notes, one of which is near 1.4 kHz and the other near 1.5 kHz. Hall-Craggs (1974) has described a similar case of a beat due to a frequency difference in the overlapping sections of duet calls of Whooper swans *Cygnus cygnus*. The somewhat irregular pulsation at the end of the main element, with a frequency of about 45 Hz, is produced by the bird itself and not by interference.

Chiroxiphia pareola

The duet of this species has been described in an earlier paper (Snow 1963a). The basic note is a simple, incisive *chup*, normally uttered in groups of 2 – 4 notes, one bird beginning a fraction of a second before the other (figure 20.2J). The interval between groups is usually less than 1 s. The *chup* is a very different sound from the musical whistle of *C. linearis* and *C. lanceolata*. It lasts only about 0.03 s. The fundamental frequency starts at about 2.6 kHz and descends sharply to about 1.9 kHz, and there is a series of harmonics (not shown in figure 20.2J) up to at least 8 kHz. This main note is followed immediately by a fainter, lower-pitched note in the form of an inverted V. The second bird usually begins 0.04 – 0.06 s after the first bird, and there is a tendency, with each group of *chups*, for the interval to become a little shorter with each successive note. Thus in the group of three pairs of *chups* shown in figure 20.2J, the successive intervals are 0.065, 0.06 and 0.05 s.

Apart from the difference in the note itself, the duet of this species differs from those of the other three species in that the duration of the note is shorter than the interval between the notes uttered by the two birds, so that there is no overlap. The interval is nevertheless too short to be perceived clearly by the human ear, but the nearly synchronous utterance has an echoing quality that was well described by Gilliard (1959) – although he did not appreciate that it was produced by two birds – as 'a resounding phrase that rang through the forest like the clicking of billard balls'.

Chiroxiphia caudata

The first impression made by the duetting of this species is of a confused gabbling. Analysis by sonagram, however, shows a definite pattern. The basic element of the duet (figure 20.2*K*) consists of two parts with rapid frequency modulation. The first part begins with a steep downward slur, followed by a slower ascending slur and finishing with a very steep downward slur. The second part, which follows after a pause of about 0.04 s, has the form of an inverted acute-angled V. The frequency range, from the lowest to the highest-pitched parts of the phrase, is about 1 – 3 kHz. Typically, these phrases are uttered in groups of three or four, with intervals of about 0.4 s between the beginning of each prase, one bird beginning to call 0.17 – 0.24 s before the other. There is a tendency, within each group of phrases, for the interval between the two birds to shorten a little with each successive phrase. Thus in figure 20.2*L*, the intervals between the two birds are 0.23, 0.20 and 0.17 s. As shown in figure 20.2*L*, the calls of the second bird tend to fill the gaps between those of the first bird, so that the rhythm may be lost and turn into a confused gabbling sound, as already mentioned.

Organisation and Coordination of Duets

In *C. pareola* and *C. caudata* duets are initiated by one male, which calls until another joins him, at which point he breaks into the duet phrases and the other accompanies him. Probably the same happens in *C. linearis* and *C. lanceolata*. The 'summoning calls', as they may be called, show some variability, and no attempt has been made to analyse them. Some at least (see, for example, Figure 20.2*A*) are abbreviated or incompletely developed versions of the calls used in the duet. In my earlier study (Snow, 1963*a*) I suggested, from an examination of slight differences in the calls of the two birds apparent in a series of sonagrams, that the bird that uttered the summoning call was the leader in the duet that followed, and it seems probable that the same holds in the other species; however, suitable recordings are not available to test this.

It is exceptional for more than two birds to take part in coordinated calling, except perhaps briefly and abortively. Once when I was watching *C. pareola*, two adults that were perched close together and calling synchronously were joined by an immature male, who perched about two feet away and called for some time in unison with them (Snow, 1963*a*). It would in fact be impossible, in view of the timing of the successive phrases, for a third bird to join in the duets of *C. pareola* or *C. caudata*, calling in sequence after the second bird, without destroying the pattern, as is apparent from inspection of figure 20.2*J* and *L*: in *C. pareola* the effect would be to destroy the rhythm by filling the gaps between successive pairs of *chups*, and in *C. caudata* each phrase uttered by the third bird would almost completely overlap the first bird's next phrase. In *C. pareola* (as in *C. linearis* and *C. lanceolata*) the point is perhaps hardly worth making, since only two males take part in the courtship dance that follows the duet, but in *C. caudata* three (and occasionally more) birds take part in the dance, so that it would not be surprising if three birds were found to take part in the synchronised calling that precedes it. Indeed, until the sonagrams showed the contrary, I had thought from its gabbling quality that the sound might be produced by three or more birds.

The interval between the calls of the two duetting birds may be related to their positions in relation to each other. Thus, to judge from the available recordings, the interval is shortest in *C. pareola* (0.04 – 0.06 s), in which the two males call side by side, facing the same way and almost touching one another. In *C. lanceolata*, the two birds may almost touch one another, or they may be up to about 30 cm apart, either facing the same way or in different directions (P. Schwartz, personal communication). In the recordings of this species analysed, the intervals between the calls of the two birds are 0.04 – 0.09 s. In *C. linearis*, B. K. Snow (personal communication) has observed duetting birds perched close together on the same branch, facing either the same way or opposite ways, and Slud (1964) gives the distance between duetting birds as about 8 inches (20 cm). The intervals between the calls of the two birds in the recordings analysed are about 0.14 s. Finally, in *C. caudata*, in which the two males may perch several metres apart in the same tree, the sonagrams show intervals of 0.17 – 0.24 s. These figures suggest that visual cues are used in the synchronisation, probably the movements of the head and neck immediately preceding the call. The same is suggested by examples of good coordination after variable intervals of silence. Thus the single duet phrase of *C. lanceolata* shown in figure 20.2*I*, with an interval of 0.09 s between the two birds, followed 3 s of silence. It is hard to believe that the second bird could have followed so closely after the first unless it had been watching its movements. On one occasion while I was watching *C. pareola* two birds came together, began to call while they were still several metres apart and continued to do so for some time (Snow, 1963*a*). At this unusual distance their calls were very poorly synchronised.

It is probably significant that it is in *C. pareola*, the species in which the males perch closest together facing the same way, that the duet is made up of the shortest and most rapidly repeated notes (the *chup*) and that the number of notes in each group of calls is variable. Although the second bird can have no means of knowing whether the leader, when he begins to call, is going to utter a group of 2, 3 or 4 *chups*, it is able to follow exactly, neither stopping short nor giving an extra call after the leader has finished. The shortening of the interval between the two birds within each group of calls, noted in both *C. pareola* and *C. caudata*, must mean that as the second bird gets into the rhythm of the duet its reaction time decreases.

Vocalisations Associated with Later Phases of Display

After a sequence of duetting, which may go on, with pauses, for many minutes, the usual sequel is for the males to fly down to a display perch and begin to jump up alternately, each bird hovering for a moment in the air before landing again on the perch. They may perform alone, or a female may join them on the perch, in which case the display is orientated towards her. Each bird, as it hovers in the air, utters a vibrant, twanging note, which may be written *myaaa* (Slud) or *ar-r-r-r*. This phase of the display has been described for *C. linearis* (Slud, 1964; B. K. Snow, personal communication), *C. lanceolata* (Friedmann and Smith, 1957; D. W. Snow, unpublished) and *C. pareola* (Gilliard, 1959; Snow, 1963*a*), and seems to be identical or nearly so in all three species. Accounts of the display of *C. caudata* are conflicting, some clearly being based on inadequate observation or faulty recollection. Observations by the author and D. Goodwin in southeastern Brazil in October 1972 agree closely with those of Holt (1925), and indicate that it is usual for three males to take part in the jumping display.

Each in turn jumps up, hovers in front of the female, and then moves back to the rear of the line while the two others shuffle forward, so that there is a constant changing of place. In its essentials the display is the same as that of the other species, with three (or occasionally more) birds taking part instead of two. In all four species as the dance proceeds it becomes more excited, the jumps become lower and more rapid, and the twanging calls take on a wavering, bleating quality. Finally, the dance is brought to an end when one bird utters a loud and very sharp *zeek* or *zeek-eek*. This way of ending the dance seems to be essentially the same in all the species, but the movements of the displaying birds are very rapid and difficult to follow at this stage and the exact sequence of events needs further study. The *zeek* has the effect of immediately inhibiting any further jumping, as already recorded for *C. pareola* (Snow, 1963a). In *C. caudata*, with three birds displaying, the effect is even more striking. The following account is based on several courtship sequences observed at very close quarters in the Boraceia Forest Reserve in southeastern Brazil. As the dance approaches its end, one of the males, as he hovers in front of the female, instead of the twanging *ar-r-r-r-r* utters a rapid *flup-flup-flup* — a noise which sounds as if it might be made by the wings but is almost certainly vocal — than a moment later, still hovering, gives a very sharp, penetrating *zeek-eek*, and immediately flies off to a perch a few feet away. At the sound of the *zeek-eek* the other males crouch absolutely motionless and remain still for a second or two, and then they too fly off to perches near by. If the female remains on the perch all the males may return after an interval and again begin to dance, or one male only, the one that uttered the *zeek-eek*, may return and perform the third and final phase of the display which leads to copulation.

This final phase of courtship is very similar in the three species in which it has been reported. In *C. pareola* the male flutters to and fro around the female, with a floating butterfly-like flight that conspicuously displays the blue on the back, crossing and recrossing her perch and every now and again alighting near her and lowering his head to display the red crown patch. Eventually he lands close beside her and mounts. A low *quaaaa* may be uttered when the male lands on an outlying perch, with a mechanical wing click at the moment of taking off to fly towards the female. Other details of this display are given in Snow (1963a). Very similar pre-mating displays have been seen in *C. linearis* (B. K. Snow, personal communication) and *C. caudata* (personal observation), but without the accompanying call and wing click.

Discussion

The system of courtship found in the genus *Chiroxiphia* may be unique, or at least is extremely uncommon; the nearest parallel may be found in the Black-and-gold cotinga, *Tijuca atra,* of which only a very incomplete preliminary study has been made (Snow and Goodwin, 1974). Such a system raises a number of points of theoretical interest, of which three will be briefly discussed: first, the predisposing conditions leading to the evolution of such a system; secondly, the stages leading to joint displays of the kind shown by *Chiroxiphia* from the more usual lek system that presumably preceded it; and thirdly, the bearing of *Chiroxiphia* on general theories concerning the evolution of secondary sexual characters.

Predisposing conditions

If reliance can be placed on the results of very incomplete field work (Snow, 1971), an essential element in courtship of the blue-backed manakins is that one male is able to maintain dominance over a small group of males. The dominant male is probably the only one of the group that regularly mates with the females visiting the group's display area. In this respect the situation in *Chiroxiphia* may have a parallel in some cotingas, such as the Calfbird, *Perissocephalus tricolor* (Snow, 1972), in which the males display in small compact groups within which probably only one or two dominant males mate with visiting females; but there is a major difference in that each male Calfbird has its own display perch. For one bird to be able to maintain dominance, it is essential that the group be small. It would not, for instance, be feasible in a manakin such as *Manacus manacus* or *Pipra erythrocephala,* in which leks may contain 20 or more males. This suggests that one of the underlying factors in the evolution of the courtship system found in *Chiroxiphia* may have to do with the density and local distribution of the population.

Over most of their range blue-backed manakins are birds of tangled secondary growth or forest edges with thick undergrowth, rather than birds of primary forest. In many areas they are now abundant, but this may not be the natural condition. One can only guess at their original abundance and local distribution before man altered the environment. To judge from the situation in relatively unaltered forest in Guyana and southeastern Brazil, and from general statements in the literature, it is likely that they were normally rather patchily distributed along forest edges and other places where there were breaks in the forest allowing the development of secondary growth. It may well be that, in addition to this habitat requirement, limitations were set for their local distribution by the nature of their food supply, but there is no information on this point.

If a group of males is small enough to be dominated by one male, which effectively prevents the others from mating, there is then no reason why the subordinate males should not support and enhance the displays of the dominant male. This may in fact bring them an advantage in two ways. First, since a subordinate bird's chance of reproductive success depends on his eventually becoming the dominant bird of the group, and since display areas are traditional, lasting for many years if the habitat is not altered, it will be to the future advantage of a subordinate male if the group's display area has been advertised as effectively as possible so that it attracts as many females as possible. Secondly, if the individuals of the group are related to one another, kin selection will make it advantageous for subordinate males to advertise the display ground even if they do not mate. Again, there may be parallels in some of the cotingas, such as the Bearded bellbird *Procnias averano,* and the Calfbird, in which subordinate males take up display perches near the perch of a dominant male, and if an opportunity arises usurp his position (Snow, 1970; 1972).

Longevity may be an additional factor. In the manakin *Manacus manacus* adults may have an annual survival rate as high as 89 per cent, and senility may not be an uncommon cause of death (Snow and Lill, 1974). If this is general, it is worth while for a younger bird to accept a subordinate position in a group, since it is likely to outlive the dominant bird. In contrast, in species with very high annual mortality rates that are independent of age, as in many northern passerine species, it would be much less profitable to wait for the death of a dominant bird since the subordinate bird would be just as likely to die first.

The transition from leks to communal displays

It may safely be assumed that the Blue-backed manakins' system of courtship is a secondary development, derived from a lek system of the usual kind in which each male occupied his own display perch. The reason for this assumption is simply that *Chiroxiphia* is a single genus among a number of related genera in which typical lek systems are highly developed, and that its closest affinities seem to be with *Pipra*, a genus of typical lek birds. Likewise *Tijuca*, whose courtship seems to show the closest parallel to *Chiroxiphia*, is also a member of a family (Cotingidae) containing many typical lek species.

The transition from a system of small leks of the usual sort to that found now in *Chiroxiphia* must have involved a marked reduction in overt aggressive behaviour between males. In *M. manacus* and *Pipra erythrocephala* males regularly come together on neutral ground between their courts or display perches. They may in fact spend a significant fraction of the day in this way (in *P. erythrocephala*, up to about 30 per cent of the daylight hours – Snow, 1962), but the two birds remain in a state of aggressive tension and no display develops. In contrast, a striking feature of the behaviour of blue-backed manakins is the absence of any overt aggressiveness between males. Even at the point in the display sequence where the dominant male has to get rid of his dancing partner or partners, before the precopulatory display, this is achieved by a ritual involving a seemingly automatic response by the subordinates to a vocal signal.

Starting from a hypothetical situation in which males are distributed in groups that are small enough for one bird to be able to maintain dominance and do all the mating, as discussed in the previous section, it is not difficult to see how overt aggressiveness between males might be progressively reduced. Aggressive behaviour would no longer be useful in helping a subordinate male to secure a display perch. Its other potential use, to usurp the dominant male's position, is also excluded, since all the evidence tends to show that a dominant bird, as long as it is fit, can normally maintain its dominance against individual rivals. If, further, a subordinate bird's best chance of reproductive success is to enhance the dominant male's displays and wait for its own turn to become dominant, as has been argued, overtly aggressive behaviour will be a disadvantage. Reduction of aggressiveness will allow males to come together without being inhibited from displaying by the aggressive tension that inhibits neighbouring males of *M. manacus* and *P. erythrocephala*.

Reduction of overt aggression seems to have resulted in a readiness on the part of *Chiroxiphia* males to adopt different subordinate roles according to immediate circumstances. Thus, it is very common in this genus for males, especially young males, to take the place of a female in the courtship display if no female is present.

In further investigating the transition from a lek system to a system of communal or joint courtship, it would be especially interesting to make detailed observations on *Pipra aureola*. In this species small groups of males appear to be usual. Neighbouring males occasionally visit each other at their display perches and perform coordinated displays, which are, however, quite different from anything found in *Chiroxiphia* (Snow, 1963c). It may be that this species is in a transitional evolutionary stage.

Secondary sexual characters in chiroxiphia

It is not the purpose of this chapter to discuss the evolution of secondary sexual characters, but joint courtship such as is found in *Chiroxiphia* raises some theoretical questions. There is no evidence bearing on the point, but it is probable that female choice is the main selective agent in the evolution of secondary sexual characters in the genus. The alternative hypothesis, that these characters have evolved through selection for inter-male dominance, seems unlikely in view of the evidence for extreme reduction of inter-male aggressiveness and the fact that males do not direct any displays at one another but pool resources in joint displays before a female. If female choice is the main agent, the female must base her choice on the characters of both dominant and subordinate males, since both play equal parts in courtship up to the stage immediately preceding mating. She has no choice at the end, since the dominant male himself dismisses his subordinates.

If subordinate males had no reproductive potential, the effect would simply be that the female's choice would be only half influenced by the characters of the male with which she mates, and its selective effect, at half strength, would be confined to this one bird. But this is almost certainly not the usual situation. If kin selection operates, and if the males are related, the subordinate male's characters will be subject to selection whether he is reproductively successful later or not. If the subordinate male later becomes dominant, his characters may be subject to selection before he actually reproduces, since he may later fertilise a female with whom he has previously danced as a subordinate.

The four species of *Chiroxiphia* have diverged to varying degrees in the secondary sexual characters of the males and in the first phase of courtship, that is, in the duetting calls the main function of which is to advertise the display ground to females. They have, however, diverged very little or not at all in the second and third phases of courtship, except that in *C. caudata* more than two males participate in the joint dance. The interspecific differences have certainly evolved in isolation. At the borders between their ranges neighbouring species approach closely without penetrating each other's ranges or, so far as is known, hybridising. It seems therefore that the different species are sufficiently similar to be ecologically incompatible, and that differences in the initial stages of the courtship serve as isolating mechanisms. The later phases of courtship are so similar that they would be most unlikely to be effective in preventing hybridisation.

Acknowledgements

I am most grateful to Joan Hall-Craggs for making the sonagrams used in figure 20.2 and for helpful criticism of the text. Jan Lindblad, Paul Schwartz and my wife, Barbara K. Snow, kindly allowed me to use tape recordings made by them in Guyana, Venezuela and Costa Rica. Derek Goodwin generously made available his field notes on *Chiroxiphia caudata*. My wife allowed me to use her field notes on *C. linearis*, and I am especially grateful to her for the benefit I have received in discussing with her some of the ideas put forward here.

References

Friedmann, H. and Smith, F. D. (1955). A further contribution to the ornithology of northeastern Venezuela. *Proc. U.S. natn. Mus.*, **104**, 463 – 524.

Gilliard, E. T. (1959). Notes on the courtship behavior of the Blue-backed Manakin (*Chiroxiphia pareola*). *Am. Mus. Novit.* 1942, Pp. 19.

Hall-Craggs, J. (1974). Controlled antiphonal singing by Whooper Swans. *Ibis*, **116**, 228 – 231.

Holt, E. G. 1925. The dance of the Tangara (*Chiroxiphia caudata* (Shaw)). *Auk*, **62**, 588 – 589.

Slud, P. (1964). The birds of Costa Rica. *Bull. Am. Mus. nat. Hist.*, **128**, Pp. 430.

Snow, B. K. (1970). A field study of the Bearded Bellbird in Trinidad. *Ibis*, **112**, 299 – 329.

Snow, B. K. (1972). A field study of the Calfbird *Perissocephalus tricolor*. *Ibis* **114**, 139 – 162.

Snow, D. W. (1962). A field study of the Golden-headed Manakin, *Pipra erythrocephala*, in Trinidad, W. I. *Zoologica*, **47**, 183 – 198.

Snow, D. W. (1963*a*). The display of the Blue-back Manakin, *Chiroxiphia pareola*, in Tobago, W. I. *Zoologica*, **48**, 167 – 176.

Snow, D. W. (1963*b*). The evolution of manakin displays. *Proc. 13th Int. Ornith. Congr.*, 553 – 561.

Snow, D. W. (1963*c*). The display of the Orange-headed Manakin. *Condor*, **65**, 44 – 48.

Snow, D. W. (1971). Social organization of the Blue-backed Manakin. *Wilson Bull*, **83**, 35 – 38.

Snow, D. W. and Goodwin, D. (1974). The Black-and-Gold Cotinga. *Auk*, **91**, 560 – 569.

Snow, D. W. and Lill, A. (1974). Longevity records for some neotropical land birds. *Condor*, **76**, 262 – 267.

Thorpe, W. H. (1972). Duetting and antiphonal song in birds: its extent and significance. *Behaviour Monogr. Suppl.* **18**.

Wetmore, A. (1972). The birds of the Republic of Panama. Part 3. *Smithson. misc. Collns*, **150**, Pp. 631.

21 Reliability in communication systems and the evolution of altruism

Dr A. Zahavi*

The evolution of a communication system depends on the existence of individuals which gain from it, that is the senders of the signals and their receivers. These two share a common interest about which they communicate. Common interests of this kind form the basis of the communication between such individuals as a sexual pair, parents and their offspring, and members of a flock or group which feed, roost or breed together. It is less often realised that individuals which are usually regarded as conflicting in their interests – for example, prey and predator, sexual rivals, a parasite and its host – may also share a common interest which may form the basis for the evolution of a communication system between them. Warning coloration, warning calls and other signals which are given by a prey species towards a predator (Smythe, 1970; Alcock, 1975) and also threat display among rivals, are some examples of signals exchanged between individuals which mainly conflict in their interest.

When two rivals communicate, for example by threatening, there exists the possibility that one of them may gain an advantage by cheating. Thus one may give a signal to show that it is a very powerful and able warrior while it is really a weakling and a coward. If that were possible, there is no reason why all individuals should not resort to cheating. If all cheat, the particular communication system concerned with threat will be worthless. Since from observations it is clear that the communication of threat is effective, there must be a reliability component in the system which guards against cheating. Although the problem of cheating seems to be peculiar to communication between rivals, which mostly conflict in their interests, there is no reason why a smaller conflict between otherwise collaborating individuals may not give some advantage to cheating. A male may try to cheat a potential female mate so as to increase its chances to get more or better females (Williams, 1966; Zahavi, 1975). An offspring may demand of his parents more than his share of care, food and so on (Trivers, 1974) to the extent that a conflict may arise between parents and their offspring. A group of territorial birds may likewise share a mixture of common and conflicting interests (Zahavi, in press). It is not easy to imagine two individuals with a common

*Born in Israel in 1928, Dr Amotz Zahavi is a bird watcher who studied the biology of birds with H. Mendelssohn, animal behaviour with N. Tinbergen and evolution with David Lack. From 1956 to 1969 he was General Secretary of the Society for the Protection of Nature. He is now Director of the Institute of Nature Conservation Research, and teaches eco-sociology, evolution and behaviour at Tel-Aviv University.

interest, that may not conflict in their interests at some time or in certain conditions. Thus keeping as free as possible from false information (cheating) is a basic
problem in the evolution of communication systems. This problem has been discussed in an indirect way (Zahavi, 1975) in relation to the evolution of characters
selected by sexual selection.

The Theory of the Handicap Principle in Sexual Selection

Certain aspects of sexual display, like the song of birds, their nuptial plumage or
various behaviour patterns, may be considered as signals aimed to inform a
potential mate about the quality of the sender. Since the sender of the signal may
gain by advertising false information, (that is, exaggerating its quality) and a
potential mate may lose if it believes a false signal, the receiver of a signal will
act in its own interests only if it considers signals that are not easily open to
cheating. The theory of the handicap principle (Zahavi, 1975) suggests that a
signal is reliable when the difficulty of its performance is related to its meaning in
quantity and quality.

A signal which means "I am very strong" should be more difficult to send than
the signal "I am strong". Furthermore, a stronger individual should find it less
costly to signal its strength than a weaker one, cost being measured in terms of
reproductive potential.

An individual that advertises its sexual quality with bright colours may suffer
a higher risk of predation than a dull-coloured one. In surviving, however, it
advertises a superior ability to avoid predators, demonstrating its quality in a way
which the dull-plumaged bird cannot match. Should the brightly-coloured individual lack the inherent ability to avoid predators which its colours advertise, then
it is particularly at hazard and likely to suffer predation; thus the reliability of the
signal is maintained by direct and rigorous selection.

Parent – Offspring Conflict – the Evolution of Signals which Threaten by Self Destruction

A fledgling which begs for food attracts its parents by sound, by movements of its
wings, and by the bright colours of its bill and gape. The signals also attract the
attention of the observer and most probably may attract the attention of a predator. Why is this apparently high cost of the communication between parents and
offspring necessary? I suggest that when parent – offspring relationships are
understood to contain an element of conflict (Trivers, 1974) the cost is a necessary component of the system which makes the signals reliable. The conflict
which Trivers identifies is that, although parents must consider their overall
breeding potential while caring for young, offspring must care only for themselves
(his statement was qualified by reference to the further effects of kin selection).
Thus, in their own interests, parents cannot always act according to the demands
of the group of offspring which they are tending at the moment. When offspring
try to induce their parents to feed them more, by calling louder, they may reveal
their whereabouts to predators and consequently stand a higher risk of predation.
The cost to the offspring of their begging call maintains the reliability of the
message to the parent. A signal for food which does not lower the chances of

survival of the fledgling (for example, in circumstances where predation is low) should not be as effective in inducing feeding behaviour as one made where predation risks are high. The strategem of threat by self-destruction, open to subordinate individuals which are of value to those whom they threaten, is probably widespread in its application. In human terms it appears among children who threaten to run away or damage themselves if their parents do not meet their wishes.

The white colour of the shell of a dove's egg may similarly be interpreted as a signal (a threat) to the parent to incubate more than its own selfish interests demand. Likewise the sparse down of a grebe chick may be an adaptation to make the chick helpless in water, so that the parent will be obliged to take it sooner on its back. The chick signals its body temperature to the parent by a bare patch on the forehead which is blue when the chick is cold and red when the chick is warm (personal observation). The signal of helplessness is an effective blackmail and functions as such only when the individual is really helpless. Consequently some individuals which threaten by evolving helplessness may do worse than they would have done otherwise. But on the average they do better, since parents respond to the threat more readily than they would have responded without it. They respond quicker because, although they sacrifice to the offspring more than they otherwise would, they stand to lose more if their offspring die. They choose the better of the undesirable alternatives which they face.

Warning Calls – as Communication between Prey and Predators

Warning calls are usually considered as a signal by which one prey individual warns other prey individuals of the approach of a predator. Warning behaviour is also considered (Maynard Smith, 1965) as an altruistic behaviour, since the warning individual may expose itself to danger when giving the warning. Perrins (1968) suggested that in certain circumstances the warning call may confuse the predator rather than help it to locate the prey. Charnov and Krebs (1975) suggested that in some cases the warning individual may gain by diverting the attention of the predator to other individuals that are easier to catch. But in many cases, for example with the Arabian babbler *Turdoides squamiceps* (personal observation), the warning individual often leaves cover to warn. The first indication that a group of babblers is present is often their warning call sounds.

Recently Smythe (1970) and Alcock (1975) suggested that warning signals may act as a communication system between predator and prey. Smythe's view is that the signals of the prey inform the predator that it has been seen. In a system in which a predator whose presence is known cannot easily catch prey, it is to the benefit of both prey and predator for the latter to move away to new ground once it has been sighted. I feel that this may explain the willingness of many prey species to give warning calls, even to leave cover and approach or mob the predator, effectively displaying confidence in the message which they deliver. Unless the prey endangered itself in displaying its confidence, the predator need not believe the message to be true. The confidence is shown by the warning call, which gives away the whereabouts of the prey to the predator, but even more by mobbing.

Similarly, it is generally accepted that a non-palatable prey may benefit by communicating, through colour and behaviour, its non-palatability to the predator. I suggest that the reliability of warning coloration depends on the danger to the prey inherent in bright coloration. Since only animals which have the protection of their poison may advertise themselves with impunity, warning coloration is generally a reliable communication system between prey and predator.

Threat Displays

Rivals often use threat display to resolve their conflicts. Threat may be considered as a signal to a rival with the aim of resolving a conflict without a clash (Alcock, 1975). Each of the rivals attempts, with threat signals, to impress its opponent with its potential as a fighter and its readiness to fight. Threat signals are effective since both rivals may profit by resolving the conflict without a fight. But if the threat signals were easily open to bluff, that is, if there were no mechanism to keep the correlation between the signal and the readiness and ability of the individual to fight, threat would lose its value as a means of settling a conflict without a fight. The use of a threat signal which endangers the threatening individual, in correlation to the magnitude of the threat signal, deters fighters of poor quality from threatening too much. Only the high quality fighters may threaten without great harm to their potential as fighters. A survey of threat displays reveals the handicap involved in all of them. The warning which the threat gives to the threatened individual decreases the efficiency of the attack, since it reduces the advantage of surprise. The side display used by many species as threat exposes a vulnerable part of the threatening individual and endangers it more than if it were facing its opponent head on. That is the reason why the 'underdog' in an encounter between dogs usually faces the opponent head on, since it cannot afford the handicap involved in a side display. Likewise a dominant Big horn sheep gives the advantage of the first strike to the subordinate (Geist, 1971). The magnitude of the horns and antlers of dominant ungulates is effective as a threat, since only very vigorous individuals may fight and survive with such a burden carried on their head.

Altruism among Babblers as a Communication System about Dominance Hierarchism

Arabian babblers live in groups of 3 – 15 birds. Each group defends its territory against its neighbours and nomadic birds. Data from population dynamics studies of babblers (unpublished observations) suggest that territory is essential for their survival and breeding. It is therefore reasonable to assume that group members have much common interest in defending their territory against all other babblers. On the other hand, since usually only one pair breeds in the territory, there is also much competition over breeding. There is very little overt fighting among adult group members; on the contrary there is much activity among them, such as allo-feeding and sentry duties, which is commonly interpreted as 'altruistic'.

In a system where the survival of an individual depends on the well-being of other individuals, there is much for all – even the strongest and most dominant individuals – to gain by avoiding fights. Strong animals gain by avoiding possible damage to themselves, and also by preventing damage to rivals which, in other circumstances, become helpers. Thus there should be considerable advantage in the evolution of a system of communication which enables conflicting group members to predict, communicate and agree about the outcome of a potential fight, and consequently avoid the need for it.

I suggest that the 'altruistic' activity of babblers may fit such a demand. Details and data of many of these activities will be published elsewhere; here I am concerned only with a reinterpretation of altruism illustrated by allofeeding among adult male babblers. A few hundred observations of the passage of food among known individuals show that the donor of the food is, with fewer than 1 per cent exceptions, practically always dominant to the recipient. It is possible indeed to draw up a dominance hierarchy from observations of allofeeding. Do babblers use allofeeding as a communication system?

If presenting food is a display of status, accepting it indicates a lowering of status. Hence it is not surprising to see babblers, especially those high in the hierarchy, which refrain from accepting food from their superiors. The very few cases where food is passed from a subordinate to a dominant become displays of rebellion against the established dominance hierarchy: it happened, in the cases observed, that fights (which are normally exceptionally rare among adult group members) immediately followed such rebellion, presumably to re-settle the changed social order. If allofeeding were simply a passage of food from one individual which has surplus to another in need of it, it would be difficult to explain situations in which an individual acts at one moment as a donor to its social inferior, and a minute later as a recipient from its superior.

My interpretation of allofeeding in this species is further supported by the proclamation of the feeding act by the donor. When allofeeding occurs in other species it is usually the receiver which makes a call — generally referred to as a begging call. In babblers the donor calls just before delivering the food, often delaying the passage as if to enable other members of the group to see the act. Furthermore, allofeeding usually occurs on top of a tree or in open spaces, where it can readily be seen by the group.

Much altruistic activity, if not all, may have originated in this way through simple individual selection among animals that cooperate with each other. It may thus be unnecessary to invoke more complicated selective mechanisms, such as group selection and kin selection, to explain its evolution.

Some General Considerations of the Handicap Principle

In the preceding paragraphs a number of characters have been reinterpreted on the basis of simple individual selection. The handicap principle suggests that every communication system has its reliability component which takes the form of a handicap. Since communication systems occur commonly, and many may not yet have been recognised as such, it is possible that several characters besides altruism may be reinterpreted as reliability components of hitherto unsuspected communication systems. In this context it is interesting to re-examine the controversy between Lack and Wynne-Edwards about the role of territorial behaviour in the regulation of animal populations to the level of their food supply.

It was Wynne-Edwards's view (1962) that territorial behaviour regulates size of population through a mechanism involving group selection. Lack (1966) did not support the concept of group selection, but took an opposing view which, in its extreme, questioned whether territorial behaviour had any role at all in regulating population numbers through the partitioning and conservation of food supplies. My own interpretation of territorial behaviour, in terms of the handicap principle, is based on an assumption that an animal advertising ownership of territory is, at the same time, advertising its own quality as a territory holder. The handicap (that is, the cost in terms of extra work) involved in defending a territory that is larger than the animal needs for feeding, is repaid by its value as an advertisement of the superior quality of the individual. Thus the phenomenon which Wynne-Edwards explained in terms of group selection, and Lack had difficulty in accounting for at all, can be resolved as a simple case of individual sexual selection. This interpretation is based on an earlier suggestion by O'Donald (1963) which involved, however, a different understanding of sexual selection.

When handicaps are considered as no more than the costs of social signals, it is not necessary to postulate a causal relationship between the cost of the signal and its meaning. This present reinterpretation, which regards handicaps as essential reliability components in communications systems, suggests that the cost is a necessary component of the signal; the more significant the signal, the higher the costs to the performer. Since it is reasonable that individuals should attempt to advertise (that is, signal) differences in their qualities, the characters important in determining quality should be affected adversely by the signal.

With this hypothesis it should be possible to predict from a study of the signal (for example its colours, sounds, movements) the nature of the selecting factors that are acting on the population. Theoretically a long-term experiment may decide between the two approaches. If it is true that a communication system simply has its costs, lowering the costs should result in increased use of the system. On the other hand the handicap principle elaborated here suggests that lowering the costs may at first increase the use of the system, but further decrease in cost should destroy it altogether. Since the lowering of cost has probably occurred often in nature, it should not be surprising to find adaptations in communications systems which maintain the value of the systems despite cost reductions. Such an adaptation should imply the understanding of a signal in relation to the circumstances – that is, whether the cost in a particular situation is high or low, whether a consequent evaluation of the particular situation is high or low, and what the signal accordingly means. This we understand intuitively; the man who dares to stand up when the bullets are *not* flying around is not considered particularly courageous.

Although I have discussed the value of the handicap principle in a restricted range of contexts, I believe it to be applicable to all communications systems. Other examples to which it may apply are relationships between host and parasite, and host and symbiont, and that between two interacting unicellular organisms which communicate by chemical means. With adaptations which will be discussed elsewhere, relationships among true social insects and even among cells of a multicellular organism may also be explained in this way.

Acknowledgements

I owe my inclination to reinterpret data with the orthodox approach of individual selection to the time I was privileged to spend with David Lack. If the theory is true it is credit to his teaching. If it fails, the mistakes are mine. Many thanks are due to my wife Dr A. Kadman Zahavi for fruitful discussions and many improvements in the manuscripts. Thanks are also due to Drs J. Terkel, R. Trivers, H. Kruuk and J. Krebs, and Professor Barends and Dr Bernard Stonehouse for comments on earlier drafts. The research on babblers is supported by a grant from the Israel Commission for Basic Research.

References

Alcock, J. (1975). *Animal Behaviour. An Evolutionary Approach.* Sinauer Associates Inc., Sunderland, Massachusetts.

Charnov, L. E. and Krebs, J. R. (1975). The evolution of alarm calls; altruism or manipulation. *Am. Nat.*, **109**, 107 - 112.

Geist, V. (1971). *Mountain Sheep. A Study in Behaviour and Evolution.* University of Chicago Press, Chicago.

Lack, D. (1966). *Population Studies of Birds.* Clarendon Press, Oxford.

Maynard Smith, J. (1965). The evolution of alarm calls. *Am. Nat.*, **99**, 59 - 63.

O'Donald, P. (1963). Sexual selection and territorial behaviour. *Heredity*, **18**, 361 - 364.

Perrins, C. (1968). The purpose of the high-intensity alarm call in small passerines. *Ibis*, **110**, 200 - 201.

Smythe, N. (1970). On the existence of 'pursuit invitation' signals in mammals. *Am. Nat.*, **104**, 491 - 494.

Trivers, R. L. (1974). Parent offspring conflict. *Am. Zool.* **14**, 249 - 264.

Williams, G. C. (1966). *Adaptation and Natural Selection: A Critique of Some Current Evolutionary Thoughts*, Princeton University Press, Princeton, New Jersey.

Wynne-Edwards, V. C. (1962). *Animal Dispersion in Relation to Social Behaviour,* Oliver and Boyd, Edinburgh.

Zahavi, A. (1975). Mate selection – a selection for a handicap. *J. Theor. Biol.*, **53**, 205 - 214.

Zahavi, A. (1976). Cooperative nesting in Euro-asiatic birds. *Proc. 16th int. Ornith. Congr.*, 685 - 693.

22 The adaptive significance of variations in reproductive habit in the Agavaceae

Dr W. M. Schaffer and M. V. Schaffer*

Introduction

Of considerable interest to students of life history phenomena is the identification of circumstances that favour the evolution of life cycles in which a single episode of reproduction is followed by rapid degeneration and death of the reproductive individual. Among non-annual plants and animals, this kind of life history is relatively uncommon, but curiously has arisen independently in several groups (for example, palms, periodic cicadas, bamboos, salmon). In this chapter, we summarise evidence implicating pollinator foraging behaviour as the factor which has led to the evolution of "Big Bang" reproduction in yuccas and agaves. In addition we also consider the effects of variation in plant reproductive expenditure on the pollinators, and finally the way in which plants and pollinators may have coevolved.

Variations in Reproductive Expenditure in Yuccas and Agaves

Yuccas and agaves are semisucculent plants that grow in the deserts and chapparal regions of western North America. Traditionally (see, for example, Lawrence, 1951), the two genera have been placed in different families (Lily and Amaryllis), but karyotypic studies (for example, Gómez-Pompa, *et al.*, 1971) suggest a much closer affinity. Gentry's (personal communication, 1972) morphological studies of *Agave dasylerioides* point to a similar conclusion.

With regard to structure and ecological requirements, the two genera are similar in several respects. They share a common basic morphology which consists of a rosette of stiff, often armoured leaves from which emerges a flower-bearing stalk. Often, they occur in the same locality, although agaves are more usually found in rockier, better-drained soils. Despite these similarities, there are important differences in their reproductive biology. Whereas all yuccas appear to be pollinated exclusively by moths of the genus *Tegeticula* (Riley, 1892; Trelease, 1893; Powell and Mackie, 1966) with which they practise the marvellous symbiosis that has become in

*William M. Schaffer is a student of the late Robert MacArthur and received his doctorate in biology from Princeton University in 1972. His principal interests are in the areas of life history theory and community co-evolution. At present, he is an Assistant Professor of Ecology and Evolutionary Biology at the University of Arizona (Tucson).

*M. Valentine Schaffer is a student of French history. Her major interest is the reign of Louis XIV. This chapter is the result of three summers' collaboration.

standard textbook fare (for example, Proctor and Yeo, 1972), agaves receive benefit from a variety of animal pollen vectors (Baker, 1961). These include pollen-
and nectar-eating bats (Porsch, 1936; Van der Pijl, 1936; Howell, 1972), carpenter
bees and bumblebees (Schaffer and Schaffer, 1976), and probably also hummingbirds in the case of *A. polianthaflora*. A second major difference relates to expenditure on sexual reproduction. In most agaves, the flowering rosette dies shortly
after fruit set, although it may have previously duplicated its genotype by pro-
pagating vegetatively (Gentry, 1972, personal communication; personal observation). Most yuccas, on the other hand, do not die after flowering. Because of the
aforementioned tendency of agaves to reproduce asexually, the distinction in
reproductive habit is not quite that of semelparity *versus* iteroparity, although it
should be noted that some agaves (such as, *A. ochaui*, *A. multifilifera*, *A. chryso-
glossa*) reproduce exclusively by seed. In any event there is a qualitative
difference between the two genera in expenditure on sexual reproduction by
functionally independent rosettes. This dichotomy is summarised in table 22.1
where we have listed the post-flowering half life (PFHL) for several species. PFHL

Table 22.1   Post-flowering rosette survival in yuccas and agaves

	Yucca			*Agave*	
Species	N†	Half-life (months)	Species	N†	Half-life (months)
whipplei	265	3.7‡	*utahensis*	86	1.5‡
glauca	266	~48*$	*deserti*	85	2.0‡
utahensis	243	~56*$	*chrysantha*	79	2.5‡
elata	357	~70*$	*toumeyana*	124	3.0‡
			palmeri	74	5.4$
			schottii	136	7.8$
			parviflora	237	29.7$

*Estimated by extrapolation.
†Number of individuals.
‡Survivorship data calculated from a single cohort.
$Survivorship data calculated from multiple cohorts.

is defined as the time taken for half the rosettes in a population to die after
flowering. Where possible, post-flowering survival was determined by following a
single cohort. In long-lived species, however, it was necessary to use multiple
cohorts, in which case the year of flowering was inferred from the condition of the
woody stalk. Note that in each genus, there is one species which is exceptional to
the generic norm. Thus, *Yucca whipplei* has a PFHL of about 4 months, whereas
in the remaining species, the corresponding figure is much greater. Similarly, most
agaves have PFHL's of less than a year, but rosettes of *A. parviflora* survive for up
to 36 months. During this time, they propagate asexually by means of suckers.

How do we account for the evolution of this variation? Clearly, differential
ancestry is not the answer, for if this were the case, we would not expect exceptional species like *Y. whipplei* and *A. parviflora*. Similarly, PFHL does not appear
to vary consistently with the intensity of vegetative reproduction. Within
Y. whipplei, for example, there are some populations which produce an abun-

dance of suckers and others which propagate largely by seed (Haines, 1941; Hoover, 1973). In both cases, however, the flowering rosettes die shortly after fruit set. Similarly, vegetative reproduction is entirely lacking in *A. dasylerioides* and *A. ochaui*, but whereas the former appears capable of flowering in successive years (Gentry, personal communication), the latter is a true monocarp (PFHL ≃ 3 months, personal observation). Nor do the *names* (as opposed to the foraging behaviour) of the pollinators appear to be the decisive factor. *Yucca whipplei*, for example, is pollinated by moths of the same genus (albeit a different species) that pollinate the perennial yuccas. Finally, there do not seem to be any clear-cut correlations of PFHL with climate or edaphic factors.

If the obvious hypotheses fail, what then are the selective agents producing the variation in reproductive habit summarised in table 22.1? Several possible explanations are suggested by the theory of optimal reproductive strategies developed by Cole (1954), Williams (1966), Gadgil and Bossert (1970) and Schaffer (1974; see also Schaffer and Gadgil, 1975; Schaffer and Rosenzweig, 1977). The principal relevant conclusion from this literature is that an all or nothing pattern of reproduction will be selected for if at every age, i, the relationship between current fecundity, b_i, and post-breeding survival multiplied by subsequent reproductive value, $p_i(v_{i+1}/v_0)$, is of the kind shown in figure 22.1a — that is, if the second derivatives,

$$\frac{d^2 b_i}{d\left(p_i \frac{v_{i+1}}{v_0}\right)^2}$$

are positive*. By way of contrast, we observed that sub-maximal reproduction will be optimal if these derivatives are negative (figure 22.1b). Among the circumstances that might cause $b_i [p_i(v_{i+1}/v_0)]$ to have positive second derivatives, three may apply to the present situation. The first of these is that the percentage of seed set might increase with increasing reproductive expenditure if seed predators were able to totally destroy the crops on small inflorescences (before dehiscence), but were swamped by the sheer number of seeds or flowers on larger flower stalks. Janzen (1976) has proposed a variant of this explanation to account for massive and synchronous flowering in bamboos, and Gentry (personal communication) reports at least one instance of near-total destruction of flowers in the bud by an unspecified insect in the case of *A. falcata*. Nevertheless, after three years of studying agaves in Arizona, Utah, and Sonora, we have not personally encountered such a situation and are therefore not inclined to believe that it is of general importance.

A second possible explanation also supposes that seed set increases convexly with reproductive output, but in this case the proposed mechanism involves pollinator selectivity for large flower stalks. If, within a population of plants, the number of pollinators relative to the number of flowers is not too large, one might imagine that for energetic reasons (Heinrich, 1972; 1975; Heinrich and Raven, 1972),

---

*The reader is reminded that the theoretical studies cited do not deal explicitly with the problem of asexual reproduction. An extension of the theory to cover organisms capable of vegetative as well as sexual reproduction is at present in preparation.

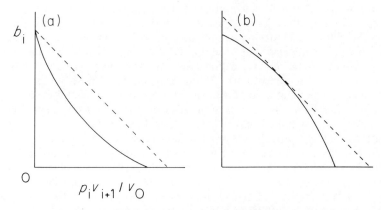

Figure 22.1 Trade-off between current fecundity, b_i, and subsequent survival and reproductive value, $p_i(v_{i+1}/v_0)$. Life history theory suggests that selection maximises for all ages, i, the sum of these two variables. (a) If $b_i[p_i(v_{i+1}/v_0)]$ is concave, an all or nothing pattern of reproduction is favoured. (b) if $b_i[p_i(v_{i+1}/v_0)]$ is convex, intermediate reproductive rates will be optimal.

pollinators will more often choose to visit the larger inflorescences that would provide the greatest nectar and pollen rewards. Recent studies of avian foraging behaviour (Krebs *et al.*, 1974; Gill and Wolf, 1975a) suggest that nectivorous birds are capable of foraging in a manner that comes close to maximising the rate of caloric intake. Our own studies support a similar conclusion for *Xylocopa* and *Bombus* visiting *A. totumeyana* and *A. schotti*. (For relevant discussions of foraging behaviour in *Bombus*, see Brian, 1957; Manning, 1956.) In addition, Howell (personal communication) reports observations on nectar feeding bats coming to *A. palmeri* that are also consistent with optimal foraging. The hypothesis of pollinator selectivity thus predicts that within populations of Big Bang (short PFHL) Agavaceae, the percentage of flowers fertilised should increase with inflorescence size, whereas this should not be the case for non-Big Bang species. Note that if this hypothesis is correct, it has another aspect, namely that plant reproductive strategy should reciprocally influence pollinator foraging behaviour. We will return to this co-evolutionary theme in the discussion.

A final possibility suggested by the theory is that the cost (Schaffer and Gadgil, 1975)* of heightened reproductive expenditure levels off with increasing effort in Big Bang species but continues to rise at a constant or increasing rate in non-Big Bang species. There is some circumstantial evidence to support this idea, and it comes from observations on root structure. As mentioned previously, agaves and *Y. whipplei* are often restricted to rocky situations and, as might be expected, have superficial root systems (Gentry, personal communication; personal observation). The roots of perennial yuccas, on the other hand appear to go much deeper (Campbell, 1932), and these plants are typically found in deeper soils. From this, one could argue that any water expended by an agave on flowering could not be

---

*These authors show that the cost of increasing reproductive expenditure from E_i to E_i' is $pV(E_i) - pV(E_i')$ where $pV = p_i(v_{i+1}/v_0)$.

replaced until the next rain which might not occur in an arid environment for some time. Conceivably, the cost of even a small reproductive expenditure could be severe. Deep roots, on the other hand, might permit a more continuous uptake of water with the consequence that the cost of flowering might increase more gradually with reproductive expenditure. This hypothesis is appealing and deserves to be tested. Lacking the manpower and funds to do so, we have set it aside and concentrated out efforts on the matter of pollinator selectivity. Fortunately the two explanations are not mutually exclusive, and in fact may both be pertinent.

Evidence for the Pollinator Selectivity Hypothesis

To test the hypothesis proposed in the previous section, we first studied the manner in which the percentage of flowers developing into fruits varied with stalk height within populations of 13 species. In each instance, flower stalks were divided into several categories according to size. On each stalk, we counted the number of fruits set and also the number of scars left by flowers which did not develop into mature capsules. The mean percentage of flowers developing into fruits was then calculated for each size category and the slope, m_f, of the regression line: percentage flowers developing into fruits against stalk height, computed. For purposes of comparison, the stalk heights in all species were standardised relative to the most common size category. Assuming that pollination is the prime determinant of whether or not a given ovary matures, the slopes so calculated are an index of pollinator preference for large inflorescences.

The data from which these putative indices were determined are presented in the Appendix (table A1). The indices themselves are given in table 22.2. On the whole, they seem to support the hypothesis very nicely. Thus, for the iteroparous yuccas, m_f ranges from -0.02 to $+0.01$, whereas for *Y. whipplei*, $m_f = +0.08$.

Table 22.2 Putative pollinator selectivity based on percentage fertilisation data

Species	N‡	m_f§	r¶	Species	N	m_f	r
		Big Bang			Big Bang (PFHL < 8 months)		
whipplei	23	0.08	0.88	*utahensis*	86	0.24	0.95
		Repeat Bloomers		*deserti**	122	0.15	0.80
elata	84	0.01	0.23	*chrysantha*	20	0.18	1.00
glauca	117	0.00	0.00	*toumeyana**	91	0.20	0.93
standleyi	100	0.00	0.02	*palmeri*	48	0.31	0.75
*utahensis**	161	−0.02	−0.10	*schottii**	182	0.17	0.81
					Non-Big Bang (PFHL = 29.7 months)		
				parviflora†	211	−0.03	−0.32

*Average of two samples.
†Average of three samples.
‡Total sample.
§Slope of the regression line: percentage flowers developing into fruits against standardised stalk height.
¶Average correlation coefficient; calculated from grouped data.

Again, for the Big Bang agaves, m_f ranges from +0.15 to +0.32, but for *A. parvi-flora*, $m_f = -0.03$. Analysis (*t* test) indicates that these differences are significant within and between genera. Thus, *Y. whipplei* and *A. parviflora* differ significantly from their congeners. Also, the difference in m_f values for all iteroparous species is significantly different from that of all Big Bang species ($t = 5.97$, $P < 0.005$ one-tailed). In addition, if we take all species together, we find that m_f is inversely, and significantly, correlated with the PFHLs given in table 22.1 (figure 22.2).

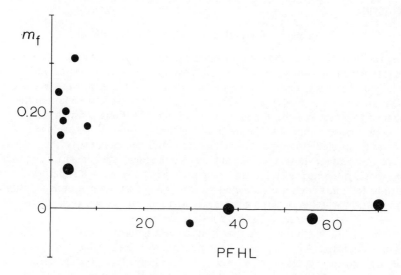

Figure 22.2 Slope, m_f, of the regression line: percentage flowers developing into fruits versus stalk height, plotted against post-flowering half-life, PFHL, for 11 species of yuccas, ●, and agaves •, $r = -0.77; P(r) < 0.005$, one-tailed.

A second line of evidence supporting the hypothesis comes from direct observations of the pollinators themselves. As mentioned above, yuccas are pollinated exclusively by moths of the genus *Tegeticula* whereas the agaves treated here derive benefit from large bees (*Bombus* and *Xylocopa*) and nectar-feeding bats (chiefly *Leptonycteris*; see Howell, 1972; and Schaffer and Schaffer, 1977 for details). Thus far, we have restricted our attention to the bees that visit *A. parviflora* (*Bombus sonorus*), *A. schottii* (*B. sonorus*). *A. toumeyana* (*Xylocopa arizonensis*) and *A. utahensis* (*X. arizonensis*). Pollinator selectivity was estimated in the following way: (1) The flower stalks in a population were measured for total height and the length of stalk (cm) in bloom. These data were recorded and the plants tagged. (2) A census schedule was established – generally every 2 – 3 h from dawn to dusk. (3) The plants were then censused for several days by an observer who would walk the study site along a pre-arranged route and note the number of pollinators present on each flower stalk. An effort was made by the observer not to look at the stalk until he had approached closely enough to make out the number on the tag. (4) Stalks were assigned to size categories and for each census the number of pollinators per cm of stalk in bloom per observation were calculated for each size category. (5) These numbers were averaged for each stalk size category and the slope, m_p, of the regression line: pollinators per cm bloom per observation against stalk height was calculated. As for m_f, stalk heights were

Table 22.3 Pollinator selectivity in agaves (direct observation)

Agave species	Pollinator	Sample*	r†	m_p‡
A. parviflora	*Bombus sonorus*	108/274	0.13	0.01
A. schottii (1973)		42/351	0.99	0.13
A. schottii (1974)	*Bombus sonorus*	63/837	0.78	0.09
A. schottii (1975)		227/1418	0.77	0.07
A. toumeyana (1974)	*Xylocopa arizonensis*	122/178	1.00	0.17
A. toumeyana (1975)		661/1890	0.66	0.10
A. utahensis	*Xylocopa arizonensis*	7/45	0.64	0.15

*Pollinator observations/plant observations.
†Correlation coefficient of grouped data; see table A2.
‡Slope of regression line: pollinators/cm in bloom/plant observations against standardised stalk height.

standardised relative to the most common size class. Table 22.3 gives the values of m_p obtained for two populations of *A. schottii* (PFHL = 7.8 months), 3 populations of *A. toumeyana* (PFHL = 3 months), and one population each of *A. parviflora* (PFHL = 29.7 months) and *A. utahensis* (PFHL = 1.5 months). The data from which these values were obtained are given in table A2 in the Appendix. Once again, we find that our observations are in harmony with the notion that when visiting populations of Big Bang species, pollinators prefer the larger flower stalks,

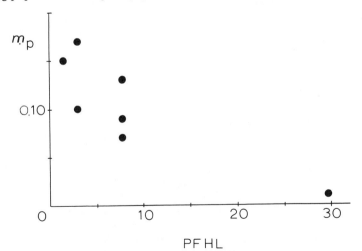

Figure 22.3 Slope, m_p, of the regression line: pollinators/cm stalk in bloom/plant observation *versus* stalk height, plotted against PFHL for four species of agaves. $r = -0.85$; $P(r) < 0.01$, one-tailed.

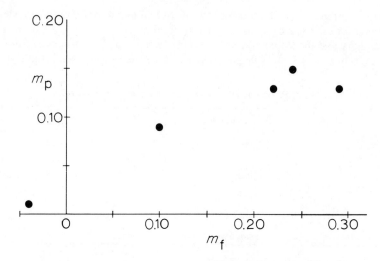

Figure 22.4 m_p plotted against m_f in five populations for which both estimates of pollinator selectivity were available. $r = 0.95$; $P(r) < 0.01$, one-tailed.

whereas in non-Big Bang species, such preference is not exercised. Thus, for the Big Bang agaves, m_p ranges from 0.07 to 0.17, whereas in *A. parviflora*, $m_p = 0.01$. Comparing *A. parviflora* with the other populations, we find that the difference in m_p values is significant ($t = 2.77$; $P < 0.025$ one-tailed), and again, m_p is significantly correlated with PFHL (figure 21.3; $r = -0.89$, $P(r) < 0.005$, one-tailed). Moreover, in those populations for which we have values of both m_f and m_p in the same season, we note that the two estimates of pollinator selectivity are well correlated (figure 22.4; $r = 0.95$; $P(r) < 0.01$, one-tailed). This suggests that the m_f values, of which we have a far greater number, are good estimators of pollinator behaviour.

Discussion

An interesting point about the pollinator selectivity data presented in table 22.3 is that the same pollinator, *Bombus sonorus*, appears to prefer the taller flowers stalks when working *A. schottii* ($m_p = 0.07 - 0.13$), but not when visiting *A. parviflora* ($m_p = 0.01$). This brings us back to the observation made previously, that pollinator selectivity should depend, at least in part, on plant reproductive strategy. Specifically, we propose that increases in plant expenditure on reproduction should select for increasing pollinator preference for the larger inflorescences within a population. If correct, this statement gives us the basis for analysing plant – pollinator co-evolution as shown in figure 22.5. Here, optimal plant reproductive effort is plotted against pollinator selectivity and optimal selectivity against reproductive expenditure. Selection (arrows, figure 22.5a) will move a given system (represented by a point on the graph) towards the isoclines of maximum fitness. Notice that such a system can have (but does not require) two co-evolutionary

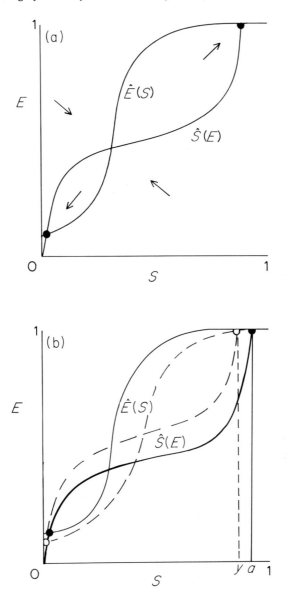

Figure 22.5 Co-evolution of plant reproductive effort, E, and pollinator selectivity for large flower stalks, s. $\hat{E}(s)$ is the optimum effort value for given selectivity; $\hat{s}(E)$ is the optimum selectivity for given effort. (*a*) Evolution moves a system toward the lines of maximum fitness (arrows). Note that there can be two co-evolutionary steady states. (*b*) The solid lines represent possible isoclines for *Agave*; the dashed lines, for *Yucca*. Note that the Big Bang equilibrium is less likely in the latter case. See text for discussion.

steady states. One of these corresponds to Big Bang reproduction in the plants and high pollinator selectivity; the other to low values of these parameters. Such a model, of course, would explain the patterns we have observed. However, the question remains, do we have evidence supporting the assumption that pollinator selectivity increases with the average reproductive effort made by plants within a population? At present, the answer is yes and no.

To begin with, we note that increasing plant reproductive expenditure results in an increase in the amount of rewards available to pollinators. If the number of pollinators does not likewise increase, the ratio of rewards to foragers will increase. So if pollinators are foraging to maximise their rate of caloric intake, they should concentrate their attentions on the more profitable, larger flower stalks. This last remark follows from almost any theory of optimal diets (for example, MacArthur, 1972; Schoener, 1974; see also Heinrich, 1975; Heinrich and Raven, 1972), and appears to apply to the bees which visit *A. toumeyana* and *A. schottii*. Two lines of evidence support this contention. First, in *A. schottii* pollen and nectar resources are most available to bees at dawn and at dusk, whereas during the middle of the day they are quite scarce (Schaffer and Schaffer, 1976). Significantly, pollinator selectivity is greatest at the times of maximal resource availability and is much lower at other times of the day. This rhythm has been observed in populations of *B. sonorus* working *A. schottii* in Molino Basin, and Patagonia, Arizona, and is based on sizeable samples. Second, pollinator selectivity can be artificially enhanced by bagging plants for a time and allowing nectar and pollen to accumulate to greater than normal levels. When the bags are removed, the pollinators exhibit greater preference for the tall stalks than is usual for that time of day. Moreover, selectivity returns to normal levels on the day after the experiment. Manipulations of this sort have been performed on two populations of *A. schottii* and one population of *A. toumeyana*. Limitations of space preclude a more detailed description of the methods and results, but the main point seems quite clear, that *short-term* increases in resource availability on all flower stalks in a population leads to a greater level of pollinator preference for the larger stalks. Unfortunately, this observation does not quite justify the analysis summarised by figure 22.5. For here we are talking about *long-term* evolutionary increases in rewards. If pollinator density can track increases in these rewards, then the ratio of resources to foragers will not increase and optimal pollinator selectivity should remain constant. Two aspects of yucca – agave biology suggest that this will not happen. First, yuccas and agaves are characteristic of unpredictable environments. To a great extent, the plants can escape the effects of drought by not reproducing in a dry year, whereas the pollinators are likely to suffer high mortality. Thus, uncertainty in rainfall probably reduces the *average* consumer/resource ratio. Secondly, in the case of the pollinators of *Agave*, population growth depends on a series of plants that bloom at different times of the year. Thus, an increase in rewards provided by one plant species would be expected to have but minimal effect on overall pollinator density and might even reduce it (Fretwell, 1972). In the case of *Yucca*, this argument does not apply — the moths depend solely on yucca plants for larval food. On this ground, we would expect the moths to be better able to track increases in plant reproductive effort. Accordingly, we expect optimal selectivity to increase less rapidly with plant reproductive effort. Another difference between the yucca and agave systems is that in *Yucca* the pollinator extracts a price for its services by injecting eggs into the plant's ovary. The resulting larvae consume a portion of the developing seeds before pupating in the ground at the base of the plant. Consequently, we might expect optimal plant effort to increase less rapidly with pollinator selectivity than in *Agave*.

These considerations are summarised in figure 22.5*b* where hypothetical isoclines for *Yucca* and *Agave* are compared. Notice that the moth–yucca interaction is the less likely to have the high effort steady state. Additionally, we note that the Big Bang equilibrium may correspond to a lower value of pollinator selectivity. Possibly this explains the relatively low value of m_f observed for *Yucca whipplei*.

In conclusion, we would like to suggest that when the pollinators are territorial, (those discussed above are not) their preference for large inflorescences may have a selective effect quite opposite to that which we have indicated. Specifically, studies of hummingbird – flower interactions suggest that dense clumps of flowers are less efficiently pollinated, or at least outcrossed, than sparser stands (Linhart, 1973; Gill and Wolf, 1975*b*; Stiles, 1975; Brown, personal communication). This follows from the observation that dense patches are more likely to be included in a territory from which other birds are excluded. So in this case maximal reproductive effort on the part of the plants would be selected against. It is perhaps therefore significant that of the 129 hummingbird flowers in western North America studied by Grant and Grant (1968), only 5 are annuals. In contrast, the annual habit is common in co-occurring bee-pollinated species.

Acknowledgements

This study owes much to the ideas of the late Drs David Lack and Robert Mac-Arthur, and was inspired largely by their divergent but complementary approaches to the study of evolutionary ecology. It is a pleasure to acknowledge our debt to these two great colonisers. In addition, we wish to thank colleagues and friends who have been a source of comment and criticism over the past three years. James H. Brown, Astrid Kodric-Brown, Howard Scott Gentry, Arthur Gibson, Lionell G. Klikoff, and Michael L. Rosenzweig deserve special acknowledgement. This work was supported generously by the Universities of Utah and Arizona and by NIH Biomedical Sciences Grant No. FR-070902 to the University of Utah.

References

Baker, H. G. (1961). The adaptation of flowering plants to nocturnal and crepuscular pollinators. *Q. Rev. Biol.*, **36**, 64 – 73.

Brian, A. D. (1957). Differences in the flowers visited by four species of bumblebees and their causes. *J. Anim. Ecol.*, **26**, 71 – 91.

Campbell, R. S. (1932). Growth and reproduction of *Yucca elata. Ecology*, **13**, 364 – 374.

Cole, L. C. (1954). The population consequences of life history phenomena. *Q. Rev. Biol.*, **29**, 103 – 137.

Fretwell, S. (1972). *Populations in a Seasonal Environment.* Princeton University Press, Princeton, New Jersey.

Gadgil, M. and Bossert, W. (1970). Life history consequences of natural selection. *Am. Nat.*, **104**, 1 - 24.

Gentry, H. S. (1972). The agave family in Sonora. *Agriculture Handbook*, 399. US Department of Agriculture.

Gill, F. B. and Wolf, L. L. (1975*a*). Foraging strategies and energetics of East African sunbirds at mistletoe flowers. *Am. Nat.*, **109**, 491 - 510.

Gill, F. B. and Wolf, L. L. (1975*b*). Economics of feeding territoriality in the Golden-winged sunbird. *Ecology.*, **56**, 333 - 345.

Gómez-Pompa, A., Villalobos-Pietrini, R. and Chimal, A. (1971). Studies in Agavaceae. I. Chromosome morphology and number of seven species. *Madroño*, **21**, 208 - 221.

Grant, K. A. and Grant, V. (1968). *Hummingbirds and their flowers.* Columbia University Press, New York.

Haines, L. (1941). Variation in *Yucca whipplei. Madroño*, **6**, 33 - 45.

Heinrich, B. (1972). Energetics of temperature regulation and foraging in a bumblebee, *Bombus terricola* Kirby. *J. comp. Physiol.*, **77**, 49 - 64.

Heinrich, B. (1975). Beeflowers: A hypothesis on flower variety and blooming times. *Evolution*, **29**, 325 - 334.

Heinrich, B. and Raven, P. H. (1972). Energetics and pollination ecology. *Science*, **176**, 597 - 602.

Hoover, D. A. (1973). Evidence from population studies for two independent variation patterns in *Yucca whipplei* Torrey. M.S. Thesis. California State University, Northridge.

Howell, D. (1972). Physiological adaptations in the syndrome of chiropterophily with emphasis on the bat *Leptonycteris* Lydekker, Ph.D. Thesis, University of Arizona.

Janzen, D. H. (1976). Why bamboos wait so long to flower. *Ann. Rev. Ecol. Syst.*, **7**, 347 - 391.

Krebs, J. R., Ryan, J. C. and Charnov, E. L. (1974). Hunting by expectation or optimal foraging? A study of patch use by chickadees. *Anim. Behav.*, **22**, 953 - 964.

Lawrence, G. H. M. (1951). *The taxonomy of vascular plants.* Macmillan, New York.

Linhart, Y. B. (1973). Ecological and behavioral determinants of pollen dispersal in hummingbird-pollinated *Heliconia. Am. Nat.*, **107**, 511 - 523.

Manning, A. (1956). Some aspects of foraging behavior of bumblebees. *Behavior*, **9**, 164 - 201.

MacArthur, R. H. (1972). *Geographical ecology.* Harper and Row, New York.

Porsch, O. (1936). Das Bestaubungsleben der Kakteenblute II. *Gesell. Jb. Cact.*, **1**, 119 - 133.

Powell, J. A. and Mackie, R. A. (1966). Biological interrelationships of moths and *Yucca whipplei. Univ. Cal. Publ. Ent.*, **42**, 1 - 59.

Proctor, M. and Yeo, P. (1972). The pollination of flowers. Taplinger Publishing Company, New York.

Riley, C. V. (1892). The yucca moth and yucca pollination. *Ann. Rep. Mo. Bot. Grdn.*, **3**, 99 - 159.

Schaffer, W. M. (1974). The evolution of optimal reproductive strategies: the effects of age structure. *Ecology*, **55**, 291 - 303.

Schaffer, W. M. and Gadgil, M. V. (1975). Selection for optimal life histories in plants, *in Ecology and Evolution of Communities*. (ed. M. L. Cody and J. M. Diamond) Belknap Press, Cambridge, Massachusetts.

Schaffer, W. M. and Rosenzweig, M. L. (1977). Selection for optimal life histories. II: Multiple equilibria and the evolution of alternative reproductive strategies. *Ecology,* **58** (in press).

Schaffer, W. M. and Schaffer, M. V. (1977). The reproductive biology of Agavaceae: I. Pollen and nectar production in four Arizona agaves. *S. W. Natur.* **22** (in press).

Schoener, T. W. (1974). The compression hypothesis and temporal resource partitioning. *Proc. natn. Acad. Sci. U.S.A.*, **71**, 4169 – 4172.

Stiles, F. G. (1975). Ecology, flowering phenology, and hummingbird pollination of some Costa Rican *Heliconia* species. *Ecology.*, **56**, 285 – 301.

Trelease, W. (1893). Further studies of yuccas and their pollination. *Ann. Rep. Mo. Bot. Grdn.*, **4**, 181 – 226.

Van der Pijl, L. (1936). Fledermause und Blumen. *Flora*, **131**, 1 – 40.

Williams, G. C. (1966). Natural selection, the costs of reproduction and a refinement of Lack's principle. *Am. Nat.*, **100**, 687 – 690.

Appendix

The following tables summarise the data from which the values of m_f and m_p presented in tables 22.2 and 22.3 were calculated.

Table A.1 Proportion of flowers fertilised in yuccas and agaves

	Yucca whipplei (1972, Pinon Flat, California, PFHL = 3.7 months)					
R.S.H.*	−0.50	−0.30	−0.10	0.10	0.30	
N†	3	5	5	5	5	
Prop. Ft.‡	0.01	0.01	0.03	0.03	0.08	
s.e.m.§	−	−	−	−	−	
	m_f¶ = 0.08, r¶ = 0.88					
	Y. elata (1973, Akela Flats, New Mexico, PFHL ≃ 70 months)					
R.S.H.	−0.85	−0.57	−0.29	0	0.29	0.57
N	1	4	26	33	11	9
Prop. Ft.	0.00	0.07	0.06	0.06	0.04	0.04
s.e.m.	−	−	0.01	0.01	0.02	−
	m_f = 0.01, r = 0.23					
	Y. glauca (1973, Las Lunas, New Mexico, PFHL ≃ 38 months)					
R.S.H.	−0.44	−0.22	0	0.22	0.44	
N	9	29	55	20	4	
Prop. Ft.	0.07	0.08	0.06	0.06	0.08	
s.e.m.	−	0.02	0.01	0.02	−	
	m_f = 0.00, r = 0.00					

Y. standleyi (1973, Vermillion Cliffs, Arizona)

R.S.H.	−0.36	−0.18	0	0.18	0.36	0.55
N	6	33	38	17	5	1
Prop. Ft.	0.08	0.06	0.05	0.08	0.07	0.07
s.e.m.	−	0.01	0.01	0.02	−	−

$m_f = 0.00, r = 0.02$

Y. utahensis (1972, St George, Utah, PFHL ≃ 56 months)

R.S.H.	−0.52	−0.26	0	0.26	0.52
N	4	29	31	22	2
Prop. Ft.	0.03	0.07	0.10	0.10	0.09
s.e.m.	−	0.02	0.03	0.01	−

$m_f = 0.06, r = 0.80$

Y. utahensis (1973, St George, Utah)

R.S.H.	−0.60	0	0.40
N	12	43	18
Prop. Ft.	0.16	0.10	0.06
s.e.m.	0.09	0.02	0.02

$m_f = -0.10, r = -1.00$

Agave utahensis (1973, St George, Utah, PFHL = 1.5 months)

R.S.H.	−0.78	−0.44	−0.22	0	0.22	0.44
N	9	9	16	30	16	6
Prop. Ft.	0.10	0.29	0.29	0.35	0.37	0.43
s.e.m.	−	−	0.05	0.05	0.06	−

$m_f = 0.24, r = 0.95$

A. deserti (1973, Kingman, Arizona, PFHL = 2.0 months)

R.S.H.	−0.80	−0.40	0	0.40
N	5	11	53	19
Prop. Ft.	0.00	0.18	0.19	0.19
s.e.m.	−	0.08	0.03	0.04

$m_f = 0.15, r = 0.80$

A. deserti (1974, Kingman, Arizona)

R.S.H.	−0.40	0	40
N	3	25	6
Prop. Ft.	0.05	0.19	0.17
s.e.m.	−	0.04	−

$m_f = 0.15, r = 0.79$

A. chrysantha (1972, Tortilla Flat, Arizona, PFHL = 2.5 months)

R.S.H.	−0.22	0	0.22
N	4	10	6
Prop. Ft.	0.25	0.29	0.33
s.e.m.	−	0.12	−

$m_f = 0.18, r = 1.00$

A. toumeyana (1973, Tortilla Flat, Arizona, PFHL = 3 months)

R.S.H.	−0.33	0	0.33
N	19	30	25
Prop. Ft.	0.63	0.63	0.70
s.e.m.	0.07	0.04	0.06

$m_f = 0.11, r = 0.87$

A. toumeyana (1974, Tortilla Flat, Arizona)

R.S.H.	−0.33	0	0.33
N	4	8	5
Prop. Ft.	0.50	0.56	0.69
s.e.m.	−	−	−

$m_f = 0.29, r = 0.98$

A. palmeri (1972, Portal, New Mexico, PFHL = 5.4 months)

R.S.H.	−0.22	0	0.22	0.44
N	9	21	12	6
Prop. Ft.	0.23	0.19	0.22	0.45
s.e.m.	–	0.05	0.06	–

$m_f = 0.31, r = 0.75$

A. schottii (1973, Molino Basin, Arizona, PFHL = 7.8 months)

R.S.H.	−0.29	0	0.29	0.72
N	9	47	28	11
Prop. Ft.	0.25	0.41	0.47	0.50
s.e.m.	–	0.04	0.04	0.11

$m_f = 0.23, r = 0.90$

A. schottii (1974, Molino Basin, Arizona)

R.S.H.	−0.29	0	0.29
N	18	54	15
Prop. Ft.	0.42	0.50	0.48
s.e.m.	0.11	0.05	0.06

$m_f = 0.10, r = 0.22$

A. parviflora (1972, Patagonia, Arizona, PFHL = 29.7 months)

R.S.H.	−0.46	−0.23	0	0.23	0.46
N	2	31	45	17	5
Prop. Ft.	0.30	0.23	0.25	0.28	0.30
s.e.m.	–	0.04	0.03	0.04	–

$m_f = 0.02, r = 0.25$

A. parviflora (1973, Patagonia, Arizona)

R.S.H.	−0.40	0	0.40	0.80
N	16	59	16	2
Prop. Ft.	0.32	0.32	0.28	0.28
s.e.m.	0.06	0.03	0.06	–

$m_f = -0.04, r = -0.89$

A. parviflora (1974, Patagonia, Arizona)

R.S.H.	−0.40	0	0.40
N	5	11	2
Prop. Ft.	0.45	0.29	0.40
s.e.m.	–	0.13	–

$m_f = -0.06, r = -0.31$

---

*Relative stalk height, standardised to the most common size category.
† Sample size.
‡ Mean proportion of flowers developing into fruits.
§ Standard error of the mean (95% confidence), calculated only for sample sizes > 10.
¶ Calculated from grouped data.

Table A.2 Bee selectivity in agaves

Agave parviflora ($N* = 4$; Pl.O.† = 274; B.O.‡ = 108)

R.S.H.	−0.40	0	0.40	0.80
BB/cm/P.O.	0.08	0.19	0.09	0.13

$m_p = 0.01, r = 0.13$

A. schottii (1973; $N = 3$; Pl.O. = 351; B.O. = 42)

R.S.H.	−0.29	0	0.29	0.71
BB/cm/P.O.	0.00	0.05	0.07	0.14

$m_p = 0.13, r = 0.99$

A. schottii (1974; N = 12; Pl.O. = 1195; B.O. = 63)

R.S.H.	−0.29	0	0.29
BB/cm/P.O.	0.00	0.06	0.05
s.e.m.	0.01	0.04	0.04

$m_p = 0.09, r = 0.78$

A. schottii (1975; N= 17; Pl.O. = 1418; B.O. = 227)

R.S.H.	−0.22	0	0.22	0.44
BB/cm/P.O.	0.03	0.03	0.03	0.08
s.e.m.	0.03	0.01	0.02	0.04

$m_p = 0.07, r = 0.77$

A. toumeyana (1974; N = 13; Pl.O. = 178; B.O. = 122)

R.S.H.	−0.27	0	0.27
CB/cm/P.O.	0.04	0.08	0.11
s.e.m.	0.03	0.02	0.05

$m_p = 0.13, r = 1.00$

A. toumeyana (1975; N = 24; Pl.O. = 1890; B.O. = 661)

R.S.H.	−0.22	0	0.22	0.44
CB/cm/P.O.	0.05	0.04	0.13	0.09
s.e.m.	0.02	0.01	0.04	0.03

$m_p = 0.10, r = 0.66$

A. utahensis (N = 1; Pl.O. = 45; B.O. = 7)

R.S.H.	−0.80	0.40	0.00	0.40
CB/cm/P.O.	0.00	0.16	0.01	0.25

$m_p = 0.15, r = 0.64$

---

*Number of censuses.

†Number of plant observations.

‡Number of bee observations.

Publications by David Lack

Books

1934 *The Birds of Cambridgeshire*. Cambridge Bird Club, Cambridge.

1943 *The Life of the Robin*. Witherby, London. 4th edn. (revised) 1965.

1947 *Darwin's Finches*. Cambridge University Press, Cambridge. Harper Torchbook paperback edn. 1961. Japanese edn. 1974.

1950 *Robin Redbreast*. Oxford University Press, Oxford.

1954 *The Natural Regulation of Animal Numbers*. Clarendon Press, Oxford. Paperback edn. 1970. Russian edn. 1957.

1956 *Swifts in a Tower*. Methuen, London.

1957 *Evolutionary Theory and Christian Belief*. Methuen, London. Reprinted with new chapter 1961.

1965 *Enjoying Ornithology*. Methuen, London.

1966 *Population Studies of Birds*. Clarendon Press, Oxford.

1968 *Ecological Adaptations for Breeding in Birds*. Methuen, London.

1971 *Ecological Isolation in Birds*. Blackwell Scientific Publications, Oxford and Edinburgh.

1974 *Evolution Illustrated by Waterfowl*. Blackwell Scientific Publications, Oxford and Edinburgh.

1976 *Island Biology. Illustrated by the Land birds of Jamaica*. Blackwell Scientific Publications, Oxford and Edinburgh.

Papers

1930 Double-brooding of the Nightjar. *Br. Birds*, **23**, 242 - 244.

1930 Some diurnal observations on the Nightjar. *Lond. Nat.*, 1930, pp. 47 - 55.

1930 Spring migration, 1930, at the Cambridge Sewage Farm. *Br. Birds*, **24**, 145 - 154.

1931 Coleoptera on St. Kilda in 1931. *Entomologists' Monthly Mag.*, **47**, 1 - 4.

1931 Observations at sewage farms and reservoirs, 1930: Migration at Cambridge, autumn, 1930. *Br. Birds*, **24**, 280 - 282.

1932 Some breeding habits of the European Nightjar. *Ibis*, (13) **2**, 266 - 284.

1932 Birds of Bear Island. *Bull. Br. Ornith. Club*, **53**, 64 - 69.

1932 With John Buchan and T. H. Harrisson. The early autumn migration at St. Kilda in 1931. *Scott. Nat.*, 1932, pp. 1 - 8.

1932 Further notes on insects from St. Kilda in 1931. *Entomologists' Monthly Mag.*, **48**, 139 - 145.

1933 With G. C. L. Bertram. Bear Island. *Geog. J.*, **81**, 45 - 53.

1933 Trichoptera, Lepidoptera and Coleoptera from Bear Island. *Ann. Mag. nat. Hist. Ser.* 10, 12, 205 - 210.

1933 Nesting conditions as a factor controlling breeding time in birds. *Proc. zool. Soc. Lond.*, 1933, pt. 2, pp. 231 - 237.

1933 Habitat selection in birds with special reference to the effects of afforestation on the Breckland avifauna. *J. Anim. Ecol.*, 2, 239 - 262.

1933 With G. C. L. Bertram. Notes on the birds of Bear Island. *Ibis*, (13) 3, 283 - 301.

1933 With L. Lack. Territory reviewed. *Br. Birds*, 27, 179 - 199.

1934 Some insects from the Scoresby Sound region, East Greenland, with an account of the fauna of a nunatak. *Ann. Mag. nat. Hist. Ser.* 10, 14, 599 - 606.

1934 Habitat distribution in certain Icelandic birds. *J. Anim. Ecol.*, 3, 81 - 90.

1934 With T. H. Harrisson. The breeding birds of St. Kilda. *Scott. Nat.* 1934, pp. 59 - 69.

1934 With B. B. Roberts. Notes on Icelandic birds, including a visit to Grimsey. *Ibis*, (13) 4, 799 - 807.

1934 With G. C. L. Bertram and B. B. Roberts. Notes on East Greenland birds, with a discussion of the periodic non-breeding among Arctic birds. *Ibis*, (13) 4, 816 - 831.

1934 With L. S. V. Venables. Territory in the Great Crested Grebe. *Br. Birds*, 28, 191 - 198.

1935 Territory and polygamy in a Bishop-bird (*Euplectes hordeacea hordeacea* Linn.). *Ibis*, (13) 5, 817 - 836.

1935 The breeding bird population of British heaths and moorland. (Together with an account of a census over seven years on Skokholm, Pembrokeshire, by R. M. Lockley.) *J. Anim. Ecol.*, 4, 43 - 57.

1936 On the pugnacity at the nest of a pair of *Onychognathus walleri walleri. Ibis*, (13) 6, 821 - 825.

1936 With H. L. Lack. Review. Territory. Some recent American work. *Br. Birds*, 29, 255 - 258.

1936 With L. S. V. Venables. Further notes on territory in the Great Crested Grebe. *Br. Birds*, 30, 60 - 69.

1937 The function of the Goldcrest's crest. *Br. Birds*, 31, 82 - 83.

1937 The psychological factor in bird distribution. *Br. Birds*, 31, 130 - 136.

1937 A review of bird census work and bird population problems. *Ibis* (14) 1, 369 - 395.

1937 With L. S. V. Venables. The heathland birds of South Haven Peninsula, Studland Heath, Dorset. *J. Anim. Ecol.*, 6, 62 - 72.

1938 With G. C. L. Bertram. Notes on the animal ecology of Bear Island. *J. Anim. Ecol.*, 7, 27 - 52.

1938 With R. M. Lockley. Skokholm Bird Observatory homing experiments. 1. 1936 - 37. Puffins, Storm-Petrels and Manx Shearwaters. *Br. Birds*, 31, 242 - 248.

1938 With J. Fisher. Field observations on birds in 1937. *Bull. Anim. Behav.*, 1, 8 - 17.

1939 The behaviour of the Robin. Part I. The life-history, with special reference to aggressive behaviour, sexual behaviour, and territory. Part II. A partial analysis of aggressive and recognitional behaviour. *Proc. zool. Soc. Lond.*, **109A**, 169 -219.

1939 The display of the Blackcock. *Br. Birds.* **32**, 290 – 303.

1939 Further changes in the Breckland anifauna caused by afforestation. *J. Anim. Ecol.*, **8**, 277 - 285.

1939 With J. T. Emlen. Observations on breeding behaviour in Tri-coloured Redwings. *Condor*, **41**, 225 - 230.

1939 With L. S. V. Venables. The habitat distribution of British woodland birds. *J. Anim. Ecol.*, **8**, 39 - 71.

1940 The Galapagos Finches. *Bull. Br. Ornith. Club,* **60**, 46 – 50.

1940 The releaser concept in bird behaviour. *Nature, Lond.*, **145**, 107 - 108.

1940 Observations on captive Robins. *Br. Birds*, **33**, 262 - 270.

1940 The behaviour of the Robin. Population changes over four years. *Ibis*, (14) **4**, 299 - 324.

1940 Courtship feeding in birds. *Auk*, **57**, 169 - 178.

1940 Habitat selection and speciation in birds. *Br. Birds*, **34**, 80 - 84.

1940 Evolution of the Galapagos Finches. *Nature, Lond.*, **146**, 324 - 327.

1940 Variation in the introduced English Sparrow. *Condor*, **42**, 239 - 241.

1940 Pair-formation in birds. *Condor*, **42**, 269 - 286.

1940 With L. S. V. Venables. Migratory Limicoline birds in the Galapagos. *Ibis*, (14) **4**, 730 - 731.

1941 Some aspects of instinctive behaviour and display in birds. *Ibis*, (14) **5**, 407 - 441.

1941 Notes on territory, fighting and display in the Chaffinch. *Br. Birds*, **34**, 216 - 219.

1941 With W. Light. Notes on the spring territory of the Blackbird. *Br. Birds*, **35**, 47 - 53.

1941 With others. Courtship feeding in birds. *Auk*, **58**, 56 - 60.

1942 Ecological features of the bird faunas of British small islands. *J. Anim. Ecol.*, **11**, 9 - 36.

1942-
43 The breeding birds of Orkney. *Ibis*, (14) **6**, 461 - 484; **85**, 1 - 27.

1943 The age of the Blackbird. *Br. Birds*, **36**, 166 - 175.

1943 The age of some more British birds. *Br. Birds*, **36**, 193 - 197, 214 - 221.

1944 Early references to territory in bird life. *Condor*, **46**, 108 - 111.

1944 The problem of partial migration. *Br. Birds*, **37**, 122 - 130, 143 - 150.

1944 Birds perching on reptiles. *Ibis*, **86**, 222.

1944 Ecological aspects of species-formation in passerine birds. *Ibis*, **86**, 260 - 286.

1944 Correlation between beak and food in the Crossbill, *Loxia curvirostra* Linnaeus. *Ibis*, **86**, 552 - 553.

1944 With W. B. Alexander. Changes in status among British breeding birds. *Br. Birds*, **38**, 42 -45, 62 - 69, 82 - 88.

1945 The Galapagos Finches (Geospizinae). A study in variation. *Occ. Pap. Calif. Acad. Sci.*, **XXI**.

1945 The ecology of closely related species with special reference to Cormorant (*Phalacrocorax carbo*) and Shag (*P. aristotelis*). *J. Anim. Ecol.*, **14**, 12 - 16.

1945 With G. C. Varley. Detection of birds by radar. *Nature, Lond.*, **156**, 446.

1946 Clutch and brood size in the Robin. *Br. Birds*, **39**, 98 – 109, 130 – 135.

1946 The balance of population in the Heron. *Br. Birds*, **39**, 204 – 206.

1946 Do juvenile birds survive less well than adults? *Br. Birds*, **39**, 258 – 264.

1946 The taxonomy of the Robin *Erithacus rubecula* (Linn.). *Bull. Br. Ornith. Club*, **66**, 55 – 65.

1946 The names of the Geospizinae (Darwin's Finches). *Bull. Br. Ornith. Club*, **67**, 15 – 22.

1946 Competition for food by birds of prey. *J. Anim. Ecol.*, **15**, 123 – 129.

1946 With R. Parkhurst. The clutch-size of the Yellowhammer. *Br. Birds*, **39**, 358 – 364.

1946 With J. H. Owen. The eggs of the Red-backed Shrike. *Ool. Rec.* **20**, 1 – 12.

1947 A further note on the taxonomy of the Robin, *Erithacus rubecula* (Linn.). *Bull. Br. Ornith. Club*, **67**, 51 – 54.

1947 The significance of clutch-size in the Partridge (*Perdix perdix*). *J. Anim. Ecol.*, **16**, 19 – 25.

1947–
48 The significance of clutch-size. *Ibis*, **89**, 302 – 352; **90**, 25 – 45.

1947 With H. Arn. Die Bedeutung der Gelegegrosse beim Alpensegler. *Orn. Beob.*, **44**, 188 – 210.

1948 Further notes on clutch and brood size in the Robin. *Br. Birds*, **41**, 98 – 104, 130 – 137.

1948 Natural selection and family size in the Starling. *Evolution*, **2**, 95 – 110.

1948 Notes on the ecology of the Robin. *Ibis*, **90**, 252 – 279.

1948 The significance of litter-size. *J. Anim. Ecol.*, **17**, 45 – 50.

1948 With A. Schifferli. Die Lebensdauer des Stares. *Orn. Beob.*, **45**, 107 – 114.

1949 Comments on Mr. Skutch's paper on clutch-size. *Ibis*, **91**, 455 – 458.

1949 The significance of ecological isolation. in *Genetics, Paleontology and Evolution* (ed. G. L. Jepsen *et al.*), Princeton University Press, Princeton, New Jersey, pp. 299 – 308.

1949 The apparent survival-rate of ringed Herons. *Br. Birds*, **42**, 74 – 79.

1949 Vital statistics from ringed Swallows. *Br. Birds*, **42**, 147 – 150.

1949 The position of the Edward Grey Institute, Oxford. *Bird Notes*, **23**, 231 – 236.

1949 Family size in certain Thrushes (Turdidae). *Evolution*, **3**, 57 – 66.

1949 An ornithological examination paper. *New Nat.* no. 6, 49 – 53.

1949 With E. Lack. Passerine migration through England. *Br. Birds*, **42**, 320 – 326.

1949 With E. T. Silva. The weight of nestling Robins. *Ibis*, **91**, 64 – 78.

1949 With H. N. Southern. Birds on Tenerife. *Ibis*, **91**, 607 – 626.

1950 Breeding seasons in the Galapagos. *Ibis*, **92**, 268 – 278.

1950 The breeding seasons of European birds. *Ibis*, **92**, 288 – 316.

1950 Family-size in titmice of the genus *Parus*. *Evolution*, **4**, 279 – 290.

1950 Birds and the Field Centres. *Rep. Counc. Promot. Fld. Stud.* 1949 – 50, pp. 24 – 33.

1951 Migration through the Pyrenees. *Bull. Br. Ornith. Club*, **70**, 59 – 61.

1951 Obituary. B. W. Tucker (1901 - 1950). *Ibis*, **93**, 300 – 305.

1951 Geographical variation in *Erithacus rubecula*. *Ibis*, **93**, 629 – 630.

1951 Ornithological investigations in forests. *Report on Forest Research for the year ending 1950*. Forestry Commission, London, pp. 125 – 126.

1951 Population ecology in birds. A review. *Proc. Xth Int. ornith. Congr.* Uppsala 1950, pp. 409 - 448.

1951 Old brecks or new forests? *Bird Notes*, 24, 215 - 216.

1951 With E. Lack. Migration of insects and birds through a Pyrenean pass. *J. Anim. Ecol.*, 20, 63 - 67.

1951 With E. Lack. The breeding biology of the Swift *Apus apus. Ibis*, 93, 501 - 546.

1951 With E. Lack. Further changes in bird-life caused by afforestation. *J. Anim. Ecol.*, 20, 173 - 179.

1952 Special review. Reproductive rate and population density in the Great Tit: Kluijver's study. *Ibis*, 94, 167 - 173.

1952 With E. Lack. Visible migration at Land's End. *Br. Birds*, 45, 81 - 96.

1952 With E. Lack. The breeding behaviour of the Swift. *Br. Birds*, 45, 186 - 215.

1952 The nesting of titmice in boxes, 1950. *Report on Forestry Research for the year ending 1951*. Forestry Commission, London, pp. 138 - 139.

1953 Darwin's Finches. *Scient. Am.*, April 1953, pp. 67 - 72.

1953 With H. Arn. Die mittlere Lebensdauer des Alpenseglers. *Orn. Beob.*, 50, 133 - 137.

1953 With E. Lack. Visible migration through the Pyrenees: an autumn reconnaissance. *Ibis*, 95, 271 - 309.

1954 The evolution of reproductive rates, in *Evolution as a process*, (ed. J. S. Huxley), London, pp. 143 - 156.

1954 Visible migration in S. E. England, 1952. *Br. Birds*, 47, 1 - 15.

1954 The stability of the Heron population. *Br. Birds*, 47, 111 - 119.

1954 Two Robin populations. *Bird Study*, 1, 14 - 17.

1954 Call-notes, *Erithacus* and convergence. *Ibis*, 96, 312 - 314.

1954 Cyclic mortality. *J. Wildl. Mgmt*, 18, 25 - 37.

1954 With E. Lack. The home life of the Swift. *Scient. Am.*, July 1954, pp. 60 - 64.

1955 The summer movements of Swifts in England. *Bird Study*, 2, 32 - 40.

1955 Summary report on nesting tits. *Bird Study*, 2, 199 - 201.

1955 British tits (*Parus* spp.) in nesting boxes. *Ardea*, 43, 50 - 84.

1955 Visible migration through the Pyrenees. *Acta XIth Congr. Int. Orn.*, Basel 1954, pp. 176 - 178.

1955 With D. F. Owen. The food of the Swift. *J. Anim. Ecol.*, 24, 120 - 136.

1955 With M. G. Ridpath. Do English Woodpigeons migrate? *Br. Birds*, 48, 289 - 292.

1955 The mortality factors affecting adult numbers, in *The Numbers of Man and Animals*, (ed. J. B. Cragg and N. W. Price), Oliver and Boyd, London, pp. 47 - 55.

1955 With E. Weitnauer. Daten zur Fortpflanzungsbiologie des Mauerseglers (*Apus apus*) in Oltingen und Oxford. *Orn. Beob.*, 52, 137 - 141.

1956 Seaward flights of Swifts at dusk. *Bird Study*, 3, 37 - 42.

1956 Spine-tailed Swifts of the Old World. *Bull. Br. Ornith. Club*, 76, 72 - 73.

1956 A review of the genera and nesting habits of Swifts. *Auk*, 73, 1 - 32.

1956 The species of *Apus. Ibis*, 98, 34 - 62.

1956 Further notes on the breeding biology of the Swift *Apus apus. Ibis*, 98, 606 - 619.

1956 Variations in the reproductive rate of birds. *Proc. R. Soc. Series B*, **145**, 329 -333.

1957 The Chaffinch migration in North Devon. *Br. Birds*, **50**, 10 - 19.

1957 Notes on nesting Nightjars. *Br. Birds*, **50**, 273 - 277.

1957 The first primary in Swifts. *Auk*, **74**, 385 -386.

1957 With J. Gibb and D. F. Owen. Survival in relation to brood-size in tits. *Proc. zool. Soc. Lond.*, **128**, 313 -326.

1958 The significance of the colour of turdine eggs. *Ibis*, **100**, 145 - 166.

1958 Special review. Recent Swiss and British work on watching migration by radar. *Ibis*, **100**, 286 -287.

1958 The return and departure of Swifts *Apus apus* at Oxford. *Ibis*, **100**, 477 -502.

1958 A quantitative breeding study of British tits. *Ardea*, **46**, 91 - 124.

1958 Swifts over the sea at night. *Bird Study*, **5**, 126 - 127.

1958 Weather movements of Swifts 1955 - 1957. *Bird Study*, **5**, 128 - 142.

1958 A vision of Rome, 1960, by Cassandra Lark. *Bird Notes*, **28**, 426 -428.

1958 Migrational drift of birds plotted by radar. *Nature, Lond.*, **182**, 221 - 223.

1958 A new race of the Whiterumped Swift. *J. Bombay nat. Hist. Soc.*, **55**, 160 - 161.

1958 *Apus pallidus* in Northern Rhodesia. *Ostrich*, **29**, 86.

1958 Watching migration by radar. (BBC broadcast.) *Birds Ctry Mag.* **11**, 112 - 116.

1958 With E. Lack. The nesting of the Long-tailed Tit. *Bird Study*, **5**, 1 - 19.

1958 With J. G. Tedd. The detection of bird migration by high-powered radar. *Proc. R. Soc. Series B*, **149**, 503 -510.

1959 Some British pioneers in ornithological research, 1859 - 1939. *Ibis*, **101**, 71 -81.

1959 Migration across the North Sea studied by radar. Part 1. Survey through the year. *Ibis*, **101**, 209 -234.

1959 Migration across the sea. *Ibis*, **101**, 374 -399.

1959 Watching migration by radar. *Br. Birds*, **52**, 258 -267.

1959 With K. Williamson. Bird-migration terms. *Ibis*, **101**, 255 -256.

1959 With M. J. Goodacre. Early breeding in 1957. *Br. Birds*, **52**, 74 -83.

1960 The height of bird migration. *Br. Birds*, **53**, 5 - 10.

1960 Autumn 'drift-migration' on the English east coast. *Br. Birds*, **53**, 325 -397.

1960 Migration across the North Sea studied by radar. Part 2. The spring departure 1956 -59. *Ibis*, **102**, 26 -57.

1960 Hints on research for bird-watchers. *Bird Study*, **7**, 9 - 20.

1960 Robins for Christmas. *New Scient.*, **8**, 1639 - 1641.

1960 The influence of weather on passerine migration. A review. *Auk*, **77**, 171 -209.

1960 A comparison of "drift-migration" at Fair Isle, the Isle of May and Spurn Point. *Scott. Birds*, **1**, 1 -33.

1960 A survey of the conflict between evolutionary theory and Christian belief. *Proc. R. Inst.*, **38**, 303 -310.

1960 The implications of evolution. *Christian Evidence Society.* pp. 11.

1960 The conflict between evolutionary theory and Christian belief. *Nature, Lond.*, **187**, 98 - 100.

1960 Bruterfolg: Verhältnis von Eizahl, Schlüpfzahl and flüggen Jungen, in *Biologische Studien am Alpensegler*, (ed. H. Arn-Willi), Solothurn, Switzerland, pp. 119 - 132.

1962 Radar evidence on migratory orientation. *Br. Birds*, **55**, 139 - 158.

1962 Migration across the southern North Sea studied by radar. Part 3. Movements in June and July. *Ibis*, **104**, 74 - 85.

1962 With E. Eastwood. Radar films of migration over Eastern England. *Br. Birds*, **55**, 388 - 414.

1962 With J. L. F. Parslow. Falls of night migrants on the English east coast in autumn 1960 and 1961. *Bird Migrat.*, **2**, 187 - 201.

1963 Migration across the southern North Sea studied by radar. Part 4. Autumn. *Ibis*, **105**, 1 - 54.

1963 Migration across the southern North Sea studied by radar. Part 5. Movements in August, winter and spring, and conclusion. *Ibis*, **105**, 461 - 492.

1963 Weather factors initiating migration. *Proc. XIIIth Int. ornith. Congr.*, Ithaca 1962, pp. 412 - 414.

1963 Cuckoo hosts in England. *Bird Study*, **10**, 185 - 202.

1963 An undiscovered species of Swift. *Bird Notes*, **30**, 258 - 260.

1963 With H. E. Axell, J. L. F. Parslow and J. Wilcock. Migration at Minsmere, seen and unseen. *Bird Notes*, **30**, 181 - 186.

1964 A long-term study of the Great Tit (*Parus major*). *J. Anim. Ecol.*, **33** (Suppl.), 159 - 173.

1964 Significance of clutch-size in Swift and Grouse. *Nature, Lond.*, **203**, 98 - 99.

1965 Evolutionary ecology. *J. Anim. Ecol.*, **34**, 223 - 231.

1965 With R. E. Moreau. Clutch-size in tropical passerine birds of forest and savanna. *Oiseau Revue fr. Orn.* **35** (no. spécial), 76 - 89.

1965 Natural selection and human nature, in *Biology and Personality*, (ed. I. Ramsey), Blackwell, Oxford, pp. 40 - 48.

1966 Wilfred Backhouse Alexander, 1885 - 1965. *Rep. Oxf. ornith. Soc. for 1965* pp. 2 - 5.

1966 W. B. Alexander. (Obituary.) *Nature*, **209**, 759 - 760.

1966 Special review — links between breeding behaviour and ecology. *Ibis*, **108**, 290 - 291.

1966 Are bird populations regulated? *New Scient.*, **31**, 98 - 99.

1966 Special review — an authorised taxonomy? ('The birds of the Palearctic fauna — non-passeriformes', by C. Vaurie. London, 1965.) *Ibis*, **108**, 141 - 142.

1966 With P. Lack. Passerine night migrants on Skokholm. *Br. Birds*, **59**, 129 - 141.

1967 The significance of clutch-size in waterfowl. *Rep. Wildfowl Trust*, **18**, 125 - 128.

1967 Interrelationships in breeding adaptations as shown by marine birds. (Presidential address.) *Proc. XIVth Int. ornith. Congr.* Oxford 1966, pp. 3 - 42.

1968 Bird migration and natural selection. *Oikos*, **19**, 1 - 9.

1968 The proportion of yolk in the eggs of waterfowl. *Wildfowl*, **19**, 67 - 69.

1968 The sequence in European bird-lists. (Letter,) *Ibis*, **110**, 107 - 113.

1969 Subspecies and sympatry in Darwin's Finches. *Evolution*, **23**, 252 - 263.

1969 Drift migration: a correction. *Ibis*, **111**, 253 - 255.

1969 Tit niches in two worlds; or Homage to Evelyn Hutchinson. *Am. Nat.* **103**, 43 - 49.

1969 Population changes in the land birds of a small island. *J. Anim. Ecol.*, **38**, 211 - 218.

1969 Of birds and men. *New Scient.*, **41**, 121 - 122.

1969 The numbers of bird species on islands. (Witherby Memorial Lecture) *Bird Study*, **16**, 193 – 209.

1970 Island birds. (Paper presented at Symposium on Adaptive Aspects of Insular Evolution, Puerto Rico, 1969.) *Biotropica*, **2**, 29 – 31.

1970 Animal populations in relation to their food resources. *Symp. Br. ecol. Soc.* **10**, xiii – xx (introduction).

1970 The endemic ducks of remote islands. *Wildfowl*, **21**, 5 – 10.

1973 The numbers of species of hummingbirds in the West Indies. *Evolution*, **27**, 326 – 337.

1973 With A. Lack. Birds on Grenada. *Ibis*, **115**, 53 – 59.

1973 With P. Lack. Wintering warblers in Jamaica. *Living Bird*, **11**, 129 – 153.

1973 With E. Lack, P. Lack and A. Lack. Birds on St. Vincent. *Ibis*, **115**, 46 – 52.

1974 Population, Biological. *Encyclopaedia Britannica*, 15th edn., **14**, 824 – 838.

Author index

Subject index